PUBLICATIONS

OF THE

NAVY RECORDS SOCIETY

VOL. 124

THE POLLEN PAPERS

THE NAVY RECORDS SOCIETY was established in 1893 for the purpose of printing rare or unpublished works of naval interest.

Any person wishing to become a Member of the Society is requested to apply to the Hon. Secretary, c/o Royal Naval College, Greenwich, London, SE10 9NN. The annual subscription is £10, the payment of which entitles the Member to receive a free copy of each work issued by the Society in that year, and to purchase earlier volumes at reduced prices. The subscription for Libraries and Institutions is £12.

Subscriptions and orders for earlier volumes should be sent to the Hon. Treasurer, c/o Barclays Bank, 54 Lombard St., London, EC3P 3AH.

THE COUNCIL of the NAVY RECORDS SOCIETY wish it to be clearly understood that they are not answerable for any opinions or observations that may appear in the Society's publications. For these the responsibility rests entirely with the Editors of the several works.

Arthur Hungerford Pollen

THE
POLLEN PAPERS

*The Privately Circulated Printed Works
of Arthur Hungerford Pollen, 1901–1916*

Edited by
JON TETSURO SUMIDA

PUBLISHED BY GEORGE ALLEN & UNWIN
FOR THE NAVY RECORDS SOCIETY
1984

© The Navy Records Society, 1984.
This book is copyright under the Berne Convention. No reproduction without permission. All rights reserved.

George Allen & Unwin (Publishers) Ltd,
40 Museum Street, London WC1A 1LU, UK

George Allen & Unwin (Publishers) Ltd,
Park Lane, Hemel Hempstead, Herts HP2 4TE, UK

Allen & Unwin Inc.,
9 Winchester Terrace, Winchester, Mass 01890, USA

George Allen & Unwin Australia Pty Ltd,
8 Napier Street, North Sydney, NSW 2060, Australia

British Library Cataloguing in Publication Data

Pollen, Arthur H.
  The pollen papers.—(Publications of the Navy Records Society; v.124)
  1. Naval gunnery
  I. Title   II. Sumida, Jon Tetusuro
  III. Series
  628.3′251   VF346
  ISBN 0-04-942182-4

Library of Congress Cataloging in Publication Data

Pollen, Arthur Joseph Hungerford, 1866-1937.
  The Pollen papers.
(Publications of the Navy Records Society; vol. 124)
Includes bibliographical references and index.
1. Fire control (Naval gunnery)—Addresses, essays, lectures. 2. Naval strategy—Addresses, essays, lectures. 3. Naval tactics—Addresses, essays, lectures. I. Sumida, Jon Tetsuro, 1949-
II. Navy Records Society (Great Britain) III. Title.
IV. Series: Publications of the Navy Records Society; v. 124.
DA70.Al vol. 124 [VF520] 359′.00941s 83-15592
ISBN 0-04-942182-4 [623′.553]

Photoset in 10 on 12 point Times by Grove Graphics, Tring
and printed and bound in Great Britain by
William Clowes Limited, Beccles and London

THE COUNCIL
OF THE
NAVY RECORDS SOCIETY
1983

PATRON
H.R.H. THE PRINCE PHILIP, DUKE OF EDINBURGH, K.G., O.M., F.R.S.

PRESIDENT
THE RT HON. THE LORD CARRINGTON, P.C., K.C.M.G., M.C.

VICE PRESIDENTS
Sir John LANG, G.C.B.
Richard OLLARD, F.R.S.L., M.A.
R. J. B. KNIGHT, M.A., Ph.D.
Capt. A. B. SAINSBURY, V.R.D., M.A., R.N.R.
Prof. B. McL. RANFT, M.A., D.Phil., F.R. Hist.S.

COUNCILLORS
J. GOOCH, B.A., Ph.D., F.R.Hist.S.
P. M. KENNEDY, M.A., D.Phil., F.R.Hist.S.
H. C. TOMLINSON, M.A., Ph.D., F.R.Hist.S.
P. M. H. BELL, B.A., B.Litt.
Prof. J. S. BROMLEY, M.A., F.R.Hist.S.
D. K. BROWN, M.Eng., C.Eng., F.R.I.N.A., R.C.N.C.
J. D. BROWN
Admiral Sir James EBERLE, G.C.B.
Lieut. J. V. P. GOLDRICK, B.A., R.A.N.
Prof. D. M. SCHURMAN, M.A., Ph.D.
Geoffrey TILL, M.A., Ph.D.
Cdr. A. R. WELLS, M.A., M.Sc., Ph.D., R.N.
Jonathan COAD, M.A.
Miss P. K. CRIMMIN, B.A., M.Phil.
C. I. HAMILTON, M.A., Ph.D.
Richard HOUGH
A. P. McGOWAN, M.A., Ph.D.
Capt. R. H. PARSONS, M.A., C.Eng., M.I.E.R.E., R.N.
R. W. A. SUDDABY, M.A.
The Hon. David ERSKINE, M.A.
Prof. Paul G. HALPERN, Ph.D.
Prof. C. C. LLOYD, M.A.
A. W. H. PEARSALL, M.A.

HON. GENERAL EDITOR
A. N. RYAN, M.A., F.R.Hist.S.

HON. SECRETARY
N. A. M. RODGER, M.A., D.Phil., F.R.Hist.S.

HON. TREASURER
H. U. A. LAMBERT. M.A.

CONTENTS

Preface		page xi
General Introduction		1
I	The Pollen System of Telemetry (February 1901)	8
II	Memorandum on a Proposed System for Finding Ranges at Sea and Ascertaining the Speed and Course of any Vessel in Sight (July 1904)	14
III	Fire Control and Long-Range Firing: An Essay to Define Certain Principia of Gunnery, and to Suggest Means for Their Application (December 1904)	23
IV	A.C.: A Postscript (Mid-1905)	55
V	The *Jupiter* Letters: Extracts from Letters Addressed to Various Correspondents in the Royal Navy principally from HMS *Jupiter* (May 1906)	70
VI	Note on the Possibility of Demonstrating the Principle of Aim Correction Without the Use of Instruments Designed for the Purpose (Sent to the Director of Naval Ordnance, July 1906)	101
VII	Some Aspects of the Tactical Value of Speed in Capital Ships (November 1906)*	105
VIII	Notes on a Proposed Method of Studying Naval Tactics (Spring 1907)	114
IX	An Apology for the A.C. Battle System: Being Notes for a Lecture to the War Course College, Portsmouth (August–December 1907)	131
X	Notes, Correspondence, Etc., on the Pollen A.C. System, Installed and Tried in HMS *Ariadne* (December 1907–January 1908)*	156
XI	Extracts from a Letter to Capt. Reginald H. S. Bacon, CVO, DSO, Royal Navy. Dated 27 February 1908 (February 1908)	172
XII	Reflections on an Error of the Day (September 1908)*	178
XIII	Notes, Etc., on the *Ariadne* Trials (April 1909)	194

XIV	Memoranda and Instructions Introductory to the Use of Pollen's Tactical Instrument (May 1909)*	237
XV	[Pollen] To Rear-Admiral the Hon. Stanley C. J. Colville, CVO, CB (July 1910)	243
XVI	The Quest of a Rate Finder (November 1910)	255
XVII	Of War and The Rate of Change (December 1910–January 1911)	271
XVIII	The Gun in Battle (February 1913)	291
XIX	The Necessity of Fire Control (September 1913)	326
XX	[Untitled Paper] (May 1916?)	333

Appendices

I	Biographical Notes	357
II	Danger Space	360
III	Range Finding by Triangulation	361
IV	Triangulation by Coincidence	363
V	Reflecting Surfaces in Coincidence Range Finders	366
VI	Change in Range	367
VII	Virtual Course	370
VIII	Trigonometric Computation of Change of Range and Bearing Rates, and Deflection	371
IX	Disc-Ball-Roller Mechanism	373
Notes		374
Index		395

* Abridged; see note 9 to the Preface.

PREFACE

Arthur Joseph Hungerford Pollen, one of the ten children of John Hungerford Pollen, the prominent Tractarian turned Catholic and noted artist, was born in 1866. As a boy, Pollen attended the Oratory School in Birmingham where he proved himself to be a gifted student, musician and athlete. He then matriculated at Trinity, Oxford, studied history, and graduated with honours. While at Oxford, Pollen made his mark as a leading speaker at the Union, the famous university debating society. In 1893, he entered Lincoln's Inn to begin a career in law, and in 1895 he unsuccessfully contested a parliamentary seat as a Liberal at Walthamstow, Essex. The earnings of a legal novice were small, and Pollen found it necessary during these years to supplement his income by writing as a critic of art, literature, drama and music for the highly respected *Westminster Gazette*. In addition, he tutored the sons of wealthy families, and in the course of his duties travelled in 1895 and 1896 to the United States, Canada and India where he was able to hunt big game, a favourite pastime. In 1898, Pollen assumed the managing directorship of Linotype and Machinery Ltd (henceforward referred to as the Linotype Company), following his marriage to the daughter of Joseph Lawrence, who was the firm's chairman.

Pollen quickly established himself as a businessman of great ability, but in 1900 he became interested in the problem of aiming naval artillery after viewing a practice firing at sea. Between 1900 and 1906 Pollen formulated his ideas for a system of firing accurately when ranges were long, and in the autumn of 1906 he persuaded the Admiralty to co-operate in the development of his proposals. The interference, however, of an influential group of naval officers whose views differed from those of Pollen led to the repeated rejection of his instruments between 1908 and 1912, in spite of performances in trials that were either completely successful, or indicative of the probability of complete success with further development. In 1913 Pollen finally abandoned all hope of persuading the Admiralty to adopt his system and, through the Argo Company – a firm that he had founded to manufacture his equipment – offered it to foreign navies. Pollen's instruments were by this time perfected, but very promising negotiations with prospective foreign buyers were disrupted by the outbreak of war in 1914. The Admiralty refused to accept the Argo

Company's offer to supply the Royal Navy with fire control gear for war service and, as a consequence, Pollen spent the years from 1915 to 1918 writing for the well-known journal *Land and Water* on naval events, in which capacity he achieved great success. Following the war, Pollen resumed his career as a businessman, but with firms whose work was unconnected with naval matters. He died in 1937.

The capability of the Pollen system was far superior to that of the alternative fire control system adopted by the Royal Navy for service during the First World War. This superiority was in large part attributable to the disc-ball-roller calculating mechanism of the Pollen mechanical analogue computer, or clock, as it was called, which could solve instantaneously the second order differential equations that often described the relative movement of two ships that were converging or diverging, through the process of what is known as differential analysis.[1] The perfection of this technique in 1911 by Pollen and his design engineer, Harold Isherwood, was in advance of those who have long been considered to be the developers of this important method of mechanical computation: Hannibal Ford, the American inventor of fire control instruments, whose differential analyser was supplied to the United States Navy during the First World War, and H. W. Nieman and Vannevar Bush, whose differential analyser was not constructed at the Massachusetts Institute of Technology until 1930.[2]

Pollen's work had a great impact on the development of the Royal Navy's fire control equipment in spite of the Admiralty's refusal to adopt his system. Several of Pollen's instruments were ordered for service in small quantities before the Admiralty severed its relations with the Argo Company in 1913. In 1911, moreover, certain aspects of Pollen's clock design – though not the differential analyser mechanism – were incorporated in the fire control system that was issued to the fleet without Pollen's knowledge or permission. This was proved in 1925 to the satisfaction of the Royal Commission on Awards to Inventors, which ruled that Pollen should be paid £30,000 in compensation. Finally, the unsatisfactory performance of the service fire control system in naval engagements during the First World War, notably at the battle of Jutland in 1916, prompted the Admiralty after the war to order the redesign of the Royal Navy's fire control gear. This work was carried out with the assistance of Harold Isherwood and D. H. Landstad, Pollen's draughtsman, and resulted in the design of a fire control system that incorporated a differential analyser. The new system and its derivatives became standard in all major British warships beginning with the battleships *Nelson* and *Rodney*, which were commissioned for service in 1927.[3]

Pollen's influence on naval affairs was also exerted through the medium of memoranda, reports, essays, and selections from his correspondence which he printed privately for limited circulation at the Admiralty and in the fleet from 1901 to 1916. Pollen's advocacy of his fire control system in these works was so persuasive that the announcement of the Admiralty's decision in late 1912 to abandon their monopoly rights over his instruments prompted no fewer than ten admirals and thirty captains on the active list to register their sense of dismay in letters to the author-inventor.[4] Pollen's propaganda success was in large part attributable to the merit of his technical proposals, but there can be little doubt his readers were also attracted by the cogency of his views on broader naval matters. In a number of his printed papers Pollen provided a synthesis of strategy, tactics, and new technology that addressed directly issues of great concern to the naval officers of the day. Pollen's work was thus regarded highly by many within the Royal Navy, while the writing of Sir Julian Corbett – a better known naval theorist – on general principles of naval strategy was all but ignored by most naval officers, who considered it to be of little practical significance.[5] Admiral Sir John Fisher – the First Sea Lord from 1904 to 1910 – was particularly impressed by Pollen's ideas, which were in many respects similar to his own, and as a consequence Pollen's writing played a critical role in the development of Fisher's conception of the battle cruiser.[6]

This edition of Pollen's writing has been produced from the only known complete set of printed papers written from 1901 to 1916, which has been preserved by the Pollen family.[7] Of the twenty-two papers that constitute the set, two have been omitted,[8] four have been abridged,[9] while the remaining sixteen have been reproduced in their entirety. In the case of one paper,[10] diagrams that are referred to in the text, but which appear to have been lost, have been reconstructed from the discussion in the text and provided for the reader in the form of an editorial appendix placed immediately after the text. A brief history of naval gunnery has been provided in the General Introduction, while there is an editorial introduction to each paper in which matters relating to the writing of it are described. The notes to each editorial introduction and paper have been placed after a biographical and several technical appendices. In the original texts, ships' names were either italicised, enclosed within quotation marks or undifferentiated. For the sake of clarity and consistency ships' names have been italicised in the text in every instance as well as in the editorial material.

A number of printed papers include the texts of letters that Pollen either addressed to or received from naval officers whose names are not given. The original letters, according to Anthony Pollen, the younger son of Arthur Hungerford Pollen, were destroyed by his father in order to preserve absolutely the anonymity of those service correspondents who were associated with papers that were often critical of higher authorities. It has thus not been possible to trace the identities of the unnamed naval officers concerned. A list of the recipients is included in the notes to each editorial introduction whenever the surviving evidence has allowed such a list to be compiled. The biographical appendix gives information on persons who played an active role in British naval affairs during the period 1901–13 and who are mentioned in either the editorial introduction or the text of each paper. Prominent non-naval historical figures mentioned in the texts such as Galileo or Harvey are identified in the notes. For prominent naval historical figures such as Nelson, Drake or Togo, or battles such as Trafalgar or Tsushima, or technical descriptions of warships of the late nineteenth and early twentieth centuries that are mentioned in the editorial introduction or the text of each paper, the reader is referred to standard works of reference.[11] Persons mentioned only in the notes are not identified.

Unless stated otherwise, the Pollen correspondence cited in the notes is to be found in the collection of papers currently in the possession of the Pollen family. In most of these cases no reference has been made to the Pollen Papers as the source. For the sake of clarity, however, a reference to the Pollen Papers has been provided whenever a note contains references to material from other collections. An attempt was made to trace the source of all references in the papers to literary and naval writing, but a number have remained elusive and are not, as a consequence, cited in the notes. Descriptions of the calculating mechanisms of the Pollen clocks have been included in the editorial introduction to 'The Gun in Battle' in order to make the technical issues clear to the specialist and the sceptic. The less mechanically inclined, and more trusting general reader may pass over this section without remorse.

My thanks are due to the following persons for their assistance during the course of completing this volume: Mr David Brown of the Naval Library of the Ministry of Defence; Dr Donald Gordon of the University of Maryland; Dr Joseph Leedom of Hollins College, Virginia; Mr Richard Ollard of Collins; Mr Alan Pearsall of the National Maritime Museum; Mr Anthony Ryan of the University of Liverpool, General Editor of the Navy Records Society; and Mr John Newth of Allen & Unwin. I am deeply indebted to the late Captain Stephen Roskill, whose support, counsel and hospitality contributed so much to the making of this work, and to Professors Emmet Larkin and William McNeill of the University of Chicago, who guided me through the completion of the doctoral dissertation upon which much of the commentary in this volume has been based.[12] Finally, I must express my profound gratitude to Mr Anthony Pollen, my close colleague in the field of Pollen studies. His orderly arrangement of his father's many papers greatly eased my task of research, while I have, since the beginning of my serious study of his father and

PREFACE

the Royal Navy, benefited greatly from his advice and company. I have, of course, received his permission to reprint the works of his father included in this volume.

College Park, Maryland   Jon Tetsuro Sumida
January 1983

The Council of the Navy Records Society is once again pleased to record its thanks to the British Academy for generous financial help.

GENERAL INTRODUCTION

For three hundred years, from the mid-sixteenth to the mid-nineteenth centuries, English warships were armed with smooth-bore, muzzle-loading cannon that fired solid iron balls with black powder charges. The effective range of such ordnance was limited in part by the inaccurate performance of the guns themselves, attributable to the inherent limitations in the design and manufacture of the guns, projectiles and charges. Naval guns were, moreover, laid at a fixed angle of elevation, which meant that the elevation of the gun in relation to the target changed as the ship, upon which the gun was mounted, rolled. Gunners could thus aim and fire only when the rolling motion of the ship in one direction had slowed to a stop before starting its return. Given the difficulty of judging the timing of the rolling cycle exactly − a requirement to the setting of the gun sight − and the complication of anticipating the appearance of the target in the gun sight to allow for the delay between the visual act of sighting and the physical act of pulling the trigger or lanyard to discharge the gun, which was inherent in the slowness of human reflexes, aim was approximate rather than precise. Naval engagements, therefore, were generally fought at ranges of a hundred yards or less, where projectile trajectories were flat, and the targets were so large that many hits could be made in spite of eccentricities in aim.

During the nineteenth century, the technological advances of the Industrial Revolution produced great improvements in the design and manufacturing of naval ordnance. By 1890 breech-loading rifled guns, built from steel rather than iron, used nitro-cellulose charges to fire ballistically shaped projectiles fitted with explosives to distances of several miles with great accuracy. In Royal Navy gunnery exercises, however, few hits were made because the unsteadiness of the gun platform continued to prevent accurate aim, even though the practice range did not exceed 1,600 yards.

The development of methods which would enable guns to be aimed accurately from on board a ship that was under way began in the late 1890s. In 1898 Captain Percy Scott, a particularly inventive officer in the Royal Navy, abandoned the practice of laying guns at a fixed angle of elevation, and instead altered the manually operated elevating mechanism of a medium-calibre, quick-firing gun to facilitate rapid gun elevation and depression. He trained his gunners to use the modified mechanism to keep the gun sight, which moved with the gun, on the target despite the rolling motion of their ship by quickly elevating and depressing the gun. Scott then instructed his men to swing the barrel from side to side quickly to compensate for changes in deflection caused by their ship's yaw – an important factor because of the greater ranges at which he intended to fire. By freeing the gunner from having to aim and fire on the roll and by eliminating the effect of yaw on deflection, Scott's system of 'continuous aim' gun-laying enabled gunners to lay their guns with precision, and thus to increase greatly the proportion of hits to shots fired at ranges of 1,500 yards or less. After demonstrating the effectiveness of his methods in umpired gunnery exercises, Scott was appointed Captain, in April 1903, of HMS *Excellent*, the Royal Navy's gunnery school at Portsmouth, with a mandate to train a new generation of naval gunners.

At ranges that were greater than 1,500 yards, the holes made in a target by medium-calibre, quick-firing projectiles were too small to be seen even with the assistance of a telescopic sight, and while the splash of a projectile was easily visible at much greater ranges, its proximity to the target was impossible to establish with any degree of precision. At ranges in excess of 1,500 yards, therefore, gunners were unable to assess the effect of their shots, and so could not make the necessary adjustment in the setting of their sights when misses occurred. But in firing experiments that were carried out at ranges of from 5,000 to 6,000 yards in 1899 and 1900, gunnery officers of the Mediterranean Fleet found that the effect of naval artillery fire could be accurately evaluated when several guns were fired together at the target in a salvo. The projectiles of a salvo fell at about the same time and within a relatively small area. A salvo whose splashes were all either short or over the target indicated that the gun sights were incorrectly set for elevation, while splashes that fell entirely to the left or to the right of the target showed that gun sights were not correctly set for deflection. A salvo that resulted in splashes that were both short and long, and to the left and to the right of the target, suggested that there were projectiles in between these extremes that were probably scoring hits, and that the gun sights were thus correctly set.

The combination of continuous aim gun-laying and the salvo system enabled medium-calibre, quick-firing guns to make a larger number of hits in proportion to the number of shots fired at ranges that were much greater than 1,500 yards. On the other hand, continuous aim methods could not be used to lay heavy-calibre naval artillery because the power, as opposed to manual, elevation and training gear required by heavy-calibre guns could not be made to work rapidly enough to keep the gun and sight on the target through a ship's roll and yaw. The fall of a salvo's projectiles over an area, however, to some extent reduced the necessity of absolute precision in gun-laying, and thus in the spring of 1904 trials in the Mediterranean demonstrated that heavy-calibre guns that were fired at fixed angles of elevation and train, but that used the salvo system, were capable of making a significant number of hits at ranges that were as high as 8,000 yards. Although medium-calibre quick-firers could still fire more rapidly than heavy-calibre guns, the larger number of splashes that they produced tended to obscure the target, making it more difficult to evaluate the effect of the firing than was the case with the fewer number of splashes made by slower firing heavy-calibre guns. Medium-calibre, quick-firing projectiles were, moreover, ballistically less accurate than those of heavy-calibre guns when ranges were long because they were much lighter and were fired at a comparable muzzle velocity. In addition, they had far less effect on impact because their explosive capacity was much less than a heavy-calibre projectile.

The evident advantages of heavy-calibre naval artillery over medium-calibre quick-firers at longer ranges led to the Admiralty's decision, in early 1905, to build the battleship *Dreadnought* and the battle cruisers of the *Invincible* class, which were armed with big guns only in place of the combination of heavy and medium calibre guns that had previously been standard. The adoption of the all big gun battleship and battle cruiser by the Royal Navy did not signify that firing salvos with heavy-calibre naval artillery was fully practicable. Big guns were then fired in salvos at fixed angles of elevation and train at the end of a roll following the signal from a bell, the goal being a simultaneous discharge of several guns that would produce a closely bunched group of projectiles upon which the effectiveness of the salvo system depended. But because the judgment of the ending of a roll was likely to vary from gunner to gunner, the firing of several guns after the bell signal did not occur at exactly the same instant, with the result that the roll produced by even a moderate sea-way was enough to cause the projectiles of a salvo to scatter over a wide area, which greatly reduced the chances of achieving hits. The smoke from adjacent guns, or sea spray thrown up by a ship steaming at high speed, could momentarily obscure a gunner's view of the target through his sight and prevent him from firing at the end of the roll after the bell signal. This caused a reduction in the number of projectiles in a salvo and also decreased the probability of scoring hits.

For these reasons Percy Scott, in August 1905, proposed that the elevation, train and moment of firing of a ship's heavy-calibre guns be determined by a device called a 'director', which was to be located high on a mast and operated by a single observer. The director was to consist of a master sight whose setting for the target's range and bearing would be translated into appropriate instructions for the elevation and train of each big gun by an instrument that was called a 'converger'. It was also to include some means of firing the guns simultaneously at the end of a roll by either an electrical signal triggered by the director observer, or an order of the director observer that would be transmitted to the various gun stations and immediately carried out. The employment of such a system of 'director firing', Scott maintained, would place the sighting process above any interference from sea spray or gunsmoke and also ensure that firing was simultaneous, or at least very nearly so. In 1908, while Scott was still developing his director firing system, power-driven elevating and training mechanisms were improved to the point that heavy-calibre naval artillery could be laid accurately by continuous aim whenever rolling and yawing were minimised by very moderate seas. But following the success of Scott's perfected director in rough weather trials in 1912, continuous aim gun-laying with heavy-calibre artillery was superseded by director-controlled firing on the roll, although this process was by no means complete at the outbreak of war in 1914.

The development of continuous aim and director firing addressed the problem of laying guns accurately in a sea-way. But there was also the question of setting the gun sights for the correct elevation so that hits could be made. The precision with which this had to be accomplished depended upon the length of what was known as the 'danger space'. The danger space was defined as the difference between the minimum and maximum ranges used to set the gun sights for elevation that would enable a hit to be made. At short ranges, where projectile trajectories were practically flat and the area between an enemy ship's water-line and superstructure presented a relatively large vertical target, the danger space was great because a gun aimed at the enemy water-line could be given too much elevation on the basis of an overestimate of the target range of several hundred yards, but could still score a hit on the superstructure. At long ranges, where projectile trajectories curved sharply and the target was thus the horizontal deck rather than the vertical sides of the hull and superstructure, the danger space could not be much greater than the 30-yard width of the deck of a battleship.[1]

At short ranges, the rough estimates of the target range that could be made by eye were good enough for hits to be made because the danger space was long. At long ranges, however, a mechanical means of determining the range for setting the elevation of the guns with precision was vital for making hits because the danger space was short. The need for an instrument capable of measuring the distance between the firing ship and its target was perceived as early as 1891, when the War Office, on behalf of the Admiralty, issued an advertisement for a range finder that was suitable for naval use. Four instruments were submitted for consideration, and after sea-going trials in 1892 the range finder manufactured by the Glasgow firm of Barr and Stroud was selected for the service, although it does not appear to have reached the fleet in quantity until around 1900.

The oblong Barr and Stroud range finder was pivot-mounted on a pedestal, which allowed it to be trained on the target so that the line of sight from one end formed a right angle with respect to its longitudinal axis. The oblique angle formed by the line of sight to the target at the opposing end of the instrument was then measured by an arrangement of mirrors, lenses and a moving prism, which gave the range to the target by means of a trigonometric calculation.[2] A range finder of this type was operated by a single observer and was said to be self-contained because the train of the entire instrument could be altered to set up the right triangular range-finding problem.[3]

The optical and mechanical system that enabled the single observer to establish the two separate lines of sight simultaneously worked on the principle of coincidence.[4] With such a system, the relative alignment of the mirrors that reflected the target images from the ends of the range finder had to be maintained exactly in order to avoid serious instrumental errors. This required a rigid framework that was capable of withstanding both the shock of heavy artillery firing in close proximity and the rigors of service at sea. The distance between the ends of the instrument from which the lines of sight were established comprised the base of the range-finding triangle, and hence was called the base. The base of the Barr and Stroud range finder was 4½ feet long, which allowed accurate readings at up to 4,000 yards. Accurate observation to greater distances could only be achieved by extending the base length.[5] It was, however, impracticable to build a framework that could support a coincidence range finder whose base was longer than 4½ feet with the rigidity that was required to maintain the near-perfect alignment of the mirror reflectors.

By 1904, the experiments with continuous aim and salvo firing had indicated that engagements might begin at ranges that were considerably in excess of 4,000 yards, and the Admiralty thus issued an advertisement for a range finder of much greater range capability than the existing 4½-foot base instrument. By 1906, two instrument firms, Barr and Stroud of Glasgow, and Thomas Cooke and Sons of York, had developed range finders that satisfied the Admiralty requirement. The instruments of both companies replaced the mirrors, which had been used beforehand as reflectors, with pentagonal prisms, whose two surfaces of reflection cancelled minor shifts of alignment in the coincidence plane.[6] The use of pentagonal prisms meant that it was no longer necessary to keep the range finder perfectly rigid along its longitudinal axis, which allowed the extension of the range finder base beyond the previous 4½-foot limit, thus enabling accurate measurements to be made at much greater ranges. By 1914, coincidence range finders with a base of 15 feet were coming into service with the Royal Navy and were capable of performing accurately to ranges that were as high as 15,000 yards.

The development of range finders that were accurate at long ranges did not, however, provide a complete solution to the problem of setting gun sights accurately for elevation. When the courses and speeds of the firing ship and target were such that the range was changing rapidly, ranges measured by the range finder would be obsolete by the time they had been taken and transmitted to the guns. In an actual battle, moreover, the observation of the target was likely to be interrupted by mist, gunsmoke or shell splashes, which would prevent ranges from being taken with the range finder. And finally, account had to be taken of the fact that at long ranges the duration of flight of the projectile might be as great as half a minute or longer, during which time both the target range and bearing could change by a critical amount. Setting sights for elevation and deflection, therefore, required a means of rapidly generating the range and bearing independently of continuous observation of the target, and correcting those ranges and bearings for time of flight of the projectile and other ballistical factors.

In 1904, the Admiralty had begun experiments with a mechanical device − developed by Vickers, the large armaments firm − that generated ranges independently of observations of the target. The Vickers range indicator, or 'clock', consisted of a clockwork motor that drove a pointer around a circular dial that was marked with a sequence of ranges. After the pointer had been set to a starting range, and the motor set to run at a speed that corresponded to a change of range rate, the clock then indicated ranges in terms of that change of range rate. The initial setting of the pointer on the dial was determined from a single observation of the range with the range finder. The change of range rate was determined in either one of two ways. It could be calculated on paper by simply dividing the difference between two ranges obtained from the range finder by the time between the observations, which could be measured with a stop-watch. Alternatively, the change of range rate and deflection could be calculated by a trigonometric slide calculator that had been invented in 1902 by Lieutenant John Dumaresq of the Royal Navy, and which was known as a dumaresq after its inventor. The dumaresq, when set with the firing ship, target courses, speeds and the target bearing, indicated both the change of range rate and the deflection.[7]

The effectiveness of the dumaresq was seriously compromised, however, by the absence of any means of measuring the target course and speed, the settings of which, as a consequence, had to be guessed. In addition, both the stop-watch-Vickers Clock and dumaresq-Vickers Clock combinations were incapable of accounting adequately for changes in the change of range rate. Given firing ship and target courses and speeds that were constant, the range could not only remain unchanged or change at a constant rate, but could also change at a rate that was itself changing at a constant rate.[8] The change of range rate computed by the stop-watch method could thus result in a highly inaccurate setting of the Vickers Clock, because the change of range rate could alter considerably during the time interval between the two range observations that were used to compute a single change of range rate. A dumaresq, on the other hand, could be made to indicate the continuous variation in the change of range rate through the continuous adjustment of its setting for target bearing, because the target bearing varied in proportion to changes in the change of range rate. The continuous variation in the change of range rate indicated by the dumaresq could not, however, be represented by the Vickers Clock, whose motor could not be made to vary in speed continuously. The result was that as the speed of the Vickers Clock was altered discontinuously, in the attempt to repeat the continuous change in the alteration of range rate indicated by the dumaresq, the range indicated by the clock would become increasingly inaccurate.

Sight setting with the assistance of a range finder, dumaresq or stopwatch, and Vickers Clock was employed in conjunction with salvo firing, a process in which guns were aimed and fired in groups rather than independently. Sight setting thus became part of a centralised system of gunnery that was known as 'fire control', in which aiming instructions were calculated at a single location and then communicated to the separate gun stations by voice pipe or electrical signal. By 1906, sight setting had been refined by the development of slide rules on which gunners could calculate corrections to calculated target ranges and observed target bearings for air density, wind direction, variations in muzzle velocity caused by the forward movement of the firing ship, and travel of the target during the time of flight of the projectile. The value of ballistical corrections was limited, however, by both the inherent inaccuracy of the basic range data produced by the Vickers Clock because of the variation in the change of range rate problem, and the fact that manual calculation of the corrections was too slow a method when the range was changing rapidly. But in the autumn of 1906 the Admiralty decided to sponsor the development of a system of calculating sight setting data that would overcome these problems, which had been formulated by Arthur Hungerford Pollen.[9]

I
THE POLLEN SYSTEM OF TELEMETRY
(FEBRUARY 1901)

While visiting an uncle in Malta in February 1900, Arthur Hungerford Pollen by chance encountered a distant cousin, Lieutenant William Goodenough, who invited him to witness a sea-going gunnery exercise. Pollen accepted the offer and observed that firing took place at a range of about 1,400 yards. That very morning, he had read in *The Times* that naval artillery of the same calibre had been effective at 8,800 yards when used on land by British forces who were fighting the Boers in South Africa, which prompted him to ask his naval officer hosts why practice was restricted to ranges that were evidently far less than those to which naval guns were capable of firing accurately. Pollen was then informed that the range limitation was imposed largely by the lack of an efficient range finder. Upon his return to England, Pollen had his engineers at the Linotype Company analyse the change of range problem, and formulated a two-observer system of range-finding.[1]

Pollen first approached the Admiralty with his proposals for a system of range finding at sea for gunnery purposes at the beginning of 1901. On 26 January 1901, he wrote to Lord Walter Kerr, the First Sea Lord and a family friend, describing his two-observer range finding system and offering it to the Admiralty to develop.[2] Kerr referred Pollen to the Earl of Selborne, the First Lord, to whom Pollen restated his case in a letter written on 4 February 1901 after he had been refused an interview.[3] On Selborne's advice, Pollen submitted his proposal formally to the Admiralty via the Admiralty Permanent Secretary[4] on 25 February 1901, in the form of a prospectus, which was entitled 'The Pollen System of Telemetry'.[5]

On 7 February 1901, the Admiralty Permanent Secretary informed Pollen that his offer had been declined.[6] Pollen's prospectus did, however, impress Rear-Admiral Lord Charles Beresford, then Second-in-Command of the Mediterranean Fleet, who wrote on 17 March 1901, 'The advantages claimed for your system, provided they can be realised, are undoubtedly of the greatest value. . . . I think you have made out a very good case,' Beresford added, 'for being granted a trial of your system.'[7] Although the Pollen papers contain no other responses to the prospectus, it was undoubtedly circulated among Pollen's naval officer acquaintances in the Mediterranean Fleet.

(1) The following is an account of the new system of telemetry at sea submitted for the consideration of the Admiralty. It is claimed that this system will confer upon the users of it two overwhelming advantages.

(2) First, a captain using this system would know the exact speed and direction of an enemy as soon as his top-masts appeared on the horizon, and could shape his course accordingly. There is no need to point out that this would mean a revolution in nautical strategy.

(3) Secondly, all guns larger than the 4.7 being virtually as accurate and effective at 10,000 yards as at 5,000 – the present limit of practicable range – this system would enable fire to be opened on the enemy at distances at which he could not possibly reply, and would enable a ship equipped with it to keep out of the enemy's range until he was destroyed or put out of action – a revolution in naval battle tactics.

(4) The system consists in this. Two observers placed at a suitable distance apart each read of the angle of the object whose distance is to be ascertained. The results of their observations are transmitted to a third operator who works a calculating machine (hereinafter called 'the machine'.) The operator sets the machine according to the results communicated to him, whereupon a pointer travelling along an evenly divided scale stops where it will record virtually the exact number of yards making up the distance of the object.[8]

(5) The whole operation of ascertaining the distance of an object for the first time should not occupy two minutes. The variations of distance due both to the object and observing stations moving could be checked and altered without difficulty every five or ten seconds. Thus the distance as given by this system would be the distance that separated the observed object and the observing station at the time of the observation. To make the results useful either for navigation or calculating ranges, the course of the enemy and of the users would, of course, be plotted on a chart.[9]

(6) The system suggested is virtually a method of rapid, continuous, and accurate surveying, rendered possible by a machine that solves triangles automatically and instantaneously.

(7) The machine will measure about 30 inches by 15 by 20. It is constructed throughout of steel, and encased in solid brass. No vibration or shock due to screws or gun-fire will injure it. If exposed to the weather it will be quite unaffected.

(8) The machine can be located at whatever point in the ship is considered the most desirable centre from which to announce the range to guns.

(9) The machine has no moving gears or parts, nor plumb lines, and consequently can be placed in any position or on edge, whether screwed to the overhead deck, underfoot deck, or on a bulkhead or side of conning tower or chart-house.

(10) The machine will be accurate in its solutions to within one-tenth of one per cent – viz., suppose the angles at either end of a given base, when worked out mathematically, result in one side of the triangle being 20,000 yards long, the solution given by the machine would be within 20 yards of this result.

(11) The value of the results to be obtained from this machine, and hence

the practical utility of the system submitted, depend on the four following problems being successfully solved:

(a) First, the performance of the machine must be as claimed.
(b) Secondly, as the results given by the machine depend upon the observations transmitted to it, these must be made with sufficient accuracy to render the machine results practically accurate.
(c) Thirdly, the observing points being on a moving body the observations must be made synchronously.
(d) Fourthly, there being a short interval between the observation being taken, and the result being obtained from the machine, the actual moment of the observations must be recorded.

(12) As to the first (a), our experience in the manufacture of similar philosophic instruments enables us to guarantee an accuracy in the machine of one-tenth of one per cent in its solutions.

(13) (b) We have designed devices whereby we reasonably believe it will be possible to make and read observations accurate at least to five seconds (5″) with comparative ease. It will be remembered that the results on the Barr and Stroud machine[10] depend on an accuracy of one second (1″) in the observations.

(14) If the inaccuracy in the observations is even as great as eight seconds (8″) in the case of *each* observer (making a total aggregate error of sixteen seconds [16″]) the effect on the accuracy of the results given by the machine when the base between the observers is 150 feet, will be as follows:

At 20,000 yards the result will be accurate within 621 yards.
At 10,000 yards the result will be accurate within 155 yards.
At 5,000 yards the result will be accurate within 39 yards.

(15) When the available base is longer – as on ships of the new first-class cruiser type,[11] on which a 300ft. base could almost be obtained – the margin of inaccuracy would be proportionately less. Where a shorter base only is available, as on battleships and for straight-ahead and straight-astern observations both on cruisers and battleships, when bases of from 130ft. to 80ft. only will be obtainable, the margin for inaccuracy will be greater. In the worst of these cases – i.e., with a base of 80ft. only, at 10,000 yards – the probable limit under any circumstances for opening fire on an enemy – the range, always supposing the inaccuracy in the observations to be as great as eight seconds (8″) each, could be ascertained within 300 yards. This margin of error would be

gradually reduced as the range diminished. At 5,000 yards, it would be 70 yards only.

(16) (c) Simultaneity of observation can be simply and easily secured by a system of communication between the observers, whom we will call A and B. A having got 'on', would signal electrically to B that his observation was ready for record. B, as soon as he had got his observation, would know whether A's observation was made or not, and would signal in like manner to A. The moment of B's signal would be the moment of joint observation.

(17) (d) On B making his signal, a printing clock connected electrically with B's signal would record the time of the observation − e.g., 3 hr. 45 min. 22 secs.

(18) Each observer would then transmit his observation to the operator of the machine.

(19) We have devised a completely novel plan for reading these observations and transmitting them, so that what at first seems a complicated organisation would in practice prove exceedingly simple and expeditious.

(20) The machine has been designed to deal normally with distances of a minimum of 5,000 yards and a maximum of 20,000 yards.

(21) The minimum has been selected in the belief that at ranges less than 5,000 yards a ship would be under effective fire from the enemy, and under these circumstances range-finding from exposed parts of the ship would be out of the question.

(22) The maximum of 20,000 yards (or 11½ miles) has been selected because it is the practical limit of the field of sight at sea. At that distance an enemy's fighting tops would just be visible from the bridge of a battleship.

(23) The machine, however, can be instantaneously adjusted to deal with two alternative scales: (1) to deal with distances from 11½ miles to 45 miles for the purpose of approximating distant points of land; and (2) to cover distances from 1,250 to 5,000 yards, if very minute readings for short distances are considered desirable. It is assumed, however, that for these distances the Barr and Stroud machine, now in use, will be considered adequately effective.

(24) It will be observed that if the claims for this system of telemetry are substantiated, it will be easily possible to keep exactly and instantaneously informed of all the movements of any enemy in sight. His variations of speed and course would be continuously apparent, and the necessary tactics to circumvent him obvious. It is claimed that even in times of peace such an addition to the art of navigation would result in large coal savings, while in war it would mean control of the sea in a new and startling sense.

(25) As a means of range-finding, this system, by its extraordinary scope, expedition, and accuracy would actually double the effective range of every gun from 6" to 12" calibre in the fleet, and enable a cruiser's or battleship's captain to use his guns at the long ranges which position finders and other devices make practicable for similar guns in forts.

(26) At present the daily increasing efficiency of guns in accuracy and velocity is absolutely neutralised as far as any increase of practicable range is concerned by the non-existence of any telemetric system, whereby distances greater than 3,000 yards can be ascertained with approximate accuracy on shipboard.

(27) All firing beyond that distance now is therefore entirely a matter of experiment, and obviously the results obtainable are limited by the accuracy by which those results can be optically tested. This method is both ruinously costly, from the waste of ammunition involved, and of very limited utility because of the large element of guesswork it entails. It is probably no exaggeration to say that an expenditure of one hundred pounds worth of ammunition today would not enable the most experienced artillerists at sea to guess a range of 6,000 yards to a nearer accuracy than 500 yards.

(28) The two elements that go to make artillery practice effective being:

   (a) Accuracy of aim;
   (b) Correct elevation;

all navies are to-day in this position, that if it is desired to gain a superiority in skill over a rival it is necessary to spend more ammunition in solving the second problem each time the solution is necessary, than can be spent, after the solution is partly obtained, in acquiring proficiency in the first.

(29) It must be borne in mind, moreover, that this heavy expenditure only avails for ascertaining very inaccurately a given range at a given moment. The rapidity with which, under modern conditions, the relative distances of ships change, makes the entire experiment useless almost as soon as it is made.

(30) It is claimed, therefore, that in times of peace, and for practice purposes, the adoption of this system, by eliminating entirely the necessity for expending ammunition for ascertaining ranges would result in very great economies.

(31) Apart altogether from considerations of financial economics, the amount of ammunition that can be carried on a ship is so exceedingly limited that every shot wasted in feeling for a range, is so much deducted from the whole fighting efficiency of the vessel. In war, therefore, it would mean that a fleet equipped with this system could not only always keep out of the enemy's range, and, while in

complete immunity itself, destroy him at leisure, but so husband its ammunition as to render constant resort to arsenals less necessary.

(32) In conclusion, practice would doubtless show that some of the advantages of this system, though theoretically demonstrable, might not be obtainable in their entirety. On the other hand, experience in the working of the system might equally possibly result in even greater advantages being realised. In any case, it is claimed that a reasonable case has been made out for a genuine enquiry and careful experiment into a system whose possibilities are in their nature so important.

A. H. Pollen,
Director of the Linotype Company Limited.

II

MEMORANDUM ON A PROPOSED SYSTEM FOR FINDING RANGES AT SEA AND ASCERTAINING THE SPEED AND COURSE OF ANY VESSEL IN SIGHT

(JULY 1904)

In spite of the Admiralty's rejection of his proposals in 1901, Pollen continued to develop his system. On 20 March 1902, Pollen, along with William Henry Lock, the Manager and Secretary of the Linotype Company, and Mark Barr, a self-employed engineer, applied for a patent for a mechanism that computed a range from two bearing observations.[1] On 19 May 1904, Pollen and Lock applied for a patent on a device that enabled two bearing observations to be taken simultaneously, and that then transmitted the bearings to the computing mechanism.[2] Encouraged, in addition, by the growing enthusiasm within the Royal Navy for improvements in gunnery that had been stimulated by the successes of Captain Percy Scott, and fortified by the counsel of Lord Kelvin, the renowned practical physicist who was also a director of the Linotype Company, Pollen wrote to the Admiralty in May 1904 to propose again that his system be developed for trials.[3] Although his proposal was rejected, the influence of his father-in-law, Sir Joseph Lawrence, a prominent Conservative MP, led to Pollen being granted an interview with Lord Walter Kerr, the First Naval Lord, and Rear-Admiral William May, the Controller, on 9 June 1904.[4] Pollen was informed at this time that no further action could be taken until the Admiralty had been provided with a written description of a complete system of obtaining the data required to set the gun sights.[5] He thus consulted a number of senior naval officer gunnery experts, and at Torbay met a committee of gunnery lieutenants chosen from the ships of the Channel Fleet who had been unofficially authorised to advise him by Vice-Admiral Lord Charles Beresford, who was then Commander-in-Chief of the Channel Fleet.[6] On the basis of these consultations, Pollen wrote his 'Memorandum on a Proposed System for Finding Ranges at Sea and Ascertaining the Speed and Course of Any Vessel in Sight' in July 1904.[7]

In addition to the two-observer range-finding system that he had proposed in his prospectus of 1901, Pollen, in his 'Memorandum', described a system of plotting simultaneously observed ranges and bearings on a chart from which either future ranges of the target or the target's speed and course could be measured. On 2 November 1904, Pollen wrote to the Admiralty to ask that his system be developed for trial.[8] Later that month, he appears to have sent copies of his paper to the Earl of Selborne, the First Lord, and to Sir John Fisher, the new First Sea Lord, along with appeals for their support for his proposals.[9] On 7 December 1904, the Admiralty appointed a special committee to consider the question of developing Pollen's system for trial.[10] The committee, which met on 16 December 1904, was dissatisfied, however, with the design of Pollen's bearing observation instrument and did not recommend a trial of the system.[11]

(1) WHAT RANGE-FINDING IS

All range-finding is the result of some form of triangulation, which consists in the ascertaining of two angles of a triangle, one side being known. The amount of time occupied in what is called computing – the method of solving triangles used in land surveying – has led to all range-finding instruments being designed on the principle of using a right-angled triangle which admits of a simple method of reading, on a tangential scale, the length of the side that is desired should be known. In other words, the mechanical adjustment of an instrument reading the angle less than a right angle can be made to actually measure the length of the subtended side. The disadvantage of this system on shipboard is that it restricts instruments of this character to an extremely short base.[12]

(2) THE POLLEN SYSTEM GENERALLY

The Pollen system of surveying and range-finding is, like all the rest, based upon arriving at the elements of a triangle and solving it. Instead, however, of being limited to right-angled triangles, the system includes the solving of triangles of all kinds – scalene, right-angled, or isosceles indifferently.

It comprises the use of two specially-designed observing instruments, situated at the two ends of a measured base. These are for the purpose of measuring the base angles, or rather, one interior base angle and one exterior base angle; electrical means of communication between the observers so that only simultaneous observations are used in the system; mechanical means of recording the two angles and of subtracting the lesser from the greater (thus arriving at the apex angle); mechanical means of computing the triangle obtained (thus giving the distance of the object observed); mechanical means of registering the least second of time at which the observations were taken; mechanical means of recording on a chart the exact position of the object observed and located by the above methods; and this recording, when made in series, affords material for deducing the mean position of the observed object if stationary, or its mean course, if a moving object, from which its speed, direction, and probable distance can be arrived at; and, finally, means of transmitting these deductions to desired stations.

In short, the system consists in the application to a moving base or a moving object of the ordinary principles of surveying, where practically all the operations, except reading of the angles and compass, the charting, and deducing mean range, are automatic and instantaneous.

(3) THE OBSERVATIONS

The diaphragm of each observing telescope has a portion of its field marked off, and it is coincidence in this portion of the field with the object which constitutes an observation.[13] It is so designed that it is a comparatively simple matter to make an observation fine enough to come within the limits laid down by the instrument; but it is also so designed that, in proportion as an observer becomes more skilful, he can define his reading more closely until, under very favourable conditions, it should be possible to get to within an accuracy of 8 seconds in each observation.

It might be noted here that, whereas in depression range-finders of all kinds, the point or line on to which an observation has to be made is the water-line – always an uncertain region; in the system under examination the part to be observed will be the mast, which is a clear and definite vertical line. Other things being equal, therefore, greater possibility of accuracy will be obtainable by reading on to a ship's mast than by reading on to an uncertain water-line.

(4) THE BASE

The base, in the case of surveying from ships, will be the longest base that can be got, consistent with keeping the instruments clear of own gun fire – e.g., from the extremities of the bridges, the port ends of the fore and aft bridges for the port side, the starboard ends of the same bridges for the starboard side, the forward bridge from end to end being the base for straight ahead, and the aft bridge from end to end for straight astern.

(5) SIMULTANEITY OF OBSERVATION AND TRANSMITTING THE RESULTS

The ship, however, is a continually moving base, not only in the direction of the length of the base, but in the direction at right angles to this, owing to the yaw, and so, of the observations taken by the observers, only those which are absolutely simultaneous can be of utility for the purpose of surveying. [Clearly, if they are not simultaneous, both the length of the base and the size of the angles will be distorted.]

The observing instruments are so connected together that it is impossible that either observer can transmit an angle unless in the same second of time the observer using the other instrument has also made an observation; consequently while each observer at each instrument is continually making observations, it is only when he is successful at the same time that his fellow is successful that an observation will be sent.

There is a communicating electric circuit between the instruments, and keys in both instruments must be pressed before the circuit is closed. On these keys being depressed simultaneously the angle of arc in each instrument is automatically and instantaneously transmitted to some convenient centre in the ship, where an automatic machine records both angles and subtracts the lesser from the greater.

(6) TIME AND COMPASS RECORD

The completion of the circuit which causes the transmission from the two observing instruments also operates a printing clock, which records on a strip of paper the exact second of time at which the observation was taken. The same circuit actuates a signal, and an observer, watching the ship's compass, makes a record of the ship's course.

(7) THE COMPUTING MACHINE AND RANGE

A computing machine is now set to one of the angles transmitted, and to the difference between the two, and automatically gives, on an evenly divided scale, the exact length of one side of the triangle; in other words, gives the range between the observing ship and the observed ship at the moment when the observation was taken.

The system might stop here if it was not desired to go further than any existing range-finding system goes. It is thought, however, that, with the data now obtained, far more valuable results than the mere knowledge of the range are within reach of the system, which is therefore carried further in the following way:—

(8) CHARTING OR PLOTTING ENEMY'S COURSE

A specially constructed charting table has an arm which swings over the table from a pivot on one side, which is set to correspond exactly with one of the angles observed. On this arm, which is divided up to correspond with the range, there is a small carrier which, after the arm is set, is moved along until its position corresponds with the range given by the computing instrument. On this carrier is a pin controlled by a spring, which is now depressed to perforate a strip of paper, and against this perforation is recorded the time at which the observation was taken. The whole table on which the paper is stretched has previously been adjusted to correspond exactly with the course of the ship, and the paper, on which the mark is made, is stretched between two rollers, actuated by clockwork, which are continually moving the paper over the table at a speed corresponding to the speed at which the observing ship is going. Thus: the paper is divided up into squares of one inch each, and each of these squares represents a knot, and the ship is going 18 knots an hour, the rollers would be set to run at a speed of running 18 of these squares over the table per hour.

(9) TIME TAKEN IN ALL OPERATIONS

The whole operation of taking the observations, transmitting them, subtracting them, getting the record of the time and the ship's compass, solving the range and recording it on this charting table, should not at the most take more than 3 or 4 seconds.

(10) DEDUCING THE MEAN RANGE AND FORECASTING ACTUAL RANGE

If this operation is to be repeated several times, say, for half-an-hour, and observations made once a minute, there would be thirty records on the strip of paper.

In taking the observations errors varying from 8 seconds to 32 seconds will be made. Now an error of 32 seconds at 10,000 yards will introduce an error of 400 yards into the range, and an error of 8 seconds produces an error of 100 yards. The instruments are so designed that 32 seconds is the maximum error and 8 seconds the finest obtainable, and consequently the records on the chart will vary between extreme points 400 yards apart, and with a sprinkling of observations between them. On a series of this kind the median line will show the exact course of the enemy, and consequently those in control of the charting will be in a position to forecast what the range will be three or four seconds ahead with extraordinary closeness.*

(11) SENDING RESULTS TO CONNING TOWERS AND TURRETS

To make this forecast of value, means must now be found of transmitting this forecast to every portion of the ship where this knowledge will be useful. It is assumed that the two conning towers, the 12 inch gun turrets, and the auxiliary gun positions will be the principal centres to which this information should be sent, and for it to be valuable it must be sent instantaneously, accurately, and in the fullest detail. Consequently, a further mechanism has been devised to be controlled from this charting centre. The mechanism consists of a device of a clock face, upon the middle of which a hand, like the hand of a clock, can rotate. On the top of the clock there are two small discs, on which the letters A and R can be respectively shown. In the event of the enemy's ship being on a course

---

*NOTE. – It will be seen that in this way the range used will always be more accurate than the range calculated. Thus the accuracy of the system is always greater than the accuracy of the observations.

towards the observing ship the letter A will be shown indicating that the ship under observation is approaching. If the ship is retreating, R will be shown. The clock hand will then be moved in a position showing exactly what the course of the ship is. Above the centre of the clock there are two sets of dials, one of which will indicate the speed of the enemy, the other the speed of the observing ship, and below the centre of the clock another set of dials, which will show the range or distance that the ship is off.

(12) A CONTINUOUS SURVEY

It will thus be seen that the whole object and purpose of this system is obtained − i.e., to maintain the most accurate possible survey of any object moving within sight of the ship. This survey would naturally begin long before there was any question of the object in sight being brought under fire. The instruments, as at present designed, contemplate the beginning of the survey at a range of about 17,000 yards.

(13) STRATEGICAL ADVANTAGES

The advantages to be gained, apart from the general advantage of arriving at the exact distance of a ship within gun range, are, first, that it would enable the Captain of a ship to ascertain, with a very close approximation to accuracy, the speed and course of any ship in sight. Thus, imagine that in the evening, when there is but an hour of daylight left, an isolated ship or a fleet is observed hull down on the horizon. By the use of this system, before darkness had made further observation impossible, the speed and direction and course of that ship or fleet would be known. This is a piece of information which should be of strategical value.

Thus, a cruiser sent out on scouting might get this piece of information in the evening, and it might be to the effect that a four-funnelled cruiser of the enemy was going at a speed of 18 knots on a course between, say, Toulon and Bizerta.

The cruiser taking the observation might be both incapable of catching up the enemy or of engaging it if it was caught up, but by means of wireless telegraphy could communicate with some faster and more powerful cruiser, which, by instantly adopting the proper angle of intersection might arrive between the enemy's cruiser and the line of her advance in time to prevent her juncture with the enemy's fleet.

(14) TACTICAL ADVANTAGES

The next obviously valuable application of the system is in the direction of enabling an Admiral to engage an enemy at a known range of his own selection, and by keeping the enemy under observation, and changing his own course according as the enemy changed his, to maintain throughout an action, a fixed range. The effective range of a gun is such as being accurately known is within the ballistic capacity of the gun. Thus 10,000 yards, if accurately known, is an effective range for a 35 calibre 12in. gun, whereas 8,000 yards is not an effective range for a 45 calibre 12in. gun if the range is not accurately known. Consequently an Admiral commanding a fleet, practised in the art of this surveying system, would always be able to engage an enemy at a range at which that enemy could not possibly make an effective reply; in other words, could make an enemy keep his distance, the Admiral's – not the enemy's.

The following notes are added to deal with points that have been raised by practical officers to whom the system has been submitted.

Space Taken by the Installation

(a) The space taken by the observing instruments would be four spaces equal to 7ft. square, and so would monopolise the four ends of the bridges if the instruments were placed there. The space taken by the recording instruments, computer, and charting mechanisms would be a space of from 8ft. to 10ft. square, which would be situated below the water-line.

The Observers Can Be Protected

(b) If thought desirable, the observing instruments could be encased in light armour at the end of the bridges, so as to give them a protection equal to such protection as those in charge of fire control would have in the fighting tops. If additional weight were not objected to, armoured hoods might be put over the corner casemates on the upper deck, which would protect the instruments absolutely. An aperture 4in. wide would be sufficient to command the entire arc.

Number of Operators

(c) One observer and one assistant would be required for each observing instrument, and two operators would be required to work the computing and charting and transmitting of results to the casemates and turrets.

Great Skill Not Necessary

(d) The skill necessary on the part of the observers would not have to be of a very high order, considering the flexibility and easy manipulation of the telescope, and the large size of the field; the standard would certainly not be comparable to that required in handling the telescopic sights of guns.

Accuracy in Results

(e) Experience only can show exactly how accurate the results to be got can be. It can be said, however, with confidence that, with this length of base, and a continuous chart being kept of the movements of both ships, a high degree of accuracy should be obtained, probably to within 20 to 50 yards, at 10,000 yards.

Difficulties from Haze, Smoke, Seaway, Weather, Blast, etc.

(f) With regard to cordite haze or smoke, difficulties arising from seaway, varying physical conditions, etc., the observers at the ends of the bridges, or in the specially constructed hoods over the casemates, would be subject to exactly the same, but no greater, disabilities than the gunners; in other words, when it was impossible to take observations, it would be impossible to shoot. With regard to blast and concussion, the mechanism is not more delicate than that of gun sights, if as delicate; consequently the observers, and their instruments, would be less affected by blast than the gunners.

Transmitting Ranges That Vary Rapidly
– Say, 20 Yards a Second

(g) Range-finding at sea differs essentially from range-finding on land. On land it consists in ascertaining the distance between two fixed points. At sea range-finding, in the sense of giving guns the distance at which they are to fire, is a forecast of what will be the distance between two moving ships at a future time. Hence an accurate knowledge of range at any one moment is no help to forecasting an accurate range at a future moment, unless the rate of change is known. To ascertain this rate of change it is necessary to know the course steered and the speed of both ships. This can be obtained by the proposed system, and apparently in no other way with certainty. Assume that the actual course and speed of an enemy is recorded in the graphic and easily understood form of a chart, a rapid rate of change instead of a slow rate would introduce no new difficulty in forecasting what the future range will be.

The charting operator would be in the position of a man looking at the map of a railway line, knowing where a train was at a particular moment, and its speed. To say where it would be at a future moment would be easy.

The range indicators in the turrets, casemates, etc., could be corrected every few seconds if necessary.

True Range V. Gun Range

(h) Some eminent authorities on gunnery go so far as to say that the actual range is not necessarily useful, as the carrying power of guns varies from day to day, and that therefore it is the gun range, and not the actual range, which is needed for effective shooting. The only method of getting gun range at present seems to be to set a gun to shoot at a certain range, and then to check by observation whether it falls short of, attains, or exceeds the range it is supposed to reach, and by this means to arrive either at a new formula for giving the ranges, or a new formula for setting the sights. This system of 'spotting' is also complicated at present by being the only means by which to check the rate of change between successive ranges.

It is submitted that if each gun could be calibrated – i.e., if the formula for variation for each day or hour, as the case may be, could be ascertained before commencing firing at a target, either the sights of the gun could be so set as to allow for this variation, or the range given to the gun could be falsified to allow for the variation.[14]

The proposed system described above affords a ready and immediate means of calibrating the sights at any desired range in the following manner. It is clear from the system that, two observations having been taken, and the elements of the triangle having been arrived at, the solving machine will give the range. This process could be reversed – viz., the solving machine having been set to a range, the angle at which the gun is to be fired could be stated and the observing instruments set to the proper angles, so that the shot, on striking the water, should fall exactly within the hairlines in the field of the telescope. If it failed to fall within these limits, an immediate adjustment of the observing instruments could be made, and the range actually made by the gun recorded, which would give the formula for the variation for the day. This might be repeated for each gun on the ship, and as often as it was necessary, and at extreme ranges, the splash of the shot being easily perceptible to both observers at distances exceeding 10,000 yards. The formula thus arrived at would afford a corrective for each gun as long as it held good.

Should this scheme prove as effective in practice as it is clearly correct *a priori*, a considerable step forward will be made in ascertaining from time to time the differences between gun range and actual range, and would afford a simple, inexpensive, and rapid method of calibrating sights.

III

FIRE CONTROL AND LONG-RANGE FIRING: AN ESSAY TO DEFINE CERTAIN PRINCIPIA OF GUNNERY, AND TO SUGGEST MEANS FOR THEIR APPLICATION
(DECEMBER 1904)

Not long after he completed the 'Memorandum' in July 1904, Pollen appears to have recognised that a more extensive analysis of the fire control problem than he had provided was required, and in the course of further study and reflection he found that it was necessary to modify his proposals for its solution. In December 1904, he was able to restate his case in a paper entitled 'Fire Control and Long-Range Firing: An Essay to Define Certain Principia of Gunnery, and to Suggest Means for Their Application'.[1] In this lengthy work Pollen produced a rigorous analysis of the fire control problem, after which he put forward the two-observer range-finding and plotting schemes that he had described in his previous paper. To this he now added 'a change of range machine' that would calculate the target range and bearing, and a mechanism that would compute and make corrections for various ballistic factors. In January 1905, Pollen sent twenty-five copies of his paper to various naval officers[2] and wrote to the Admiralty to propose that his expanded system be developed for trials.[3]

Pollen's proposal was considered in February 1905 by Captain John Jellicoe, who in that month replaced Captain Henry Barry as Director of Naval Ordnance. Although explicit evidence is lacking, Jellicoe undoubtedly received, or at least read, a copy of Pollen's paper. He was also advised by Captain Edward Harding, Royal Marine Artillery, who had been a member of the specially constituted Admiralty committee of December 1904, of both the practicability of Pollen's proposed system and of the soundness of Pollen's understanding of the fire control problem as a whole.[4] Harding's counsel was supported by the naval officer members of the committee, who reversed their earlier unfavourable judgment after they had inspected a prototype bearing observation unit in February 1905 that had been built by Pollen's Linotype Company engineers, who had taken account of the criticisms that had been made by the committee the previous December.[5] On 20 March 1905, the prototype bearing observation unit was successfully tried in the old armoured cruiser HMS *Narcissus* under the supervision of Harding, who wrote a highly favourable report on 3 April 1905.[6] On 3 May 1905, the Admiralty formally accepted Pollen's offer of 17 April 1905 to supply a complete set of instruments for trial.[7]

EXTRACTS BY WAY OF PREFACE

'Fire control . . . . constitutes a problem so intricate and extensive that it should be treated as a science in itself.' – U.S.N. Gunnery Text-Book.

'Accuracy . . (in) . . . range may be regarded as the most difficult problem in naval gunnery.' – Lieut.-Commander W. S. Sims, U.S.N.

'Our effective ships have cost us over £120,000,000. We spend between £30,000,000 and £40,000,000 yearly upon our navy. The hits we can make with our biggest guns at the longest ranges are what we get for the money. The value of any system that will give us better results may be incalculable. The monopoly of a successful system means naval supremacy.' – Letter from Captain ———, R.N.

'The British Navy cannot afford to lose the benefit of any invention that can conduce to its efficiency in any particular.' – Letter from Admiral ———.

PRELIMINARY

Some Definitions

The author uses the expression *Fire Control* to mean that part of the executive details of the system or organisation by which the captain directs the fighting powers of his ship, as are concerned in supplying the guns with data for making hits. The following expressions are used in the sense set out:

True Range or Actual Range[8] — The distance between gun and target.

Gun Range — The distance the gun shoots when elevated to shoot the true or actual range.

Error of the Day — The difference between true range and gun range.

Gun range and *error of the day* are held to be caused by the irregularity in the ballistic powers of the gun, due to the effect on the ammunition of change of temperature, etc.

Effective Range — Distance the gun is elevated to shoot when true range and gun range are known, and the error of the day allowed for.

Hitting Elevation — Elevation at which a gun is found to hit by trial shots, but arrived at without knowledge of true range or gun range.

For instance, the distance of a target is measured to be 10,000 yards.

This is true range. A shot is fired from a gun elevated to 10,000 yards, and the projectile is found to hit the water 9,800 yards off — 9,800 yards is the gun range, 200 yards the error of the day. The elevation is increased by 200 yards — 10,200 yards is the effective range.

Suppose a gun is fired at the same target, its distance not being known. After a series of trial shots the gun is found to hit at the elevation of 10,200 yards. It is not known whether 100, 200, 300, or 400 yards is the error of the day, or 10,100, 10,000, 9,900, or 9,800 yards is the true range. Hence all that can be said is that it is the hitting elevation.

Gun Error — Any irregularity in a gun that is not caused by temporary variations in the ammunition. If corrigible, such error must be measured for individual guns and allowed for; such errors would tend to be eliminated by careful calibration, and more accurate manufacture.

FIRE CONTROL

THE FUNDAMENTAL FACTS

Gun Fire without Fire Control is Impotent

1 As at anything except point blank ranges great guns cannot be used without a knowledge of range and deflection, beyond the capacity of anyone in a gun position to gather for himself, it is an accepted axiom that the gun layer must be a gun pointer only.

The striking force of all navies is their battleships, and the object of a battleship is to destroy other battleships. Shot and shell, torpedos, and the ram are its means of doing so. The torpedo is better employed by other craft. The ram is a last desperate resort. The guns are, therefore, the raison d'être of the battleship, and their hitting capacity a measure of the value of all naval expenditure. If, then, fire control is necessary to enable a gun to make hits, it is an integral part of the organisation of gunnery; those engaged in it are as truly the wielders of the gun as the gun layer, and their appliances and positions are as truly vital parts of the gun and of the ship that carries it as the sights through which the layer looks, the barbette that houses the gun, or the mountings that control its movement. Without an efficient system of fire control an efficient navy is impossible.

The true standard by which to compare ships and fleets is neither tonnage, nor speed, nor armour; neither number of guns, nor weight of guns, nor their velocity or foot ton energy, but purely their hitting efficiency. The ship that hits (1) furthest, (2) oftenest, and (3) hardest is the best. The first is the most valuable pre-eminence. It is not the weight of metal that can be shot into the sea per minute, but the weight that can be shot into the enemy *before he can hit back* that makes one ship better than another. Ranging superiority is entirely a question of scientific fire control. One 30-cal. 12-in. gun that can hit at 10,000 yards is worth twenty 46-cal. 12-in. guns that cannot.[9]

Gun Range as Important as True Range

2 To give the guns the exact range of the target in yards will not, even with perfect aiming, secure hits. Great guns do not shoot alike, or any of them successively alike in varying conditions. These irregularities may be broadly said to go into three groups:

(1) Errors due to imperfect manufacture, so that guns, mountings, and projectiles of the same type are not exact reproductions of the same model. These errors are constant.

(2) Differences in the chemical constituents in different batches of what is nominally the same explosive. These errors being due to absence of uniformity, are constant for each batch differing uniformly from the standard.

(3) Variations from hour to hour in the power of the explosive due to changes of temperature, etc.

The first group of these errors can be to a great extent eliminated by ascertaining what they are by experiment and correcting the sighting and training scales to counteract them, so that the same elevation should give approximately the same hitting pattern for all the guns of a broadside.

The second can be eliminated by greater care in the manufacture and in the classification of charges; but the third form of error can only be allowed for if there is some means of measuring the distance between the actual distance of the target and the fall of the shot, on the day when firing takes place – i.e., constant errors can be allowed for by calibration and care; varying errors only by ascertaining the extent of the variation under the conditions that cause it.

Ranges Change from Minute to Minute

3 Finally, a ship in action is a moving platform, and the enemy a moving target, and so, the distance between them is not only a constantly changing distance, but, for a majority of speeds and courses, one that changes at a different number of yards for each equal period of time. The amount by which the distance or range changes is determined by the speed of the two ships and the direction or course each is steering relatively to the other.

It must be remembered, too, that the projectile takes a considerable time to travel from the gun to the target, so that, for each shot, a future position of the target is the true alignment for making hits.

THE FIVE CANONS OF FIRE CONTROL

Must Indicate and Maintain Hitting Range

4 A system of fire control to be as sure and reliable in its results as the unavoidable errors in guns and aiming permit, must be based upon a

realisation of the foregoing fundamental facts, must provide the data they indicate as necessary, and with the speed and continuity required for maintaining effective fire.

The fire control system should, therefore:

(1) Supply the exact distance of the target.
(2) Measure the percentage error of the day or hour, both before effective fire is to begin, and,
(3) Indicate the change of range from minute to minute after effective fire has begun, and supply the data for deflection.

It is assumed that the guns will have been kept calibrated as a condition of general efficiency before war is in question; and, of course, an accurate knowledge of the distance of the target must be got, and the error of the day measured at a range exceeding that at which fire, expected to be effective, will be opened.[10] If it is supposed that aimed fire should make a serviceable percentage of hits at a maximum range of 7,000 yards, when distance and gun range are known, the true distance and error of the day must be, as a preliminary, ascertained at a greater range. Further, as the rate aimed fire may attain is almost one shot per second, it will be impossible to supply the altering range by direct observation for every shot.[11] The range once got, therefore, and the conditions of its variation known, the change of range must be automatically calculated and transmitted.

Immunity from Gunfire. Continuity of Action

5 It is not enough, however, that the system should give the data enumerated above. Remembering that these are essential to efficiency, and so must be supplied throughout the action, unless the ship is to become impotent:

(4) The fire control agents, their positions and appliances, must be just as thoroughly protected by armour as any other persons, positions, or appliances vital to obtaining the object for which battleships exist.
(5) The positions and instruments must be so placed that the operators shall not be unduly impeded by the blast of the guns, or more impeded by haze, smoke, etc., than the gun layers, navigators, and others who will have to carry on their duties in the difficult conditions of action.

MUST BE PRACTICABLE
Economical. Progressive. Centralised

6 Assuming that a fire control system fulfilled the five foregoing canons, there are certain conditions which it must further comply with before it can be a satisfactory addition to a ship's polity. A battleship is a

complicated organism, costing nearly £1,500,000 to build and equip, and about £100,000 a year to maintain in working order. It employs a personnel strictly limited in number. The system must, therefore:

(1) Make only such demands on cost and space, and necessitate only such additions to weight and means of communication, as are commensurate with the benefits obtained.
(2) It must only require such service as the personnel can supply, without neglecting equally important operations.
(3) It must not postulate an unreasonable standard of skill or aptitude in its operators. The human factor must be reduced to its simplest proportions. No higher standard of skill than is commonly attained by the gun layers should be necessary.
(4) It must permit of constant practice being made, and so perfect expertness being attained, without cost, so that as large a number of persons as possible may master the necessary operations, and thus the working of the system in war be rendered independent of casualties to a few exceptional individuals. Practice in fire control must above all not be limited to practice with guns, which is necessarily restricted, owing to the heavy cost of ammunition and the deterioration of the ship's engines of offence.
(5) As the manufacture of artillery is a progressive art, and the improvements of the future may be expected to result in extension of range, the system must contain the seeds of progress, and be capable, by increased skill in its use to give ranges at a distance and with a precision beyond what is necessary, or, indeed, useful now for artillery as at present made and used.
(6) Finally, the ultimate object being to centre the direction of fire in the captain, the system must link up with and be a part of the directing system controlled from the conning tower. The selection of the target and guns whose ranges are to be taken must rest with him.

If the foregoing canons and conditions are correctly stated, it is worth seeing how far the best system now in use conforms to them.

FIRE CONTROL AS PRACTISED TO-DAY

Unprotected Observer in Unarmoured Position

7 The present system of fire control is based on the observation of the fall of the shot by one or more skilled observers, placed in unarmoured positions at the greatest elevation that can be obtained above the level of the water. Their functions are:

(1) To judge how far short or long shots may be, and direct the elevation of the gun accordingly.
(2) To judge the speed and course of the enemy so that the rate at which the range is changing from minute to minute may be calculated.
(3) After general fire has been opened, to correct the range by observation of the effect of the fire of the ship as a whole.

Using an Unarmoured Inexact Range-Finding Instrument
8 To assist these operations, they have one-observer range finding instruments that afford an approximate indication of the distance of the target in yards. The base of this instrument is too short for it to be possible to get ranges with exact accuracy, except at comparatively small distances.[12] The instrument cannot be used for measuring the distance of the fall of the shot, and neither at the elevated position nor in any other part of the ship can it be at all protected by armour.

Ranging Gun Used till Hit Obtained
9 However, the range finder will give an indication of the distance more accurate than the observers can obtain without it. At the range so given, a series of shots from a ranging gun are fired. When they are over the target, the observers direct the elevation to be diminished, and, when they are short, increased. This procedure is continued till a hit is made.

Corrects Range and Judges Rate of Change by Watching Fire of Broadside
10 The guns of the entire broadside are now elevated to the hitting elevation of the ranging gun. The observers being no longer able to distinguish the fall of the individual shots must judge by the effect of the fire of the ship as a whole whether the range is correct for the majority of the guns or not. Meanwhile on both masts the range finding instruments have been kept in use to check the varying range, and from these readings and the observation of the effect of the broadside fire, an estimate is made of the speed and course of the enemy, from which estimate the rate of change in the range is automatically calculated. A figure having thus been got of so many yards per minute, an ingenious instrument automatically adds or subtracts this continuously from the range assumed to be correct, and, by means of another ingenious appliance, this range may be put upon the sights with great rapidity.[13]

ANALYSIS AND EXAMINATION OF THIS SYSTEM

Extreme Vulnerability. See Sec. 8

11 The first obvious criticism of the above system is its extreme vulnerability. Continuity in working order depends on the observer, the mast, the instruments and appliances that he uses and the means of communication necessary to his work, *not* coming under the enemy's fire. A lucky shot at 10,000 yards might destroy the entire organisation before a shot had been fired from the ship depending upon this organisation for its efficiency. It is certain it could not survive five minutes in action under equal conditions, and, as we have seen in Section 4, great guns without fire control are impotent. There is no need to emphasise the illogicality of armouring the ship and the guns, and leaving the brain and nerve on which fire efficiency depends exposed. When armour was first used this paradox did not exist, the gun captain being able to make all necessary allowances from the turret or port; to-day this is impossible at ranges over 3,000 yards.

The truth is that the extension of range by improvement in artillery and ammunition having been gradual, the inconsistency of unarmoured range finders was not glaring when it was thought they would only be used at extreme ranges. But by now the capacity of artillery has grown so great that fire control is necessary for even very moderate ranges. A generation has grown up, therefore, that takes exposure of the fire controllers for granted; but custom does not get rid of the fact that the system is obviously dangerous and wrong.

Apart altogether then from the question of whether this system gives good results in peace time, the fact that it could not be relied upon in war makes it most important to find a substitute for it. Its only possible defence is that it does give the best results now obtainable and that there is no alternative.

Inaccurate Data

12 An analysis of the procedure described shows us that two systems of telemetry are used. The first the short base range finder; the second the gun itself. The first is admitted to fail in accuracy by at least 3 per cent at 6,000 yards, and a succession of readings, each of which may be 180 yards out, might give us totally unreliable results from which to judge the course and speed of the enemy. The second, it is true, gives the hitting elevation which should be correct for the other guns if all were calibrated, but as it does not give the percentage of the range which is the error of the day, a shorter or longer range could not be deduced from it. Thus if, say, 8,500 yards were got as a hitting elevation, and soon afterwards the range finder gave a new target at 6,000 yards, unless the error of the day

FIRE CONTROL AND LONG-RANGE FIRING   31

is known the ranging gun would have to be set to work again to get the hitting elevation at 6,000 yards. Whereas if the 8,500 yards were known to include a plus percentage of 2 per cent, the gun could be at once elevated to 6,120 yards at the shorter range.

The crux of the matter, however, is change introduced by the movements of ship and target, and it is on the capacity of the system to supply this correctly that it must be judged.

Change of Range the Test

13   Remembering that change of range depends on the speed and course of both ships, if it is to be rightly calculated the speed and course of the enemy must be correctly known. The speed and course of own ship is, of course, available. But the slightest error in calculating the enemy's speed and course will result in the range being lost immediately. As we have seen, in the heat of action, with many guns firing from the ship itself, and with many of the enemy's shells bursting round it, it will be impossible to ascertain ranges. The range must be known and kept if effective fire is to continue. Consequently action must begin with speed, course, and distance of enemy certain.

To arrive at this we must have the exact position of the enemy at two points separated by an interval both of time and distance. The line from point to point will show his course, the time his speed. The ascertaining of these points is the crux of fire control. How far does the present system supply the necessary data; if it cannot supply them can a better system be devised? The answer involves the consideration of the merits of the telemetric methods used; a comparison of them with that which is suggested.

Short Base Instrument

14   The short base instrument has the inestimable advantage of greatly simplifying the problem of observing. The split image permits of an accuracy of reading unattainable by any other known method. The fact that one observer only is employed obviates the necessity of only synchronous readings being useful, and the instrument further lends itself to extraordinary accuracy in manufacture. But with all its merits there are obvious limits to what any observer can read to, and to the fineness of adjustment that can be maintained. Hence it is probably safe to say that with exceptionally skilful observers, capable of reading to 1 second, an almost incredible fineness, seeing that it implies being able to adjust the split images of a ship at 6,000 yards to an accuracy of about an inch of linear measurement, and an instrument with only $\frac{1}{10,000}$ in. error, a greater inaccuracy than 3 per cent could be obtained. In practice it would probably be found that 4 per cent would be nearer the usual error. However, 3 per cent would be a fatal error in getting the data for change

of range. At longer ranges the error increases as the square of the distance — i.e., it will be 6 per cent at 12,000 yards.

The Gun as a Telemetric Instrument

15   If we are thus driven to regard the gun as our measuring device, we must remember that, assuming there is no error in the gun at all, it can only be of service as far as we can read the results it gives. Reading implies being able to say how far off the shot has fallen. The readings will be made from two points — one on the mast, say 100 feet above the sea, the other, after the mast is shot away, from the conning tower. The problem is one of depression range finding.

If we suppose an observer at an elevation of 100 feet above the sea, with a glass with an horizontal line that marks a 7,000 yards range, in a perfectly smooth sea, he will have to raise the glass in the following angular intervals to bring it into coincidence with other imaginary lines at 8,000, 9,000, and 10,000 yards:

7,000 to 8,000	1 min. 48 sec.
8,000 to 9,000	1 min. 21 sec.
9,000 to 10,000	1 min. 1 sec.

Let us suppose that there are posts 10 feet high rising from the sea at these ranges, and the observer's horizontal line to be successively in coincidence with their bases, he will have to raise the glass the following angular measurements to bring the line into coincidence with their tops:

At 7,000 yards — from base to top — 1 min. 38 sec.
At 8,000 yards — from base to top — 1 min. 26 sec.
At 9,000 yards — from base to top — 1 min. 16 sec.

If we substitute for the posts waves running to a height of 10 feet above the true water level — perhaps a formidable seaway — we see that the imaginary range lines run into each other hopelessly. Thus:

Ranges	Angular distance apart of ranges	Angular Height of wave at lesser distance
7,000 to 8,000	1 48	1 38
8,000 to 9,000	1 21	1 26
9,000 to 10,000	1 1	1 16

Remembering that we must suppose wave depressions of corresponding angular measurements at each range, we might then have successive shots striking at 8,000, 9,000, and 10,000 yards, and to the observer all would seem to have struck in the same place.

With waves running to only half this height – i.e., 5 feet – there still might be no distinction to be made angularly between 8,000 and 9,000, or between 9,000 and 10,000 yards. As a matter of fact the great majority of shot impacts would, if the sea were moving at all, be invisible – viz., the rise of a wave between the impact and the observer would conceal it. Finally the impact, if seen, is not a definite line, and has no edge or distinguishable position.

It is clear then that no observer, even at the height of 100 feet – a height that can only be got on a battleship of quite new design – could distinguish and hence measure ranges with any certainty, even if he had and could use instruments of precision. With the naked eye the thing is manifestly impossible. The utmost he can do is to say when a shot is short. He cannot always distinguish a shot that is long from a hit. And yet it is upon his observations that we rely, not only for range, but for the materials for calculating error of the day and change of range.

From an elevation of 30 feet only, all spotting must be still less reliable. The diagrams illustrate the difficulties from both heights.

Change of Range Based on Guess-Work

16 It is clear then the gun is useless as a telemeter, and therefore can tell us practically nothing as to the speed and course of the enemy. The hitting elevation is the only information that can be got. The rest is guess-work – extraordinarily clever guess-work, no doubt, but still guess-work – tempered by the unreliable information of the range finder. When it is remembered that we want true range, gun range, and exact data for change of range, and that the present system only gives us the hitting elevation, which is lost as soon as the range changes, it is clear that we must try and find a better method. Further, the velocity of a projectile is continually diminishing as its flight is prolonged. To get the correct deflection to allow for the movement of the target during this flight, the exact time of flight must be allowed for, so that the projectile and target shall arrive at the same point at the same time. A battleship at 10,000 yards might go more than her own length while a 12-in. shell was travelling this range. Not only exact speed and course, but exact calculation for deflection is important.

LONG RANGE FIRING

A Long-Base Absolutely Necessary

17 The one-observer instrument having broken down owing to the shortness of its base; the vertical system owing to the impossibility of relying on the surface of the sea as a reading scale, and both because the operator cannot be protected, there remains only the long horizontal base that the ship itself affords. The introduction of two observers doubles the

chances of error in observation, the split image cannot be used, the system involves new positions, means of communication, and employs more hands. On the other hand, if the base were only 150 feet, the error permissible in each observation will be 20 times larger than with the one-observer instrument, while the mast of the enemy supplies a clear and definite mark to read to, instead of the fugitive and deceptive and often invisible impact of the shot upon the water. A comparison of the permissible error in each observation by this system with those of the one-observer system, for the long range, is sufficient to show the advantages of such a system if it can be made practicable. The instrumental error permissible will also be 40 times greater than with the short base instrument. Unlike the one-observer system, errors with this system can easily be corrected for changes of temperature, and quite simple and well known devices will make it easy to measure and compensate for any distortion set up from time to time in the structure of the ship, by wave pressure, expansion, etc.

If guns are to be used at long distances, range must be known by a very small percentage, the area represented by the danger space getting, of course, smaller as the range increases. Assuming half the danger space to be the largest permissible error in the range, the figures work out approximately as follows:

Range	Danger space	Largest permissible range error	Percentage of range
10,000 yards	70 yards	35 yards	.35 per cent
9,000 yards	75 yards	37½ yards	.38 per cent
8,000 yards	80 yards	40 yards	.5 per cent
7,000 yards	90 yards	45 yards	.63 per cent
6,000 yards	120 yards	60 yards	1.0 per cent
5,000 yards	170 yards	85 yards	1.7 per cent

Chances of Hitting at Long Range

18 The formula for the probability of hits is the total error in the range divided by half the danger space. The total error may be due to miscalculation of the range itself, or error of the day, or gun error, or to all three added together. Aiming, under the admirable system of training now in use will not cause many misses, a battleship at 10,000 yards being a larger mark than the 8 feet by 6 feet target now used in practice at 1,600 yards, and constantly hit, even when firing against time. Gun error, it is assumed, may be calibrated out. Under the existing system, in a series of successive shots that have commenced with the hitting elevation, the errors in range will be those set up by inaccurate data for change of range. An example is given in the illustration of the seriousness of a slight initial

error; it is probably safe to say, therefore, that the error by range finder − the instrument now relied on for correcting these data − will be the minimum range error in a series. Expert artillerists give one hit to 30 shots as the probable chances at 10,000 yards. This is probably because in addition to change of range there are wind and deflection errors to be considered. If this estimate is correct the figures given in the following table, in the third column, are well within the mark. The fourth column gives the figures as they would be were true range, gun range, and accurate data for change of range available, deflection exactly calculated and windage measured.

It is assumed that the range is obtained accurately to within one per cent at 10,000 yards, .75 per cent at 9,000 yards, and within half the danger space at lesser ranges. (See note at end).

Range	Error in range by range finder	Half danger space	Number of shots to a hit under present conditions, omitting error of the day	Shots to a hit when range, change and error are known
6,000 yards	180 yards	60 yards	3	1
7,000 yards	473 yards	45 yards	6	1
8,000 yards	347 yards	40 yards	8.7	1
9,000 yards	454 yards	37½ yards	12.1	1¾
10,000 yards	541 yards	35 yards	15.5	3

These figures are to some extent arbitrary, but they illustrate, probably favourably to the present system, the principles on which the possibilities of getting a high frequency of hits at long ranges depends. To apply these principles is therefore what is wanted to get near the figures shown in the fourth column. As we have seen, the problem is primarily one of telemetry, but the telemetry required is of a kind that permits of measuring the fall of the shot, both as to distance and lateral diversion. Given a scientific and accurate system of telemetry, its application to naval gunnery is a mere question of suitable mechanisms for performing the necessary calculations automatically, and suitable means of securing that the elevation and alignment of the gun is in accordance with the results. It is proposed now to explain how the system might be made to work.

THE SUGGESTED SYSTEM[14]
Telemetry the Basis

19 All telemetry being based upon surveying the elements of a triangle, accuracy in result depends on fineness in the readings of the angles or length of the base. If we can say what is the limit of accuracy in reading angles, and the percentage error permissible in a given range, it is possible to lay down the minimum length of base necessary. Assuming 10 seconds to be the maximum inaccuracy we may expect, and 1 per cent to be the largest

permissible error in a range, say, of 10,000 yards, it is clear that the base of our triangle must be 150 feet, if we assume that each observer makes a 5 second angular error. Of course, on any two readings each may make a full angular error of 10 seconds, so that the cumulative error will be 20 seconds and the error in the range 2 per cent. On the other hand, all the errors may neutralise, in which case a perfectly accurate range will be got, and, if all observations average between the two, it may be said, for practical purposes, that with a maximum error of 200 yards at one end of the scale and perfect accuracy at the other, the average total error should be something less than 1 per cent.

It so happens that 150 feet is about the largest distance apart, within the ground plan of the citadel of battleships, that two observers in protected turrets or observation towers could be conveniently put. If we had two observers there, is there any means by which they could survey the elements of a triangle, of which the enemy's ship was the apex, and obtain a computation of that triangle in time to be of practical use as a range for guns?

HOW READINGS ARE MADE

Observing Angles

20   A system of telemetry designed to achieve this end has been for some time in course of development, and commences with observing telescopes pivoted at one end and having a quarter-arc sweep. The whole quarter-arc is so subdivided that a reading at every successive 15 seconds can be made. In whatever position the telescope on the arc is arrested, a reading can be automatically and instantly transmitted to another portion of the ship.

Synchronism Secured

21   Now, bearing in mind that a ship is a moving platform, moving not only in the line of its length, but thwartwise, owing to wind and current, if two readings are to be made of the position of the same object with a view to arriving at a triangle whose computation shall give the distance of that object, the readings must be made with absolute simultaneity. At 15 knots a ship travels 25 feet per second. The smallest failure in synchronism would therefore greatly distort the base. The thwartwise movement of a ship is, of course, infinitely slower; but, as the distortion introduced into the triangle is not the distortion of the base, which causes a percentage error, but a distortion of the base angle, which causes a geometric error, increasing as the square of the distance, the importance of obtaining synchronism becomes manifest. It is proposed to secure that the observations transmitted are synchronous by preventing the transmission of any that are not. Each observer will make readings as

FIRE CONTROL AND LONG-RANGE FIRING   37

rapidly as he can, and when both make independent readings simultaneously, both readings will be transmitted. It will need the coincident action of the two to actuate the electric transmitting devices. The selection of such of the observations of each operator as are simultaneous will therefore be automatic.

Readings in Rapid Succession
22   What must next be aimed at is a continuous taking of observations by each operator, so that synchronously observed angles shall be transmitted five or six times a minute.

The ordinary form of instrument for reading angles contains a vertical wire, or hairline, defining the exact centre of the field. By bringing this hairline into juxtaposition with the clearly defined outer edge of the object to be observed, the angular bearing of the edge is indicated on the divisions of the instrument. But this, clearly, is not a method that holds out much prospect of success at sea, where the platform would be continuously moving with the motion of the ship, and, to some slight extent also, would be subject to the vibration of the ship's engines. It is proposed, therefore, that the observers shall read these angles on a somewhat different principle. Instead of a single vertical hairline, two vertical hairlines will mark off rather over 1½ minutes of arc in the field, and a third line will subdivide the space between them. A series of electrical buttons will be made to correspond with seven positions, one corresponding with the centre line, three corresponding with three positions to the right, three with positions on the left – the exterior lines marking the limit. The observer's instructions will be to keep his telescope continuously traversing the target. The amount of arc of this continuous traverse will consequently be approximately 2 minutes, a comparatively large amount of arc easily controlled by the hand, readily seen and sub-divided by the eye.

Let us suppose that the part of the enemy's ship which the observer is told to read is the top mast. As the top mast approaches the left-hand vertical line, the observer will depress his left-hand button, and, as the mast moves across the line, he will change to the next button, and so forth over the whole six, as the mast takes up successive positions from the extreme left of the left line to the right of the right line. The traverse having been made, he will repeat the operation by reversing the movement of the telescope.

It is stated that the best gun layers manipulating a heavy implement can be trusted to fire aimed shots without making a greater angular error than 1' 35" from the true aim. This is less than the angular field that the observers handling a freely moving telescope, specially designed to be under the most absolute control, have to keep upon their target. Note that the gun layer, under practice conditions, has nothing in his target to assist his eye, whereas the observer would have the sharp vertical line of the enemy's mast, and a field divided by vertical lines. Most important of all, the gun layer has to fire when sights and target are in coincidence; the observer will be independent of correct alignment. All he will have to do will be to judge how far off the true reading he is. Instead, then, of his observation being limited to moments of correct alignment, he will have a liberal margin for error in all manipulative operations, and only require skill in a comparatively simple mental operation. Skilful men should soon arrive at making successive observations at very small intervals. With two observers the taking of successive synchronous observations should be limited purely by recording. A practical test of an observing telescope made on this principle shows that continuous observation can be made without difficulty.

THE REST OF THE SYSTEM AUTOMATIC

Recording and Computing

23   In some convenient position below the water-line, where space can be obtained for the necessary instruments and operators, a compartment is set aside for the rest of the operations. First, the angle readings electrically transmitted from the observers are recorded by an automatic machine. The angles read will be one interior base angle of a triangle and one exterior base angle. The exterior angle will, of course, always be greater than the interior angle, and the difference between the two will be the apex angle. The function of the recording machine is to receive these two angles, subtract the lesser from the greater, and exhibit in two lines, one above the other, the figures representing the base and apex angles.

An operator faces these dials and has under his control a calculating machine which performs mechanically the trigonometrical operation of expressing the value of one side of the triangle in terms of the base. The triangle is, in other words, mechanically computed, and the length of one of the sides – i.e., the range – given in yards. The results given by this machine are accurate to less than 1-10th of 1 per cent. The length of time that will have elapsed between the synchronous reading of the angles and the solution of the range being given in yards should not exceed two or three seconds.

Clearly, then, if all the foregoing operations can be satisfactorily performed, we have got a very accurate and rapid system of mere range finding. It is free from the general objection to two-observer systems, in that the electric devices it embodies render any conscious co-operation between the observers unnecessary. The design of the telescope and the power of the operator to select any six positions in his field as constituting his observation does away with the objection that it is unreasonable to expect accurate observations to be taken with sufficient frequency to be useful; while the rapidity and certitude with which the range can be computed, once the angles are known, obviates the last objection to the two-observer principle in range finding − i.e., that the length of time occupied in computing the triangle makes the result, when known, however accurate it may have been at the time when the observations were taken, useless by the time it can be communicated to the guns.

Lastly, it should be noted that it is not necessary in this system that one of the angles surveyed should be a right angle. The system will give equally accurate results at any point of training of the arc, so that each observer's extreme point of training is 45 to 50 degrees forward and 45 to 50 degrees aft. The apex of the triangle may be anywhere and at any visible distance within that arc. An enemy's topmast should be visible at 15 miles from the conning tower, assuming atmospheric conditions to be satisfactory.

Averaging Ranges

24 If the results of these observations could be averaged, not in the sense of adding 10,100, 10,000, 10,025, etc. − viz., the successive ranges given, and dividing them by the number of ranges, and so getting the average range, but in the sense of placing on a chart the successive positions of the enemy indicated both by the angular observations and by the range, the *median* line of the enemy's course so made would be the exact course, and the range shown by that *median* line the exact range, accurate probably to 1-10th of 1 per cent. It now remains to be seen whether, assuming that we can get synchronous and accurate observations it is possible to convert them, not only instantly into ranges, but into a charted course of the enemy.

Charting

25 It is assumed that synchronous observations of the enemy will have been commenced a considerable distance beyond that at which it is intended to engage him. Each pair has been transmitted, the triangle computed, and his distance given within the percentage error of the angular observations. But one of the angles of his bearing and the distance are all that is needed to chart his successive positions. The next member of the system is therefore a charting table.

Over this a ribbon of paper of suitable width is run at a speed corresponding for its scale with the speed of own ship. An arm marked to different ranges carrying a movable pin is pivoted at one side of the table and can be adjusted to any selected angle on a divided arc erected above the moving paper. The arm is moved to the angle corresponding with the bearing transmitted, the carrier is run up to the calculated range and the pin is depressed to perforate the moving ribbon.

The electric circuit that transmits the observations to the recording machine actuates a chronometer governed printing device, so that the time of transmission of the angles is recorded as well as the angles themselves. The perforation on the paper has written beside it the time at which it represented the bearing and distance of the enemy.

Let these operations be repeated for some minutes and there will be a line of dots on the moving paper, and this line will be slightly irregular. The distance indicated between any two perforations will show the enemy's speed and the median line between the dots his true course and the termination of that line his actual distance at the moment. If own course has throughout been unaltered, this line will show the angle of the enemy's bearing exactly and changes in own course can be simply allowed for and the angular difference arrived at. Thus this record of the results of the observations will afford all the materials needed to satisfy the first three canons of fire control if we can automatically calculate and transmit the changes in the range.

Change of Range Machine

26 A further mechanism, provided with proper scales and dials, is then set to own and enemy's course and speed, and the true known range at the moment. The operations of this machine are governed by a chronometer and, on being set in action, a pointer automatically begins to move round the edge of a dial marked in yards. This pointer having been set to the range at the moment as a starting point, moves on to the changes of range set up by the altered positions of the ships.

The additions or subtractions made from the original range will, of course, not be a fixed amount per minute, but the actual amount of change. For instance, in Figs. 3 and 4 are shown two typical cases, and it will be seen that the initial amount of change per minute is quite materially different from the change 2 or 3 minutes later.

So long, therefore, as the courses and speeds remain unaltered, this mechanism will give the correct range at any desired intervals of time – every 10, 15, or 30 seconds; and, by simple electrical connections, can operate similar dials either in all the gun positions or in such as it is desired should engage the target whose range it announces. This mechanism will further show the altered bearing of the target as well as the altered range, and so will afford an indication to the guns of the object selected for their fire. A separate change of range and bearing machine should be available for each group of guns so that several targets could be engaged simultaneously.

Correcting Data for Range

27 After firing has commenced, it will, of course, be an object with those navigating the ship to keep her on as straight a course as possible, so as to reduce the difficulties of the gun layers. But should course or speed

for any reason be changed, the change of range machine would be corrected accordingly. But the enemy meanwhile may change his course or speed and so alter the range; and it is obvious that keeping him under synchronous observation, while the ship is wrapped in smoke and haze from its own guns, will not be easy. After firing has commenced, therefore, the synchronising circuits should be disconnected and every observation allowed to come down whether synchronous with another or not. As it is unthinkable that the enemy could be at his right bearing but at the wrong range, or at the right range but on the wrong bearing, every observation from the station, whose observation is taken as his original bearing, will be an automatic check on the range in use. So soon as it was known that the range was lost, 'cease fire' would be signalled, the synchronising circuits re-connected and the range taken before firing re-commenced.

Gun Range and Windage

28 It remains to be seen how it is proposed to measure gun range. The observing telescopes, it will be remembered, have a marked field within which the observer can select from many the observation he wishes to transmit. Let us suppose a ship ranged at 11,000 yards. It is desired to know the gun range before engaging him at 10,000 yards. It is assumed that the percentage error at 11,000 yards will hold good at lesser distances. The first gun is then elevated at say 10,500 yards, is aimed at the foremast of the enemy and fired. Both observers have received the signal 'gun range', and have the foremast within the marked off field, and each will be able to note the position of the column of water sent up by the projectile on its impact. Supposing both observers to be well practised in the art of selecting correct positions, each will be able to press the button corresponding to the position the column of water is observed to be in. Both do so and a pair of angles are transmitted (just as when ranges are being found) and, on the triangle so arrived at being computed, the distance of the fall of the shot will be known. If three shots were fired in succession and three distances got, error of the day and windage should be available, to which the elevation and deflection of the other guns will be set. The observers, of course, will have been kept practised in taking gun range during all long range firing in peace time, so that, in addition to general expertness in the use of the instrument, which could be got without firing, complete familiarity should be obtained with the form of the projectile's splash.

Deflection Machine

29 The final element that must be known to get the gun into a correct position to hit, is the proper deflection. The data necessary for calculating deflection are: drift, speed and course of own ship and enemy's ship, time of flight of projectile and wind. The taking of gun range should give

windage – the drift is known – speed and course we have already. A machine is in course of design to get these data combined correctly for different ranges, so that exact accuracy may be available should an improvement in present methods be desired.

Summary of System

30 The principles on which the proposed system is based, and the nature of the appliances it is proposed to use, has now been briefly sketched. If the results aimed at can be obtained, even approximately, a notable step forward will have been made in devising a system complying with the first three canons laid down, as it provides for true range, the ascertainment of gun range, and continuous correction. Practically the whole operation, with the exception of the taking of the observations, is automatic, and, the observations once having been taken, the results are instantaneously transmitted to the guns. It remains to be seen whether the system can be made to comply with the remaining two canons, and the conditions necessary for the system to be practicable. The problem of armouring the observing stations and the selection of suitable sites for them is a question for naval architects; but the design of the instrument, the size of the aperture necessary for using them and the space they and their operators occupy, make the problem a simple one, assuming there is no objection on the score of space and weight. Such objections as there are on this score are eminently matters for compromise, that is, that any sacrifice made must be measured by the advantage got. Various apparently practicable suggestions have been made. Subsidiary conning towers might be added either between existing conning towers and barbettes, or at either side of existing conning towers, as in the new Japanese ships.[15] Either of these would provide sheltered positions for the broadside observers. For the end-on observers fore and aft, turrets projecting from the ship's side above the main deck guns and below the upper deck guns, and turrets erected on the sides and carried above the upper guns have been suggested. Wherever placed, the observers should not be worse off than the officers in the present conning tower, or the gun layers, in the matters of haze and smoke.

The Beginning Half the Battle

31 With regard to the whole question of armoured protection, and of the use of the system during action, one important consideration should be borne in mind; and it is that of all forms of superiority in artillery – viz., greater penetration, greater speed of fire under equal conditions as to the number of shots per hit and greater ranging power, the last is so much the most important that, if it could be overwhelmingly secured, the duration of actions might conceivably be limited to a very few minutes of fire, and the question of armouring the fire control system made

negligible. Reverting, for the purposes of illustration, to the table in Section 18, let us suppose a fleet capable of making one hit with its 12-in. guns at 10,000 yards, for every three shots fired meeting another that could only score one hit in 15 shots. Let there be five ships aside, and let each gun fire three shots in two minutes. Allowing four 12-in. guns per ship, each fleet will fire 120 shots in the first four minutes of action. But, whereas the first fleet will score 40 hits in the time, the second will only score eight, and five minutes of action under these conditions would certainly decide the issue.

If the estimate of 30 shots to a hit at 10,000 yards is correct, the second fleet will only score four hits.

Other Conditions

32 The outline given of the system has already shown that it complies with the other conditions. The total necessary staff would not be larger than a battleship could easily afford; nor, even in the observers, would a high standard of skill be necessary.

The observations once got, true range, gun range, change of range and deflection are not guess-work, but are given by instruments working to absolute scientific precision and automatically. Obviously, if the skill of the observers justified it, the observing telescopes could be still more finely divided, holding out the promise of accurate ranges, gun ranges, and the material for rate of change at distances incredible to present day users of artillery. Apart from the space occupied by the observing turrets, the system is a simplification and not a complication of the organisation of a ship in action as it is to-day, and, being based on centralisation, would be naturally brought under the immediate control of the Captain, who could indicate the target by communicating its bearings as he wished it engaged.

It is perhaps worth remembering that if the base were 300 feet instead of 150 feet, one-half only of the standard of accuracy in both observation and instruments would be necessary – or if the standard stipulated could be easily obtained, an accuracy of great fineness might be got. Thus ranges might be still further increased without diminishing the chances of hits. Meanwhile the enemy's chances of hits would be zero. On a battleship a longer base, of course, means an unarmoured base. But that would be immaterial if the ship, like Dewey at Manilla Bay, never came under the enemy's fire.

Check on Enemy's Movements and Formation

33 For what it may be worth, it is apparent that the same system that at close distances will give us a chart of the enemy's positions sufficiently correct for gun fire, will at 20,000 yards or more give us – even when only his top masts are visible – a less accurate but quite serviceable chart

of his course. His formation and speed, therefore, would be familiar long before the engagement began. With this knowledge an admiral might engage with considerable tactical advantage. Note that a much longer base – viz., the whole length of a cruiser might be used for this plotting of a distant enemy; and with a base of 400 to 500 feet, an accuracy of one or two per cent might easily be attained at the extreme limit of sight. Both as a preliminary to action, and generally as a function of scouting, this capacity to lay out the course of a cruising enemy and to state his speed might give benefits quite unconnected with gunnery, but none the less valuable in war.

The following tables, diagrams and figures are marked with numbers to correspond with the sections they illustrate.

See Section 17.

NOTE ON THE THREE METHODS OF TELEMETRY

Maximum permissible total errors in base angles if range is to be correct to one per cent

Total errors in base angle must not exceed:

Range	1 per cent Range allowable error	With 4-foot base		With 100-foot base depression readings		With 150-foot base suggested scheme	
Yards	Yards	Min.	Sec.	Min.	Sec.	Min.	Sec.
10,000	100	0	.26	0	6.3	0	10.22
9,000	90	0	.28	0	7.0	0	11.32
8,000	80	0	.33	0	7.8	0	12.76
7,000	70	0	.39	0	9.0	0	14.66
6,000	60	0	.45	0	10.5	0	17.01

One-and-a-half seconds error in base angle gives an error

At a range of: Yards	Yards	With 4-foot base	With 100-foot base depression readings. Yards	Yards	150 feet
10,000	541	= 5.4 per cent	24	15	= .15 per cent
9,000	454	= 5 per cent	19	12	= .13 per cent
8,000	347	= 4.6 per cent	15	9.6	= .12 per cent
7,000	273	= 3.9 per cent	11	7.2	= .1 per cent
6,000	201	= 3.7 per cent	8	5.4	= .09 per cent

The advantage of the long over the short base is obvious. It may be measured by the relative sizes of the apex angles in the triangles surveyed.

The One-Observer Split Image Using Instrument has the advantage over any other method in the matter of taking an observation; but the shortness of the base; the fact that it cannot be used for taking gun range; is useless for assisting to get exact data for speed and course; is liable to distortion by temperature expansion, and that the smallest distortion results in enormous errors, more than counteract that advantage.

The Mast-Head Method has an advantage in a large base, but this is more than counterbalanced for taking gun range by the reading scale – *i.e.*, the surface of the sea being an unreliable scale, and the image – *i.e.*, the impact being always too indefinite or invisible. At the mast-head also, instruments of precision could not be used except with great difficulty. For these reasons the depression method may be eliminated as not being suitable to exact telemetry.

The Suggested Method seems to be the most free from objection, as the readings need not be very fine, the errors due to distortion, expansion, etc., can be corrected every minute or so if necessary without interfering with the observing operation, and both with regard to range-finding and gun range-finding, it has the advantage of observing the most definite marks – viz., the mast and projectile splash instead of impact.

See Section

FIGS. 1 and 2.

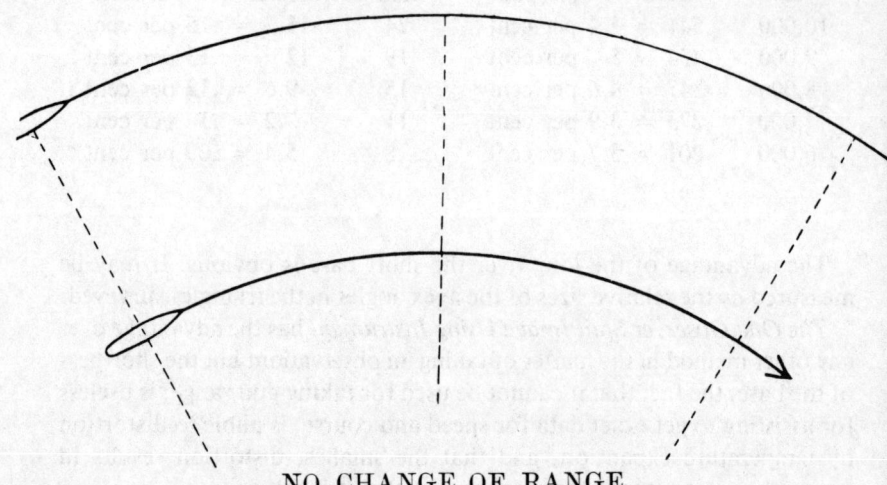

NO CHANGE OF RANGE.

COURSES CONCENTRIC. SPEEDS PROPORTIONED
TO TRAVERSE EQUAL ARCS IN EQUAL TIMES.

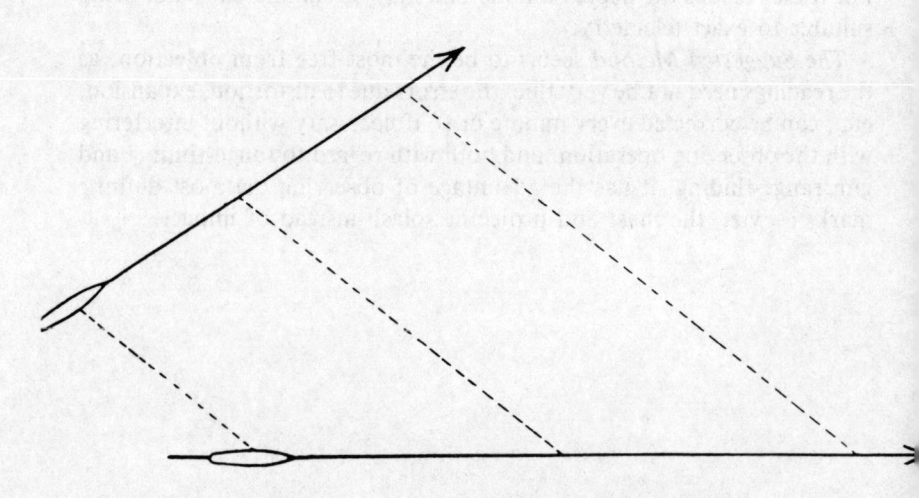

REGULAR CHANGE OF RANGE.

SPEEDS EQUAL.

FIG. 3.

Change of Range with reference to Stationary Object.

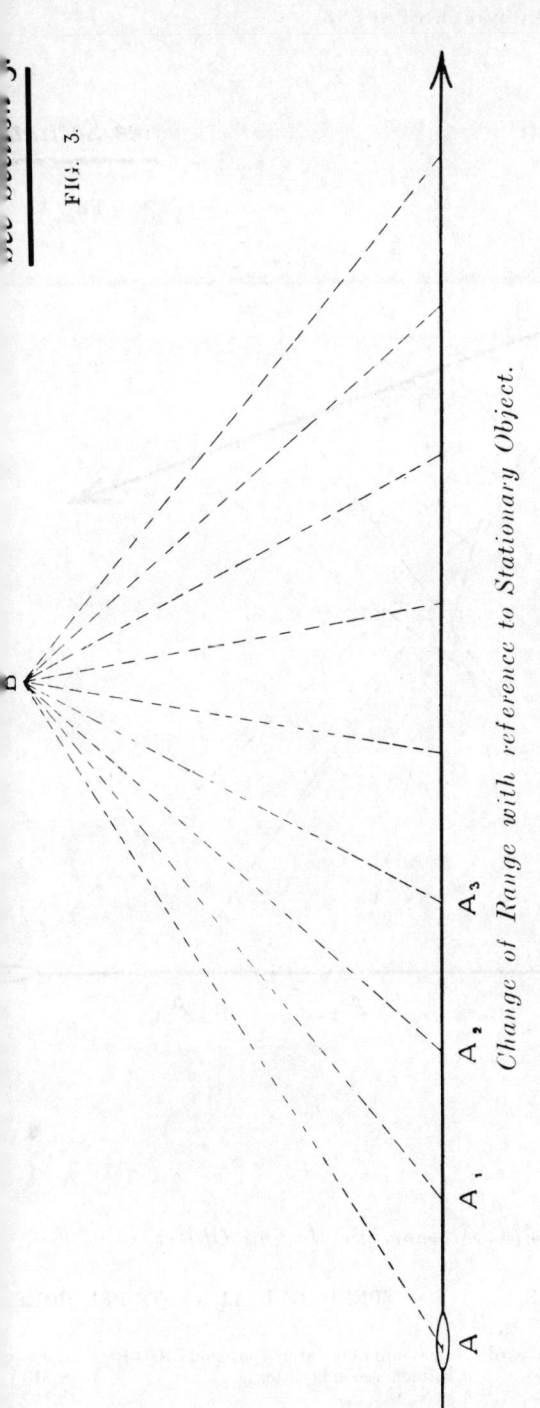

SPEED 17 KNOTS PER HOUR.

First observation (position A—B)	...	9,140 yards.
Fifteen seconds later	...	9,020 ,,
,, ,, ,,	...	8,900 ,,
,, ,, ,,	...	8,780 ,,
One minute later (position A¹—B)	...	8,660 ,,
Fifteen seconds later	...	8,540 ,,
,, ,, ,,	...	8,420 ,,
,, ,, ,,	...	8,300 ,,
Two minutes later (position A²—B)	...	8,190 ,,
Fifteen seconds later	...	8,080 ,,
,, ,, ,,	...	7,970 ,,
,, ,, ,,	...	7,860 ,,
Three minutes later	...	7,750 ,,
Fifteen seconds later	...	7,640 ,,
,, ,, ,,	...	7,530 ,,
,, ,, ,,	...	7,430 ,,
Four minutes later	...	7,330 yards.
Fifteen seconds later	...	7,230 ,,
,, ,, ,,	...	7,130 ,,
,, ,, ,,	...	7,030 ,,
Five minutes later	...	6,930 ,,
Fifteen seconds later	...	7,830 ,,
,, ,, ,,	...	6,740 ,,
,, ,, ,,	...	6,650 ,,
Six minutes later	...	6,560 ,,
Nine ,, ,,	...	5,610 ,,
Twelve ,, ,,	...	5,065 ,,
Fifteen ,, ,,	...	5,065 ,,
Eighteen ,, ,,	...	5,610 ,,
Twenty-one ,, ,,	...	6,560 ,,
Twenty-four ,, ,,	...	7,750 ,,

48 THE POLLEN PAPERS

See Section

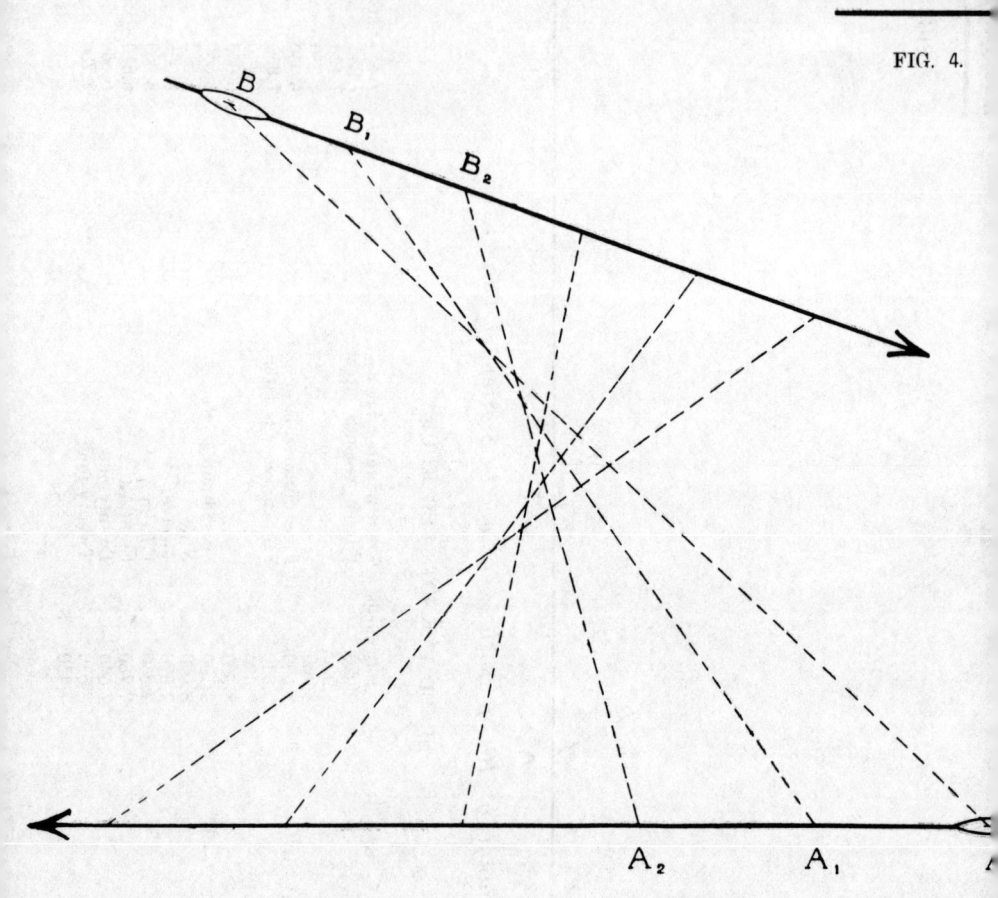

FIG. 4.

Change of Range with reference to Moving Object.

SPEED OF A, 17 KNOTS PER HOUR.　　　　SPEED OF B, 12 KNOTS PER HOUR.

First observation (positions A—B)	10,000 yards.
Fifteen seconds later	9,800 ,,
,, ,, ,, ...	9,600 ,,
,, ,, ,, ...	9,410 ,,
One minute later (positions A^1—B^1)	9,220 ,,
Fifteen seconds later	9,030 ,,
,, ,, ,, ...	8,850 ,,
,, ,, ,, ...	8,670 ,,

Two minutes later (positions A^2—B^2)	8,490 y
Fifteen seconds later	8,310
,, ,, ,, ...	8,140
,, ,, ,, ...	7,970
Three minutes later	7,800
Six ,, ...	6,150
Nine ,, ...	5,600
Twelve ,, ...	6,450
Fifteen ,, ...	8,250

FIRE CONTROL AND LONG-RANGE FIRING 49

See Section 13.

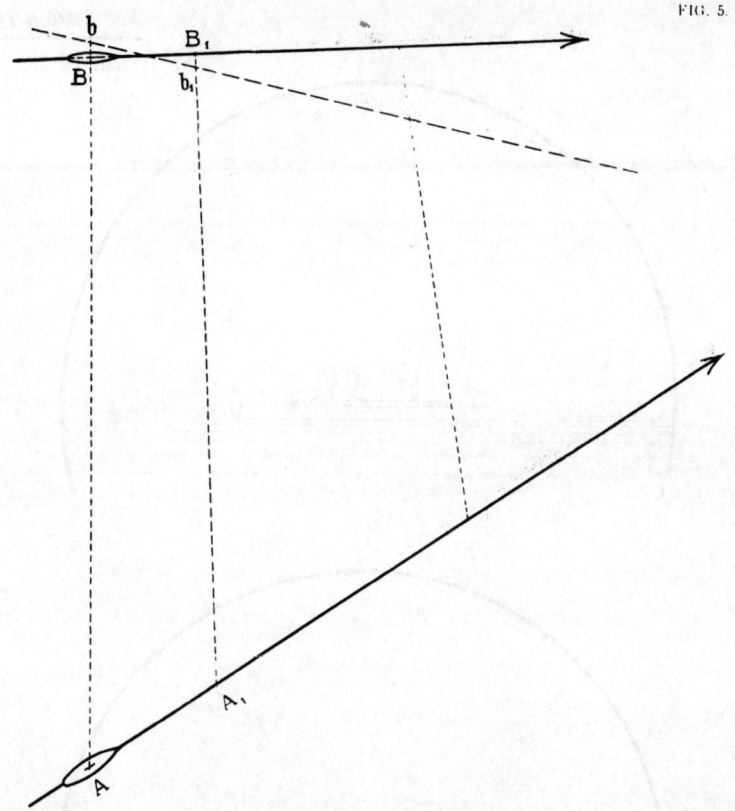

FIG. 5.

ILLUSTRATING DEGREE TO WHICH ERROR OF RANGE INCREASES IF MATERIALS FOR RATE OF CHANGE ARE NOT ACCURATELY OBTAINED.

The table gives ranges when the enemy's course shown in the straight full line, B to B¹, is assumed to be on the dotted course b to b¹; own course being in nearly the same direction upon the line A to A¹.

Own speed is 17 knots an hour, and opponent's speed is 12 knots per hour.

The diagram is drawn to scale, and three minutes has elapsed between own positions A and A¹. At the position A a positive error of 150 yards is made in taking the range, and again at the position A¹ a negative error of the same amount is made.

The true range from A¹ to B¹ is 7,000 yards, but the assumed range is 150 yards short of this.

	Real Range. Yards.	Assumed Range. Yards.	Error. Yards.
Positions A¹—B¹	7,000	6,850	150
Fifteen seconds later	6,926	6,751	175
,, ,, ,,	6,852	6,652	200
,, ,, ,,	6,779	6,554	225
One minute after A¹—B¹	6,706	6,455	251
Fifteen seconds later	6,634	6,357	277
,, ,, ,,	6,562	6,259	303
,, ,, ,,	6,491	6,161	330
Two minutes after A¹—B¹	6,420	6,063	357
Fifteen seconds later	6,350	5,966	384
,, ,, ,,	6,280	5,868	412
,, ,, ,,	6,210	5,770	440
Three minutes after A¹—B¹	6,140	5,670	470
Six ,, ,, ,,	5,320	4,520	800
Nine ,, ,, ,,	4,520	3,370	1,150
Twelve ,, ,, ,,	3,760	2,180	1,580

See Section 15.

FIGS. 6 and 7.

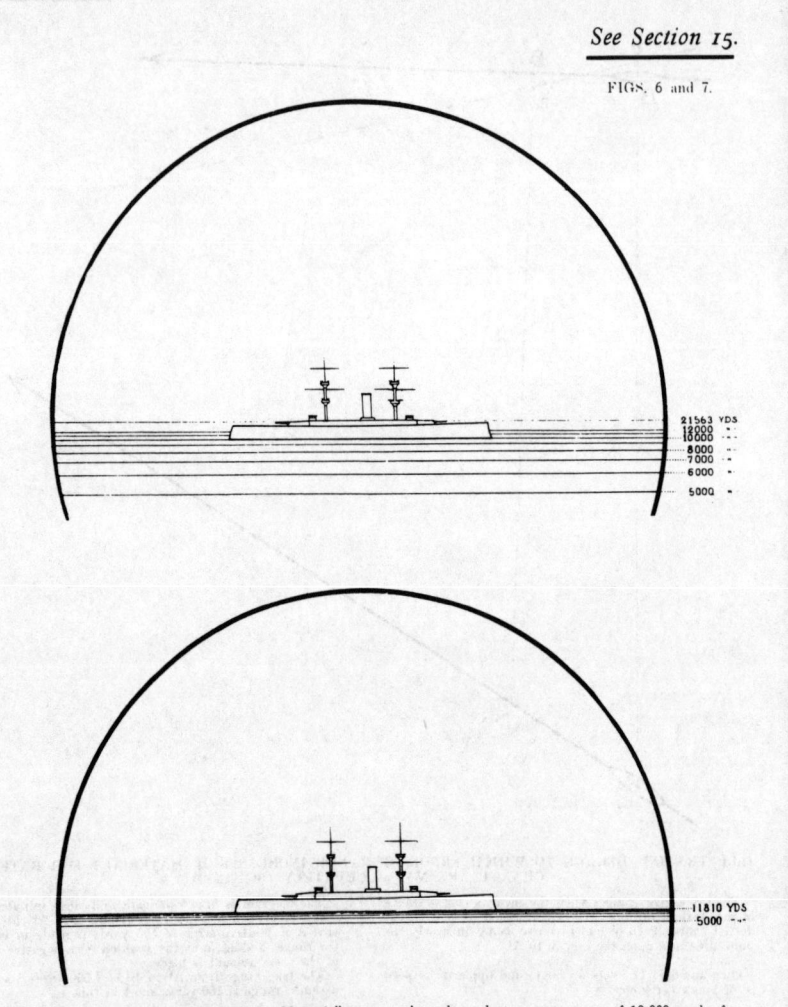

The upper diagram represents H.M.S. "Majestic" as seen through a telescope at a range of 10,000 yards, from a height of 100 feet, in a perfectly smooth sea. The ruled lines represent imaginary range marks drawn to scale for 12,000, 10,000, 9,000, 8,000, 7,000, 6,000, and 5,000 yards. It will be seen that to distinguish, even in a perfectly smooth sea, the difference between 9,500 yards and 9,700 yards would be impossible.

The lower diagram represents the same as seen from a height of 30 feet above the water. Note that it is practically impossible to distinguish any ranges above 5,000 yards from this height, even in perfectly smooth water. With any movement at all there would be no difference at all as far as the fall of the shot is concerned between 5,000 and 10,000 yards.

Note on Figs. 6 *and* 7, 10 *and* 11.—If the figure is held 9 inches from the eye the image will be exactly the size an observer would see if looking through a telescope with 1½ degrees of field and magnifying 20 diameters.

FIRE CONTROL AND LONG-RANGE FIRING 51

See Section 15.

FIG. 8.

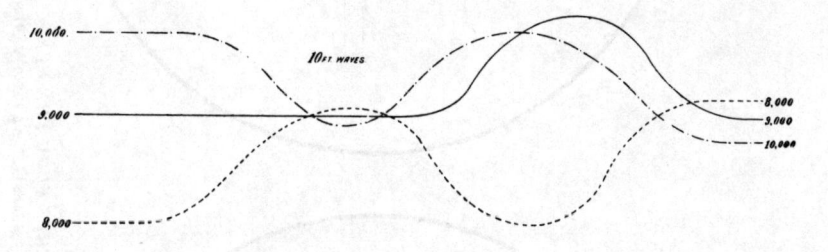

This represents an enlargement of the range marks shown on the preceding page, as seen from a height of 100 feet, and is drawn to scale. The sea is supposed to be running in waves 10 feet high. The range marks consequently take the form of curves, with the result that the 10,000, 9,000, and 8,000 marks intersect. A shot at the intersection would angularly seem to be 9,000 yards off; in reality it might be 10,000, 9,000, or 8,000 yards off.

With waves only 5 feet high the 10,000 and 9,000, and the 9,000 and 8,000 ranges intersect in a similar manner.

52   THE POLLEN PAPERS

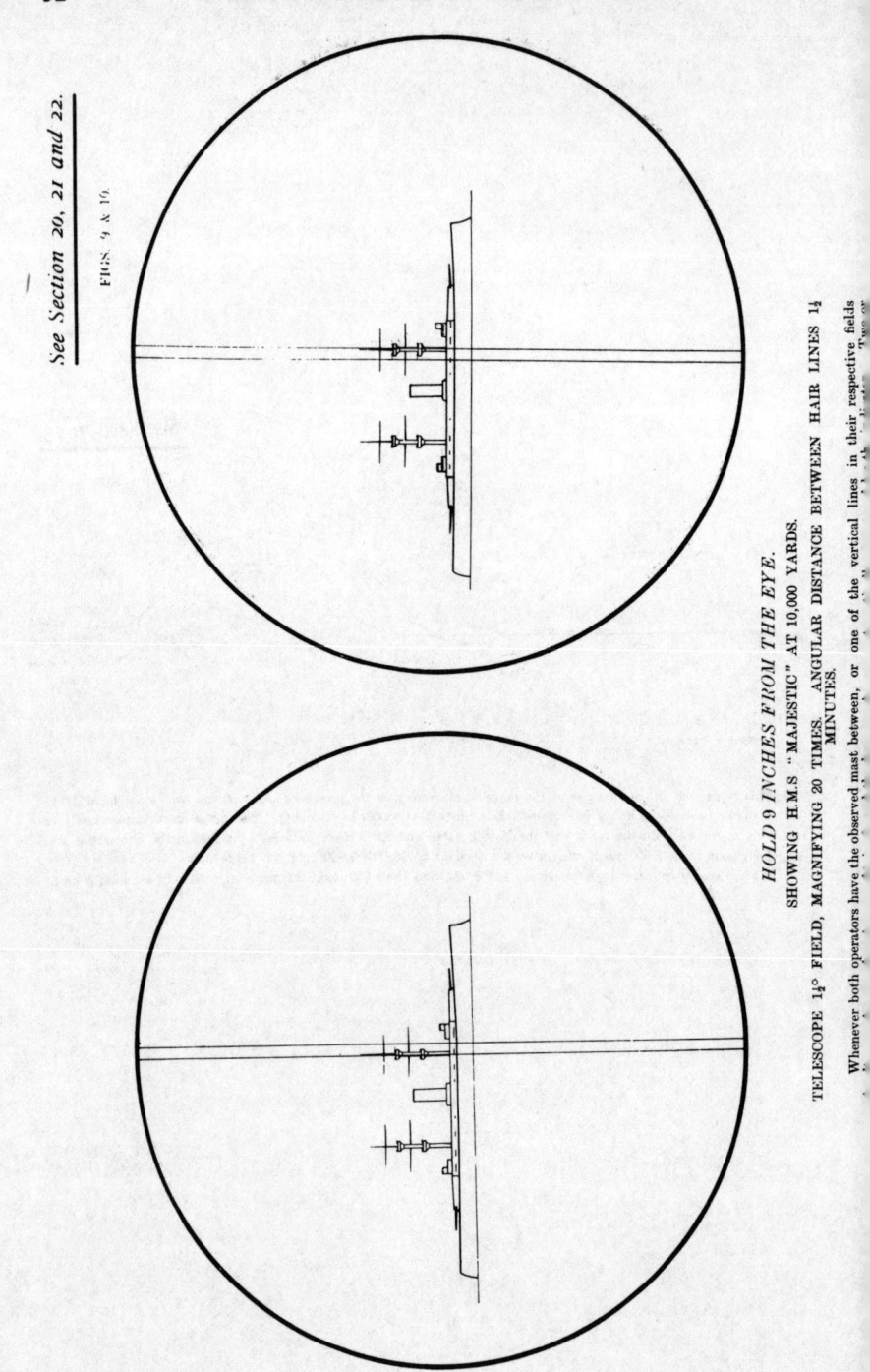

See Section 20, 21 and 22.

FIGS. 9 & 10.

HOLD 9 INCHES FROM THE EYE.
SHOWING H.M.S "MAJESTIC" AT 10,000 YARDS.
TELESCOPE 1¼° FIELD, MAGNIFYING 20 TIMES. ANGULAR DISTANCE BETWEEN HAIR LINES 1½ MINUTES.

Whenever both operators have the observed mast between, or one of the vertical lines in their respective fields

FIRE CONTROL AND LONG-RANGE FIRING 53

See Section 23.

FIG. 11.

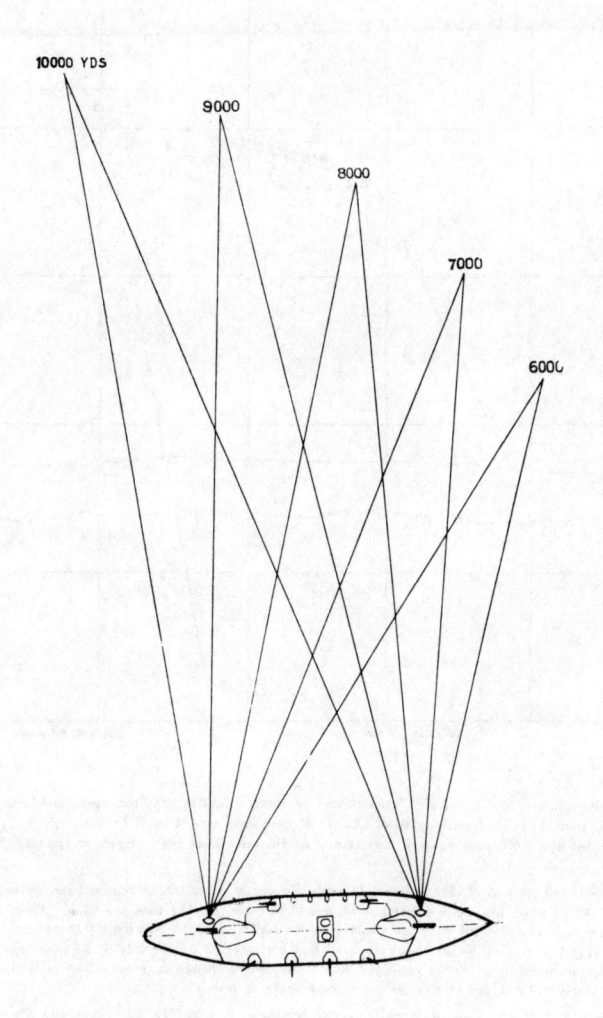

Illustrates the principle of proposed system. If drawn to scale, either the 10,000 yards range would be 225 feet away, or the base would be .03125 inch long.

See Section 25.

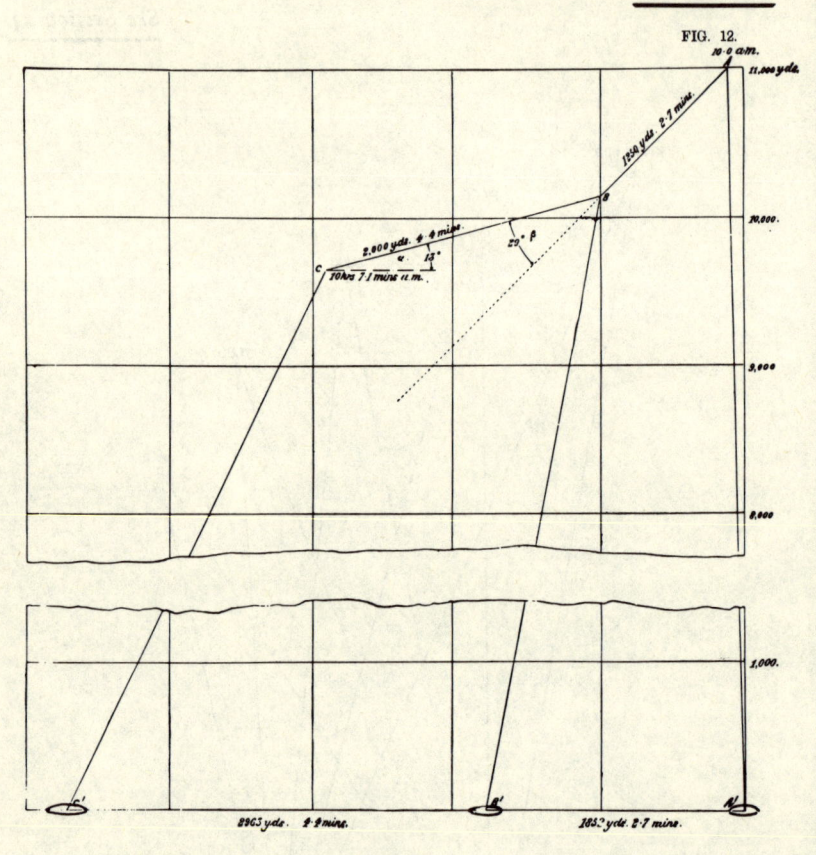

FIG. 12.

In the above figure the line A^1, B^1, C^1 represents own course, right to left; own speed is 40,000 yards per hour; own direction from A^1 to B^1, then from B^1 to C^1. At B^1 the course was changed by the angle b (29°) to port. The paper is divided into 1,000 yard squares, and travels on the table from left to right, at the rate of 40 squares per hour.

A represents the position of the enemy at 10 a.m. His course is the median line between the successive observed positions till B, at which his course apparently changes. In reality he has kept a straight course (the dotted line), but own course has been changed when the range is nearly 10,000 yards, to keep him at an advantageous distance for engaging. When C is reached by the enemy we know that his speed is 3,250 yards in 7.1 minutes—i.e., v. A to B to C, showing his speed to be 26.664 yards per hour. The rate of change machine is then set to this speed, and to the angle 13°, and will continue to give his range and bearing accurately.

The lines A^1A, B^1B, C^1C show the enemy's angular bearing at 10 a.m., 10h. 2.7m. a.m. and 10h. 4.4m. a.m.

IV

A.C.: A POSTSCRIPT
(*MID-1905*)

Pollen wrote 'A.C.: A Postscript' after the Admiralty's formal offer of 3 May 1905 to develop his instruments for trial, but apparently before the start of the preliminary trials of his instruments from the battleship HMS *Jupiter* that began in September 1905.[1] There is no record of the paper's distribution.

In 'A.C.: A Postscript', Pollen for the first time put forward the term 'Aim Correction' as preferable to the term 'Fire Control' as a description of his system's function. The sub-title 'A Postscript' applied to his correction of certain propositions that he had advanced in 'Fire Control and Long-Range Firing' which, upon further thought, he had found to be faulty. Pollen, in addition, took the opportunity to respond to criticisms of his approach that had been raised by Percy Scott and Jellicoe. Pollen's argument in his paper against the practice of determining the range solely by judging the fall of shot and making corrections accordingly appears to have been directed against the views of Percy Scott, whom Pollen had met for discussions about gunnery during the first calibration trials at Bantry Bay in May 1905.[2] The argument against 'end on' fire appears to have been directed against Jellicoe's objection to the two-observer system of range-finding on the grounds that it could not be employed against targets that were right ahead or astern, which he had made in a letter to Pollen of 14 April 1905.[3] Towards the end of his paper, Pollen argued that the rejection of 'end on' for broadside engagements and improvements in long-range gunnery must lead to certain radical changes in capital ship design beyond those attained in the new battleship HMS *Dreadnought*.

RECAPITULATION

In an essay entitled 'Fire Control and Long-Range Firing', the writer pointed out that the strength of Navies rested not on the number of ships or their weight of guns, but purely on the capacity to use their armament with effect; that the fleet that could hit farthest, hardest and oftenest must be supreme; that to hit with regularity at long range, certain principles must be recognised and certain data obtained by an organisation commonly called 'Fire Control' (or, as the writer preferably names it,* 'Aim

---

* NOTE. – The former term exactly connotes quite a different operation of great importance – viz., the tactical employment, in battle, of the ship's or fleet's armament when efficient, and not the means for securing its efficiency.[4]

Correction') and, after setting out certain basic facts of gunnery, propounded five canons of fire control, as follows:

> A system of fire control to be as sure and reliable in its results as the unavoidable errors in guns and aiming permit, must be based upon a realisation of the foregoing fundamental facts, must provide the data they indicate as necessary, and with the speed and continuity required for maintaining effective fire.
> The fire control should therefore:
>
> (1) Supply the exact distance of the target.
> (2) Measure the percentage error of the day or hour, both before effective fire is to begin, and
> (3) Indicate the change of range from minute to minute after effective fire has begun, and supply the data for deflection.
> (4) The fire control agents, their positions and appliances, must be just as thoroughly protected by armour as any other persons, positions, or appliances vital to obtaining the object for which battleships exist.
> (5) The positions and instruments must be so placed that the operators shall not be unduly impeded by the blast of the guns, or more impeded by haze, smoke, etc., than the gun layers, navigators, and others who will have to carry on their duties in the difficult conditions of action.

He went on to enquire how far the system now in use fulfilled these canons, and ended by advocating a method based on making a chart of the enemy's actual course by plotting the bearings and distances given by long base range finding, which, it was maintained, held out the promise of carrying the canons into effect.

CORRIGENDA

Shortly after that essay was written, a decision was made to experiment with this system; and, as was to be expected, certain shortcomings have come to light, while greater familiarity with the subject has convinced the writer that his first analysis and presentation of the problem were marred by ignoring certain facts, and unscientific classification of others.

The chief error seems to be the theory underlying the second canon. This may be stated: 'After every possible provision has been made to obtain the data that govern the choice of the elevation and deflection angles to secure a hit, there will remain a source of error, constant while the conditions continue, due to occult forces; that the extent of this error cannot be anticipated, as the forces causing it are immeasurable; that it is therefore imperative to ascertain the difference between the gun range of the day and the normal range; finally, that there are no means of doing this, except by ranging the splash of trial shots with absolute exactitude.

It now appears to the writer that this doctrine is unorthodox, and that a better appreciation of the truths of gunnery, and of the theory of aim correction, will show, first, that there need not necessarily be any residual error of the day unallowed for when the first shot is fired; secondly, if there is, it is not an 'error of the day', but an error of the aim correctors; thirdly, that while the solution is undoubtedly to find the difference between range attained and the target range; that, fourthly, ranging the splash is neither the only nor necessarily the best or most expeditious way of doing so, and that, lastly, the capacity to do so by direct telemetry is of no crucial consequence whatever.

He therefore proposes to partially restate the problem to render this contention clear; and, after doing so, to suggest certain tactical and economic conclusions that seem to follow from long range hitting being either attained, or attainable.

The following notes, therefore, are to supersede so much of 'Fire Control and Long-Range Firing' as take the second canon of fire control as fundamentally true, and to be in extension and emendation of the rest.

PART I – TECHNICAL

(1) The accurate statement of a problem is generally the best clue to its solution: a definition of hitting is therefore a necessary preliminary to any discussion of how to do it. It seems as if the art of hitting with a gun at any range, when both gun and target are stationary, may be almost exhaustively defined to lie in procuring such an angular relation between the sight and the axis of the bore at the moment of firing that the projectile will intersect the line of sight at the target if it was on the target at the moment of discharge; and similarly, when both or either are moving, intersect the same line at the same point if it was continually on the target from the moment of discharge till the expiration of the time the projectile takes to travel from the gun to the target.

(2) Thus, the art has two distinct parts – the choice of a line of sight, and the choice of the elevation and deflection angle. The first is the art of aiming, still called in deference to the fact that one time the gun captain was the embodiment of both arts – gun laying instead of sight laying. The second is the art of aim correction. These arts are mutually exclusive. At long range the aimer cannot use any profitable discretion in selecting

either elevation or deflection angles; at no range can any assistance be given to the layer in selecting his line of sight. It is solely with the art of aim correction, or angle calculation, that we are now concerned.

(3) If, then, the art is one of angle calculation, it must take cognisance of all the material factors; and so naturally falls into two divisions. First, starting with the information given by the range tables, and indexed on the sights, it must first obtain the data by which the elevation and deflection angles on the sights are to be selected, on the assumption that the performance of the guns will correspond with them, and next ascertain and measure any forces that will prevent the gun from achieving a normal performance, and vary the angles to counteract them.

FIRST SECTION – GEOMETRICAL

(4) The first division is concerned with the data necessary for calculating the required angles under normal conditions, and they are six in number. viz.,

I	Distance	
II	Speed	
III	Angle of Course	of Target
IV	Bearing	
V	Speed	
VI	Angle of Course	of Own Ship

These may be called the purely geometrical elements, and enable a forecast to be made of the position of the target at the end of the time of flight, and afford an index to the lateral impulse given to the projectile by the ship's motion. In 'Fire Control and Long-Range Firing' the writer has very fully explained how they may be obtained. It was proposed to take the bearings of the enemy by azimuth telescopes that could measure and transmit the angle with almost perfect fineness; to transmit these angles synchronously to a central station where the triangles surveyed could be instantly computed; to lay out the bearings and ranges thus obtained on a moving chart, and from this plotted course of the enemy so made, to deduce his distance, speed, course, and bearing. It is perhaps pedantic to insist that, as he may change both speed and course between the moment when the calculation is made and the end of the time of flight, there must always be an element of uncertainty in the forecast. For it would seem as if 30 seconds were the outside time available for these changes, and a battleship being neither able to spurt or double, it could not be often that the destined position would be evaded. Where a ship has to keep station in a fleet, it could practically never falsify the forecast at all.

(5) The author has developed an alternative method of getting the distance and bearing, and necessarily any method that obtained them with speed and accuracy would be equally useful, so long as instruments, operators, and communications were completely protected from the enemy's fire. The keystone of aim correction is not any particular means of getting bearings and range, but in the method of converting them into such exact knowledge of the enemy's speed and course that his future positions can be foretold. In other words, it is the charting table that by making the forecast of position and giving the data for change of range, makes it possible for the angles of elevation and deflection to be calculated. Automatic devices make the consequential calculations, and supply the guns with information necessary for future elevation and deflection. The changes in range are thus mechanically supplied, and the machines are corrected from time to time as any of the factors are observed to alter.

(6) So far the chart, the change of range and the deflection machines complete the purely geometrical division of the subject. Under standard conditions the range given to the gun will be that for which the sights make the appropriate allowance both for the future position of the target and diversion of the projectile from the line of the bore.

SECOND SECTION – DYNAMIC

(7) But as standard conditions are seldom known to occur, it is necessary to provide for the exceptions that are the rule, and it is the second division of aim correction that is concerned in ascertaining and counteracting these factors. These are the forces which give the projectile a variation of flight, which either in direction or length is other than that accounted for by the ship's movement or credited to the gun by the range table, and complete the list of factors necessary for arriving at the true angles for hits, of which the first six are enumerated in Section 4.

(8) They may be due to the condition of the gun itself, to the ammunition, or be purely external. They are:

VII   Variations in the natural velocity of the gun.
VIII Lateral impulse known as drift.
IX   Varying powers in the propellant.
X   Variations in projectiles and driving bands, and two principal causes purely external.
XI   Wind.
XII   Atmospheric resistance.

(9) The first of these is quite simply dealt with by recognising that the same sight is not suitable for all guns of the same nature, nor for the same gun at all stages in its life; ascertaining the peculiarities of each gun from time to time by calibration, and making the several sight notations

correspond with different angles accordingly, so that the same indicated range will give each gun its own appropriate angle for attaining it.

Drift in the present state of knowledge can only be dealt with by a constant angular set of the sight; but experiments and calculations may show a better system to be necessary.

(10) The third is not important, and can be allowed for by a nett addition to, or subtraction from, ranges within wide limits; the fourth can obviously not be allowed for at all, and remains an insoluble difficulty until perfect repetition in material and manufacture is secured. But it is fortunately not so frequent a cause of error for misses due to this alone to be serious in number.

(11) Wind presents one element of difficulty that must always prove completely baffling, namely, differing velocity in the various elevations and areas of the projectile's flight. But an anemometer at a height, connected with a velocity dial below that would give the aggregate strength of the wind made by the ship, and the free wind, and a vane giving their resultant direction, should make it easy to get a near approximation of the strength and angle of the wind that affects the projectile. A well understood formula can be embodied in the change of range machine and deflection calculator, so that the wind's influence on length and direction of flight can be added to the other elements. Of course, the strength and direction so arrived at would, even if accurate, only be an index to the wind at the firing ship's position, but it should generally be sufficiently accurate to leave the remaining error negligible.

(12) There remain the variation in the temperature and density of the air. These fortunately affect the flight in either extension or abbreviation in a manner that makes it possible to insert a cam movement in the change of range machine which with a sliding index finger on the main index finger will give with almost complete accuracy the range the gun must be elevated to meet any combination of atmospheric conditions.

ERROR OF THE DAY

(13) It is the error caused by the forces numbered 9, 11 and 12 in Section 8 that is called the error of the day in 'Fire Control and Long-Range Firing', was assumed to be inevitable and immeasurable, and only to be counteracted by ranging the splash of a ranging shot. Now a closer analysis of the problem shows that almost every factor can be very closely, if not exactly, ascertained, and the formulae for correction be embodied in the automatic machines proposed. Clearly, if the geometrical operations were correctly carried out, and the formulae were both correct themselves, and correctly made effective in the machines, error of the day would be eliminated altogether, and firing commenced with full confidence of hitting when the conditions enabled the layers to do themselves justice. But even

if so happy a consummation is never reached at all, by making these corrections the residual error must be notably diminished, and the mere fact that the corrections are possible makes any such name as 'error of the day' for this residual error unscientific and misleading. The error is not due to any mysterious cause, but to incompleteness or inaccuracy in the means taken to eliminate it. It may be necessary to ascertain it by experimental shots; but it is fatal to allow this empirical method of getting the hitting angles to dominate any theory of the aim correction. The striving must be towards avoiding it altogether, and so reducing its extent to the smallest proportions. The less the gun is used as a range finder the better for gunnery.

(14)   But once granted that with the best of instruments, the most expert of operators and the truest of calculations, the first shot is not a hit, the question arises how can we most expeditiously and surely find the missing link between our data and those we want? To the author, attractive as the idea of ranging the splash undoubtedly is, bracketting seems in many respects preferable; and as it has sometimes appeared as if bracketting under the conditions assumed is indistinguishable from the combination of spotting and bracketting now adopted in battle practice, it may be as well to imagine a case in which it is necessary, and describe the operation.

'BRACKETTING' – RANGE UNKNOWN – CHANGE KNOWN

(15)   Let us imagine that an enemy has been under observation for a certain period, that his course has been plotted, and that the change of range machine shows him to be closing in to a 10,000 yards range. The correction for temperature, density, and wind have been made, the deflection angles calculated, and the sights are being altered every five seconds as the change of range machine transmits the new ranges to the gun positions. At 10,000 yards a trial shot is fired. The change in range for the next 100 seconds are given in column form below.

(16)

Range as transmitted to guns

0'	0"	10,000*
	5"	9,925
	10"	9,950
	15"	9,075
	20"	9,900
	25"	9,870*

The first ranging shot would be fired at 10,000 yards. Let it be observed to fall short. To bracket, the range is increased by a safe amount – 400 yards. The second shot will be fired approximately 25 seconds or so after the first. But knowing that the range has changed 130 yards in the time, the sight will be raised, not to 10,400 yards, but to 10,270. This shot is observed

	30″	9,840
	35″	9,810
	40″	9,780
	45″	9,740
	50″	9,710*
	55″	9,670
	60″	9,640
1′	5″	9,500
	10″	9,560
	15″	9,520*
	20″	9,480
	25″	9,440
	30″	9,390
	35″	9,390
	40″	9,350
		9,300*

to be over. The second bracket shot is now fired 150 yards short of the first. Here again the range taken is not 10,120, but 250 yards is added to the range as shown by change of range machine, which is now − 25 seconds after the first bracket shot − 9,710. The sight is therefore raised now to 9,710 + 250, or 9,960. Let us suppose this short, and 50 yards more added. The new range will be 9,520 + 300, or 9,820, and this still being short, another 50 is added, and the last bracketting shot, which is a hit, is fired with the sights at 9,300 + 350, or 9,650 yards.

All we have now to do is to set back the index hand to this range, when future ranges will be given with virtually complete accuracy.

(17) Now compare what has been actually done with what it was intended to do.

Orders given to close bracket	Actual amount sights raised or lowered
Up 400	up 270
Down 150	down 310
Up 50	down 140
Up 50	down 170

Is it conceivable that anyone in the tops with the instruments now in use, could give the orders shown in the right hand column when his intention was to carry out the operations described by his order in the left hand column?

'BRACKETTING' – RANGE AND CHANGE BOTH UNKNOWN

(18) Combined spotting and bracketting is practised to-day with a moving ship and a stationary target, and as all the elements of change of range − i.e., speed and course of own ship is known, this is as easy as bracketting a stationary target from a stationary ship, or a moving target from a moving ship, if the movements of both, as shown above, can be accounted for. But suppose an attempt is made to bracket the case illustrated with the movements of the enemy unknown − that is to say, incorrectly judged both as to speed and course. Let it be done on the supposition that the range diminishes 100 yards per 25 seconds − or per shot − and the opening range to be guessed right within 4 per cent. The

first column gives the actual change of range per shot; the second column the ranges put on the gun; the third the alterations in the assumed rate of change, as the previous assumption was found to be incorrect; the fourth the order for closing the bracket; the fifth the net addition or subtraction to or from the previous range.

(19)

Actual range	Assumed range as varied to close bracket	Rate of change for shot	Bracket allowances
10,000	1. 9,600	−100 short	up 500 − net + 400
9,870	2. 10,000	−100 long	down 400 − net − 500
9,710	3. 9,500	−100 short	up 300 − net + 200
9,520	4. 9,700	−100 long	down 150 − net − 250
9,300	5. 9,450	−100 long	down 100 − net − 250
9,050	6. 9,250	−100 long	down 250 − net − 350
8,770	7. 8,900	−100 long − rate 200	down 300 − net − 500
8,460	8. 8,400	−200 short	up 50 − net − 150
8,120	9. 8,250	−200 long − rate 250	down 200 − net − 450
7,750	10. 7,800	−250 long − rate 300	down 250 − net − 550
7,350	11. 7,250	−300 short	up 50 − net − 250
6,950	12. 7,000	−300 long − rate 350	down 100 − net − 450
6,550	13. 6,550	−350 hit at last	

CHANGE OF RANGE THE CRUX

(20) In the operation described in Section 16, the bracket was closed with exact knowledge that the range was decreasing at the rate of 130 yards at the first shot, and 180 yards per shot for the fifth. In that set out in the above Section, it takes seven shots before the evidence is conclusive that the original rate taken is wrong. Four new rates are successively guessed – and the last, which results in a hit, is as wrong as the others. Illustrations of this character are necessarily arbitrary, but they serve to show principles. And the principle found here is that unless the change of range is accurately known, it is impossible to rely on ever making a hit without a costly expenditure of ammunition, and possibly a loss of time more valuable than the ammunition but that if it is known the thing can be done with the same ease as if both ship and target were stationary. Note also that a hit, when actual change of range is unknown, affords no data for getting future change of range – so that the range must soon be lost and the whole procedure gone through again with disastrous results to the efficiency of the ship's fire.

(21) It should be borne in mind that even if the chart were to give a range that for purposes of hitting were useless – say, with a 4 per cent error – the rate of change given, even with the same percentage error,

would be perfectly serviceable for bracketting; and in the case illustrated would necessitate only one shot more before closing. But there is no reason for assuming so large an error as 4 per cent – nor reason to suppose that any bracket could not be closed much more rapidly by firing several guns in quick succession. In any case, however, a hit would only be useful if it were the final correction for future hitting ranges. It is a fatal error to be dependent on the fall of the shot for data for calculating this rate, for this is to suppose that a hit which may be purely fortuitous, can give the data for distance, speed and course of the target. Guns, always expensive range finders, are useless for finding change of range. It is in these two particulars – i.e., reducing the final bracketting to the terms of bracketting a stationary target from a stationary ship, and being able instantly to use the result as a safe correction for future ranges that the distinction between the bracketting advocated and that in use lies. It is a distinction with a difference more of kind than degree, and seems to offer a solution of the residual error difficulty at once more rapid and more reliable than a system, however successful, for ranging the splash of a trial shot. It is impossible to guarantee the accuracy of individual ranges; but it should be easy to ensure the accuracy of change of range. To sum up: Fire control and long-range firing, assumed gun range to be the end of aim correction after the geometrical factors have been approximately ascertained, we now see that it is not on gun range, but on change of range that the whole art depends.

PART II – TACTICAL

(22) In attempting to suggest certain tactical corollaries from long-range hitting, the writer proposes to deal first with the consequences of the hitting projectile's increased angle of descent and then with those of its decreased velocity.

CHANGES IN VIRTUAL TARGET

(23) On the assumption that a battleship at point blank range presents a vertical target about 26 feet high, whether viewed broadside or end on, a calculation based on the angle of descent of a 12in. shell at different ranges, and the width and length of the ship, shows that at 10,000 yards the end-on position presents a virtual target over 100 feet high, while broadside it is only about a third of this. It must, the writer considers, be assumed that range errors will always greatly exceed deflection errors, and that consequently at any great range the probability of hitting a battleship will always be three times as great if she is end on to the guns than if she is broadside on. So true is this, that it will probably be at least twice as easy to hit at 12,000 yards,

if the enemy is end on, as at 8,000 if the broadside is presented.

(24) Now if the model of a ship be made showing only the armoured structure and the hull, and placed end on to the observer and tilted towards him, so that the end away from him is lifted by the amount of the angle of descent, at say 10,000 yards, he will see, if his eye is on a line with the end nearest him, that virtually one half of the ship is completely unarmoured – in other words, that a projectile coming in the line of his sight could, if the stern were nearest, plunge unobstructed into the propeller shafts, or, on clearing the after barbette and the stern bulkhead (if there is one) penetrate without let into the engine.

If then the end on position presents a target three times as favourable in area, it is also one enormously more favourable as to vulnerability. If the defensive properties of the ship are represented by the figure 100 in the broadside, this will have been reduced first to 33 and next to 16 by assuming the end on position.

(25) Further, however, the end on position reduces the offensive by at least half; and assuming 100 to represent the offensive figure, and 100 the defensive, or 200 the full fighting efficiency of a ship of war, the broadside position may be said to stand to the end on position in the relation of 200 to 66 or roughly 3 to 1. There can be few, if any, occasions when a commanding officer would be justified in putting himself into so hopelessly disadvantageous a situation as to assume an end on position to an enemy who is broadside to him, or remain end on to an enemy end on to him, when by maintaining his position, or turning his broadside, he could keep equality, or improve his tactical advantages from equality to 3 to 1 in his favour.

THE IMPORTANCE OF END ON FIRE

(26) The writer concludes from the foregoing reasoning that in the design of battleships, and in making provision for aim-correction, bow and stern fire may be ruled out of consideration as relatively unimportant, if by so doing any advantage is gained for broadside fire; and it has been objected to this conclusion that it ignores the unavoidable incidents of a pursuit. This objection does not seem well founded if we regard the ultimate goal of pursuit to be an engagement instead of a speed contest.

STERN CHASE THE WRONG CHASE

(27) For there can only be two cases in which the question of pursuit arises – one when the enemy seeks to terminate, the other when he strives to avoid an engagement, by flight. In the first, he has previously found the fire intolerable when he was broadside on; does he better his case by increasing the odds against him three times over? And so long as the

attacking ship does not turn after him, he is under this infinitely more destructive fire until he is well on 12,000 yards away. The time should allow of decisive damage being done to propelling gear or engines, if not to the ship's main structure, even if the attacking ship remained stationary in her position at the moment of his turning. But while a stern chase would be a wrong chase, there is no reason why the attacking ship should not pursue at such an angle as would keep him under the fire of her broadside, thus using twice the armament and keeping up superiority of defence. Very few minutes engagement under such conditions should be conclusive.

(28) Similarly, if the enemy is to be brought to engagement, we must presume superior speed in the pursuing ship or fleet. Let the pursuit be continued until the enemy's stern fire shows he is getting the range; the time has come to assume a broadside position, and with it the tactical advantages already explained.

REDISTRIBUTION OF GUNS

(29) If we then are driven to the conclusion that the broadside fire is of overwhelmingly greater importance than end on fire, it seems as if the disposition of guns in war ships of all kinds might be reconsidered. For instance, it is rumoured that the *Dreadnought* is to carry ten 12in. guns in five barbettes, of which three are in the centre line of the ship, all bearing on both broadsides, and those at the ends fore and aft respectively as well, and two form the arms of a cross on either beam and bear each on their beam and directly fore and aft, but with no training beyond the centre line.[5] Thus except when the target is in absolute line ahead or astern, she will have a broadside fire of eight guns, and a fore and aft fire of four. Now if the broadside is of overwhelmingly greater importance, two guns are being always carried that are practically useless in action – in other words, for the sake of doubling the unimportant end on fire, the ship is burdened with two guns and barbettes, etc., a matter of many tons of dead weight and many thousands of pounds expenditure, with consequent curtailment of other qualities – engine power, armour protection, coal capacity, etc.

THREE GUN BARBETTES

(30) Now suppose both the beam barbettes taken away, and each barbette in the centre line to be carrying three guns instead of two, the net addition to weight will be one gun, and the enhanced weight of three barbettes, supports, training gear, etc. But one gun, two barbettes, supports, etc., will have been taken away, and a considerable economy in weight, space, and cost would result. What is more important, while the armament has been reduced by one gun, the broadside has been increased from eight guns to nine. Meanwhile the end on fire has

only been reduced from four to three. It is possible that with three guns in a barbette instead of two there might be some slight loss of rapidity in fire. But the longer the range the more accuracy given in importance over rapidity. Whatever the potential rate then, it is doubtful if a quick rate could ever be wisely adopted, unless in salvoes, when the objection to the larger barbette would not be great. In any case the gain in other particulars must be balanced against any loss in this. Should this principle of distribution prove workable in practice, it is easy to see how the *Edinburgh* and *Natal* class of cruisers could have been greatly strengthened in offensive, while economising in weight and cost – with added speed or protection obtained by the saving.[6]

BIGGER GUNS

(31) In tabular comparisons of guns, the penetration of steel at 3,000 yards is commonly given – and there has arisen, possibly from this fact, a habit in technical papers of comparing ships by giving the net striking energy of their armament at this range. Thus, when the *Triumph* and *Swiftsure* were acquired, a well-known publication hazarded the statement that as the rapidity of fire of the new ships was so great, and the velocity of the guns so high the couple of them 'could fight the whole five ships of the *Royal Sovereign* class now in the Home Fleet with a fair chance of success.'[7] Much the same style of argument was used in Parliament to commend the purchase, and it is quoted here as the supreme illustration of the absurdities that ignoring tactical conditions inevitably involves. The comparison is of course a possible one if the range in the supposed engagement were 3,000 yards only. But change to the conditions at 10,000 yards, and the 13.5's of the *Royal Sovereigns* are overwhelmingly superior to the 10in. guns of the purchased ships.[8]

(32) And this leads to the importance of very fully considering loss of velocity, and therefore of penetration, at extreme ranges. Thus a 12in. gun would not penetrate the *Dreadnought's* armour at 10,000 yards, nor one *Dreadnought* sink another at that range. But should a 13.5 firing a 1,250lbs. shell be developed with high velocity and of serviceable accuracy, the *Dreadnought* might be at the mercy of a ship greatly inferior in size and number of guns. In other words, just as all the world is agreed that the 6in. gun, so recently fashionable, is obsolete, and that some suspect all secondary armament to be obsolete, so unquestionably will the 12in. gun become obsolete as soon as any nation has a much heavier gun equally handy and accurate.

REDISTRIBUTION OF ARMOUR

(33) But meanwhile it is a question whether, considering the increase

in the angle of descent and the loss of velocity, vertical armour might not be diminished in thickness and the amount and thickness of horizontal armour increased. The writer can see no reason, except the possibility of surprise at short range, in bad light or fog, for maintaining the proportions now used.

TWO POINTS IN POLICY – I SECRECY

(34) Obviously no progress can be made with the art of using guns in the best combination until the art of using individual guns to the best advantage is mastered. It is this art that is to-day interesting the navies of the world more than any other. Compared to it, the arts of ship-building and gun-making are insignificant. On this year's battle practice, the duplication of the fire crews of the *Exmouth's* and the *King Edward's* would add more strength to the navy than half-a-dozen new *Dreadnought's* manned as were the ships on the bottom of the list. Yet, while the most successful efforts have been made to keep the designs of our new ships secret, the means by which hits are got are blazoned abroad for the edification of foreign readers, in an orgy of competing publicity. But if it is important to preserve the secret of the designs of new ships, which can only add a score or two of 12in. guns to our strength, how much more important must it be to keep secret any method by which the efficiency of the 160 guns we already possess may be increased?

II ECONOMY

(35) When all is said and done, it is not the guns that miss but the guns that hit that make for a strong navy. Dewey beat the Spaniards and Togo the Russians because each could excel his enemy in hitting. Similarly the *Exmouth* and *King Edward* could, on the recent battle practices, take on and defeat a dozen battleships that 'on paper' should be six times as powerful. There is an obvious moral to be drawn. On the assumption that the battleships exist solely as platforms for big guns, every 12in. gun we have costs us about £265,000 apiece to put into line of battle, and we have over 160 of them. Leave out the two ships named, and the average number of hits per gun show a derisive dividend on this expenditure. Now suppose that by expending five millions on ammunition, wear and tear of guns, new instruments, experiments and training – the efficiency of the 160 guns now available could be brought up to the standard of the *Exmouth* – which would be the best investment, to spend the money doing this, or in building three more *Dreadnought's*? Only one reply is possible.

CONCLUSION

(36) To sum up, we have to recognise that, with all the world concentrated on the problem of long-range hitting, it is inevitable that a solution will be found. We ought then to be prepared for the modification of ships and fire tactics which must necessarily follow. Above all, it must be recognised that of all monopolies the secret of hitting will be incalculably the most precious. To get possession of it there should be no limit put to experiments or practice; nor can any code of secrecy be too strict for guarding every step in the hard road of its discovery. It cannot be doubted that a request to the Press not to publish any details of experiments with instruments, nor more details of competitive practices than the score of hits made, suppressing all conditions of speed of ship or target, range and rate of fire, or references to systems or instruments used, would, if made in the national interest, meet with the response desired.

V

THE *JUPITER* LETTERS: EXTRACTS FROM LETTERS ADDRESSED TO VARIOUS CORRESPONDENTS IN THE ROYAL NAVY PRINCIPALLY FROM HMS *JUPITER* (*MAY 1906*)

Although Pollen's two-observer range-finding and plotting instruments were ready for trial by September 1905, Harold Isherwood, a design engineer employed by the Linotype Company, had not been able to translate his preliminary sketches of the change of range machine into a fully practicable working design.[1] The cost of building the two-observer range-finding and plotting instruments had, moreover, greatly exceeded Pollen's estimate. The agreement between Pollen and the Admiralty embodied in the letter of 3 May 1905 had, however, made provision for a delay in the delivery of the change of range machine, and its manufacture was, therefore, suspended, pending the outcome of the trials of the completed instruments.[2]

On 25 September 1905, the Admiralty appointed a committee under the presidency of Vice-Admiral Alfred Chase Parr to carry out the trials.[3] The committee was instructed to evaluate the accuracy of range-taking with the two-observer range finder, the accuracy of range-taking off the plot, the capacity of the range finder to measure the distance to the fall of shot, the accuracy of course and speed data produced by the plot, the rapidity with which the system of range-finding and plotting could be made to function and the extent to which the instruments were affected by gun blast and exposure to sea-going conditions.[4] On the same day that the committee was appointed, preliminary experiments with a small boat from the battleship *Jupiter*, which was being fitted with the trials instruments, were begun at Portsmouth and lasted until 17 November. The sea-going trials off Glengarriff, Scotland, began in the *Jupiter* on 18 November 1905 and continued until 19 January 1906.[5]

Four major technical problems became apparent. In the first place, the faulty design of the bearings transmission mechanism resulted in the frequent sending of non-simultaneous bearings to the calculator, which thus computed ranges that were inaccurate.[6] Secondly, the manual manipulation of the plotting arm and pencil proved to be too liable to human error and too slow.[7] Thirdly, when the ship yawed a great deal due to heavy winter seas, there were many instances when the two telescopes could not be trained through a wide enough arc to keep the target continuously in sight, which greatly reduced the number of bearing observations that could be made in a given time.[8] But the most serious problem was posed, as Pollen had feared beforehand,[9] by the effect of yaw on the bearing observations required for the plot. While valid ranges could be calculated from yaw-affected bearings – so long as they were taken simultaneously – a satisfactory plot could not be obtained without bearings that were nearly perfectly accurate with respect to the mean course-line travel of the ship – that is to say, without bearings that were yaw-free. Pollen thus improvised a system of gyroscopic correction by adapting a Whitehead gyroscope of the type used in torpedoes to work a dial that would indicate the extent of deviations from mean course caused by yaw. As bearings were received, they were corrected by the amount indicated on the dial before being plotted with the simultaneously observed range. Trials with this improvisation demonstrated that plotting was feasible if provided with gyroscopically corrected bearings. The spring-wound drive of the Whitehead device, however, limited plotting runs to only four minutes.[10]

But while the trials in the *Jupiter* were in progress, Pollen was informed of two technical breakthroughs that offered solutions to most of the problems that were being experienced. He first learned that a single-observer, self-contained, coincidence range finder of a longer base than had been possible previously would soon be available.[11] The new instrument, Pollen recognised, would be sufficiently accurate to take the place of his troublesome two-observer range-finding system. Pollen was then told that Isherwood had developed a design of a compressed air drive that would keep a gyroscope running continuously at the desired constant rate.[12]

At the end of January 1906, Pollen thus cancelled further trials in the *Jupiter* and began to re-work his system in light of the latest technical developments.[13] Under Pollen's direction, Isherwood, who had by this time displaced Lock and Barr entirely, completed the re-design of the system's components between March and June 1906. Two methods of gyroscopically correcting bearings were developed. In the first, the gyroscope was to be incorporated into the plotter, where the bearings would be corrected after they had been observed by, recorded at, and transmitted from, the observing instrument. In the second, the bearing indicator was combined with the single-observer self-contained range finder in such a way that the indication of the target bearing was determined by the train of the range finder, with an electrical mechanism insuring that the taking of the range also resulted in the taking of the target bearing; bearings were corrected as they were observed because a gyroscope controlled a motor that stabilised the combined range finder and bearing indicator mounting against yaw by altering its train in proportion, an action that, in addition, relieved the range taker of the task of constantly re-laying his instrument. The plotter was improved in the light of the difficulties experienced with manual operations during the *Jupiter* trials by the design of powered mechanisms that automatically translated the electric range and bearing signals from the observing instrument into settings on the moving arm and depressed the marker. The change of range machine design remained essentially the same, but was provided with a mechanical scheme of correcting the computed ranges and bearings for various ballistical factors and the change of range and bearing during the time of flight of the projectile.[14]

In May 1906, while his system was in the final stages of re-design, Pollen collected portions of the texts of six letters which he had written between December 1905 and May 1906 and had them printed as 'The *Jupiter* Letters: Extracts from Letters Addressed to Various Correspondents in the Royal Navy principally from HMS *Jupiter*.' Copies were circulated privately in late May and June 1906,[15] and were so well received that Pollen had it reprinted the next year.[16]

In 'The *Jupiter* Letters', Pollen argued the case for the development of his reconceived system for trials by criticising the existing state of gunnery in the Royal Navy, summarising the experience of the trials in the *Jupiter*, putting forward the technical arguments in favour of his mechanised approach to fire control, and discussing the advantages of possessing a monopoly of long-range hitting with respect to capital ship design, tactics and strategy. Pollen's call for a battleship with larger guns, higher speed, and less armour may have caught the eye of Admiral Sir John Fisher, the First Sea Lord, whose views on capital ship design were quite similar.

INTRODUCTORY NOTE

The following extracts have been put together, with more additions than corrections, at the request of certain friends in the Navy, who are good enough to think that the ideas expressed in them afford food for discussion.

The writer is very far from thinking that what he has said is put into a form suitable for general circulation. In an age when every journalist is supposed to be a naval expert, there is perhaps no occasion for a layman to apologise for rushing into the discussion of highly controversial matters; nor is it his ignorance that the writer would wish to excuse. Apart from obvious defects in style, repetitions, inaccuracies of statement, etc., he is particularly conscious that a tone quite permissible in private correspondence may look stupifyingly argumentative and cocksure in the cold light of print. The consciousness is no doubt an admirable reason for not giving the letters any further publicity than the good nature of the recipients and the patient curiosity of their friends would be likely to secure. But it has been urged with a delicacy of flattery that was irresistible that their familiar and unguarded phraseology is more likely to make the ideas they convey of interest to others, than a more formal treatise.

The letters stand therefore substantially as they were written. But a word of explanation should go with them. The writer seems pretty often, he fears, to speak censoriously of the absence in the work of five years ago of the ideas of to-day. If any censure were intended it would of course be as impertinent as it must be misplaced and absurd. No art has ever been so completely in a state of flux as the art of naval tactics during the last few years. It is in a state of flux now. Those who have identified themselves with certain views and principles will naturally try to gain acceptance for those views by pointing out the defects of ships and systems embodying other views. A partisan cannot be impartial. And while the first principles of the right method of gunnery are undecided, it cannot be expected that an advocate of a novel system can be always temperate in his criticism of methods he hopes to see displaced.

This somewhat overheated style must not be misunderstood, however, to imply any want of appreciation of the value of recent work. The writer holds strongly that the fire control methods of to-day are mistaken in the sense that they are neither satisfactory now, or have the seeds of finality in them. On the other hand, he holds still more strongly that but for the loyal and devoted labour of those to whom the recent and vast revolution in gunnery is due – labour in the face of much opposition and still unrequited – no such progress, as he believes may now be achieved, would have been possible. The subject matter is too vast for any one to be entirely and solely right – the work now being done too splendid for any errors of system to make it entirely and wholly wrong. The solution to be hoped for can only be found in the unselfish co-operation of many men, and in the bringing to the test of experience of all ideas that seem to have any promise of success. The open-minded thoroughness that characterises the bulk of the Officers in the Royal Navy is an absolute guarantee that the solution will be found. The writer hopes that some of the ideas he has put forward may be of use in the process – and his ambition will be satisfied if he only succeeds in stirring the fire from which Phoenix is to rise.

He has therefore yielded to the agreeable pressure of his friends and put the letters into print; but in giving them a wider circulation must ask his readers to forgive their many crudities, on the plea that his motive, like Captain Savage's (or was it Mr Falcon's?)[17] in writing and in printing, was nothing less than 'zeal for the service' he almost abjectly admires.

I

THE THEORY OF A.C.

RIGHT GUNNERY METHODS THE GROUNDWORK OF TACTICS

THE THREE STAGES OF DIFFICULTY IN GETTING HITS

H.M.S. *Jupiter*,
Channel Fleet,
Spithead,
December, 1905.

DEAR ADMIRAL ──

... I am not surprised at your scepticism as to A.C. [i.e., the writer's system of Aim Correction] by revolutionising gunnery, revolutionising naval warfare. It is one of those sweeping statements that startle friends and delight enemies. But really there is something in it. Let us proceed logically, and begin with definitions. What is gunnery? I should say that gunnery was the art of hitting the target with the projectile fired from any gun, and that consequently perfection lay in making such hits at the greater rather than the less distance, and with greater and not less rapidity. Of course the value of a hit must depend upon the striking power of the projectile. Ideal gunnery would, therefore, be hitting at the longest possible range, with the highest possible frequency, and with the greatest possible smashing power. A.C. is concerned with the first two of these only – i.e., ensuring hits at the greatest possible range, or, rather, securing the data for getting them – and securing these data continuously. Hitting itself ultimately rests as much on good aiming as on right data, just as frequency rests on both good aiming and good drilling, and securing the greatest smashing power rests upon good previous administration. Excellence in administration – and by this I mean getting the right guns rightly distributed in ships – rests, first on the possession of right tactical ideas, and, next, on the power to carry them into effect. But it does not require arguments to show that if there is no agreement as to the best method of securing hits, there can be no certain standard of hitting to be depended upon. Unless there is some certain standard, no tactics can be built on it, and consequently no right tactical idea inspire the administration. If, then, A.C. is the best method of securing hits, it is just as vital to the right design of ships, and sound administration, as to the right use of ships and guns. In other words, if it is a tactical improvement, its effect must be felt all along the line.

The theory of Aim Correction rests on a quite simple analysis of what hitting under battle conditions connotes. There are roughly three stages of difficulty in getting hits – (A), (B), and (C). (A) If a stationary gun is firing at a stationary target – the problem that field, fortress and siege artillery have to solve – the only difficulty (apart from aiming, which henceforth we will ignore) is finding the range. The system of bracketting (see note) affords a rapid and certain way of getting this at sea, even without range finders; though of course good range finders should accelerate the process by diminishing the space to be bracketted. For the sake of illustrating my point, however, I will call the difficulty over a series of shots, ten, which is now the arithmetical coefficient of it. (B) Now, if we have a moving ship carrying a gun and firing at a stationary target, we at once complicate the process, both of getting the first range – i.e., closing the bracket – and of securing a succession of hits; for, in our first case, the range once got, hits should be secured indefinitely. The complication in (B) is caused by the ship's movement, which causes a change in the range every moment, and not only alters the range, but necessitates the gun being deflected from the line of sight, so that the lateral

NOTE – Bracketting for hits is a totally different proceeding from spotting for hits. The essence of spotting is judging the error in the range – i.e., the distance of the splash from the target. Bracketting consists in firing two shots so that the target is between them, and halving the interval by a third shot, the target must then be in the nearer or farther half. A fourth shot bisects the half the target is known to be in. The location of the target is thus found by successive shots to be in one of the halves by the third shot, one of four quarters by the fourth, one of eight eighths by the fifth, one of 16 sixteenths by the sixth, and so on until the fraction of the bracket containing the target is less than half the danger space. Of course the target may be found by the third or fourth shot. No guidance by observing the splash is necessary beyond saying whether the splash is short of, or beyond the target. This could be done from the conning tower. (See pp. 89, 90, 91.)

momentum given by the advancing ship may be allowed for. Theoretically, therefore, not only the range but the deflexion, must change every moment. And, finally, this change is seldom an evenly recurring change – in other words, the number of yards to be added or deducted, or, up to a point added, and then deducted, will not be the same for each successive minute.

Considering, then, these difficulties alone for the moment, the problem of hitting a stationary target from a moving ship is very greatly harder of solution than the problem of hitting a stationary target from a stationary ship. The solution of this problem is annually attempted in the Navy at what is sonorously called 'battle practice'[18] – and a comparison of the results obtained in this exercise with those obtained in tests where there is no range problem, practically justifies one in saying that the difficulties have been multiplied at least by ten. If, then, our first illustration represents a problem valued at 10, 'battle practice' must be rated at its square, and the coefficient of B is 100.

But there is a third stage of intricacy (C), the solution of which has not yet occupied the gunnery experts of the Navy – the stage when the target is no longer stationary, but travelling at an unknown distance, at an unknown speed, and along an unknown course. In our second problem, difficult as is the solution in practice, as is shown by the small number of hits and more by the slow rate of fire, at least all the elements that make it difficult – i.e., that cause change of range, are known – viz., the speed and angle of course of the firing ship. But under real battle conditions – which must always be the third stage – two new unknowns are added, that not only make the ascertainment of change of range difficult, but add enormously to the difficulty of getting an initial hit.

Now, reverting to our ideal of gunnery – viz., hitting at the longest range with the highest frequency, what chance is there of frequency when every moment the range is being altered according to a formula into which two unknowns enter? Am I extravagant if I say that, if 'battle practice' squares the difficulties in the gun layers' tests, that battle will square the difficulties of 'battle practice'. If not extravagant, the arithmetical coefficients of the three stages of difficulty are 10, 100, 10,000.

Remember that long range and high frequency are the ideals. Ten hits a minute – i.e., as fast as a gun can be loaded – is done in mere aiming tests – i.e., problem A. If 'battle practice' – i.e., problem B – could be carried out with equal efficiency in the ten minutes allowed, a *Duncan* could make 80 hits with her 12in. guns, and 1,200 with her 6in. As a fact, the best of the class makes 15 with her 12in. and 49 with her 6in. What would she do in battle – i.e., problem C?

If we are to suppose that we can by Aim Correction, or any other system, get real knowledge of change of range, we at a stroke have robbed problem C of all its terrors, and brought it to the simple proportions of problem A – viz., antecedent knowledge of change of range makes hitting a moving target from a moving ship as easy as hitting a stationary target from a stationary ship.

I will write again in a week or two and carry the argument further.

II

H.M.S. *Jupiter*,
Channel Fleet,
Spithead,
January, 1906.

DEAR ADMIRAL——

The master problem of gunnery is, then, ascertaining the formula for change of range, under action conditions, and it is the paramount necessity to do this that has led to the development of the Aim Correction system. It consists in taking ranges and bearings simultaneously and charting the results, so that, *as a preliminary to firing*, the nearest possible approximation to the enemy's range, speed, and course may be on record. The instruments necessary for doing this are means for measuring ranges, either by one-observer short base instruments [such as the Cooke, or Barr and Stroud], or a two-observer system, such as we are trying in the *Jupiter*. Both methods ensure the bearing and range being taken at once; both are completely protected by armour, and can be used either alternatively, or preferably in combination — i.e., two-observer for the broadsides and one-observer for fore and aft. (Continuous gyroscopic correction keeps the zero of bearing constant, in spite of yaw) and, the chart being based on successive observations, there is a natural tendency for both instrumental and personal errors to be averaged out. Ultimately I hope to get a change of range machine perfected which will automatically take care of atmospheric, temperature, and wind corrections, and, when fed with range, speed, course of enemy, speed and course of own ship, and time of flight, will give both range and deflection every ten seconds, and, perhaps, if required, continuously set the sights of all the guns of one nature to such range and deflection.

At any rate, without going into any of the details of the machinery, you will see that the theory of the system is to this extent not only scientific, but unique, in that it is based upon an analysis of all the difficulties and aims at surmounting them all. There will be immense difficulties in getting the thing to work. Perfecting instruments and gear, training operators, justifying experimental expenditure to non-technical people who must officially endorse its being incurred. But all these things have got to be done — and for this simple reason — that, when there is no flaw in the theory of a system, and that system is a necessity, that system *must* be made to work. For my own part, my six years' work, and now my practical experience of the instruments on ship-board, have left me with no doubts at all; in spite of the fact that practical success in this ship has been made impossible by an extraordinary miscalculation that neither I nor my assistants, nor any of the experts, detected until too late.[19]

All this, however, is potential fact, to be proved or disproved; and what you have asked me is not to prophesy but to explain. Well, the reason I said that the success of the A.C. would necessarily revolutionise naval warfare was this. The gun is the unit of offence in naval warfare. With the system and appliances now in use, no fleet could count on destroying a fleet of equal strength without coming within 5,000 yards range. By this I mean that a fleet of *Exmouth's* and *King Edward's* could not destroy an enemy beyond this range, because with an enemy going at anything between 14 and 18 knots on a course that can only be guessed, it would be impossible to count on keeping the range continuously, or being able to concentrate fire with certainty. The reason for this is that 'spotting' is the only method we have for range and change of range finding. There are no means for ascertaining the factors in the formula for change beforehand. There are, or are to be, nice little adding and subtracting clocks – very handy if you know what to add or subtract – but nothing to guide you in judging the amount, except watching your misses.[20] Now, watching misses is a very pretty business; but it is terribly difficult to ensure that you are right; and the results, if you are wrong, are awful. Look at the *Bulwark* – top in gun layers' test, nowhere in battle practice, and all because of the spotters' very pardonable initial miscalculations. Now if under the same kindergarten conditions of battle practice an acknowledged mastership in gunnery can be reduced for ten minutes to utter impotence and inefficiency, and a ship that does not hit with the guns that are its justification for existing is worse than impotent, what is going to happen when every aspect of 'spotting' is made tenfold harder and more confusing, first by several ships firing at the same target; and, next, by that target firing back? For this not only causes smoke and haze round the target ship, but fills the air with projectiles (which, as was found by the Japanese, make the watching of one's own shells exceedingly difficult or impossible).[21]

But it does not need much argument to show that, if spotting is unreliable at a stationary target at a middle range, when only one ship is firing, it certainly can afford no help towards the first crying need of any tactical employment of a fleet – i.e., concentration of all the ships' armaments on one of the enemy. How are we to watch the 'effect of the ship's fire as a whole', as the admirable Rapidan[22] has it, if half-a-dozen ships are churning up the sea all round the enemy? And yet it is on concentration of fire that we must lean if we are ever to destroy.

So you see I come back to where I started from. At present I can see no standard of technical efficiency in the use of their armaments that we can predicate of the units of a fleet. You cannot say that there is a definite fire zone of a battleship, or a line of battleships, such as you can say of a line of infantry, or a battery of field guns, or a machine gun party on land. Shore tactics have been built up on the known efficiency of the units of offence – the magazine rifle, the machine gun, field and horse artillery, guns of position. Their value is known, and can be reckoned with. It can be said quite definitely what is, and what is not, possible in the way of advance over the open in face of such fire, and the knowledge is a determining factor, both in tactics and strategy. But, without some such axiomatic and fundamental data to go on, how can naval tactics be worked out?

If, then, I am to answer your question, how does A.C. revolutionise naval warfare, I reply by analysing the operations necessary to good gunnery, and showing that A.C., if successful, will make certain effects certain – that is to say that, at each range, a fixed proportion of hits are to be confidently expected. Next I remind you that all tactics, with all arms, under all conditions, are based upon the efficiency of the units of offence, and thus show how naval tactics can only begin as an art when these proportions of hits are known to be certain. And, finally, if certain tactical principles emerge clear, why, there is no difficulty in showing how these principles will make new sub-divisions of fleets possible and thus multiply the means for bringing the enemy to battle. And thus the chain of reasoning as to a better gunnery system revolutionising both tactics and strategy is complete.

Without attempting to go into all technical details, let us assume that A.C. will make it possible to range and plot an enemy's course, and so start firing with the range approximately known and the data for change of range certain. A few bracketting shots should correct range error. Let each ship in the squadron make this correction on its opposite number of the enemy's fleet. There will then be no confusing of one ship's bracketting shots by another's. It will not be necessary to observe these shots for exact distance from the target. The bracket will be closed by set 'up's' and 'down's' of the requisite amount, the observer's functions being merely to say 'long' or 'short'. There is no necessity here for masthead positions or judging range. The plotting system has given the enemy's line of bearing – the range of one ship gives the range of any in the line as soon as its bearing is taken – and the change of range will be found automatically. The brackets once closed, therefore, each ship is ready to direct its fire on any target selected by the Admiral. So long, therefore, as the enemy maintains his course and speed, so long will each of our ships have the exact range for every shot.

The formula for hits will be easily obtained in peace practice – the factors being the known average aiming ability, and the known average errors in the guns and ammunition. Suppose, for instance, that some such formula as the following were arrived at as the percentage of hits to be expected from 12in. guns:

At 10,000 yards	12 per cent
At 9,000 yards	16 per cent
At 8,000 yards	22 per cent
At 7,000 yards	35 per cent
At 6,000 yards	55 per cent
At 5,000 yards	80 per cent
At 4,000 yards	100 per cent

Suppose, further, that experience has shown the extent to which sudden changes of course, etc., vary these figures, or delay the time at which these figures may be realised. Well, with these to guide him, an Admiral could choose his time, range, and method of attack, with full consciousness of the effect his artillery will produce. A reliable figure of hits at each range, and their smashing power being known, why, here is a ready made revolution in tactics.

Perhaps evolution is the happier word. Revolution implies the transition from rule to anarchy. What we are seeking is progress from chaos to order.

If both sides had the same tactical and technical skill, certain relative positions would become impossible. Thus no one at any range over 6,000 yards could ever turn end on to an enemy broadside on, as doing so would expose a target so large from a ranging point of view, and so vulnerable, as to constitute a grave disadvantage; and if this became an axiom of tactics, it would not be necessary to build ships with a view to straight ahead and straight astern fire (see note). I enclose you a sketch that illustrates the difference in the angular target presented by a ship end on and broadside on at different ranges.

Well, if we are never to be end on, our broadside fire is all we need consider, and then we shall want it always as powerful as possible. But speed, armour, coal and ammunition carrying capacity, convenience in space, etc., are all vital conditions. We shall not want in action to be burdened with anything useless. Obviously, then, it will be most desirable that all our guns should be available for either broadside, or some of them are mere passengers. It comes, then, to this: That certainty of hitting at 6,000 yards will have made a change in the distribution of guns desirable – i.e., all will be in turrets in the centre line, so that they can be used on either broadside. Possibly this will restrict the number carried, unless three can be put into a turret, but that only emphasises the second tactical

NOTE. – See last letter and diagram.

axiom that A.C. makes inevitable – viz., that, instead of multiplying the number of guns, we must increase their power.

Now, if we look at certain recent ships in the light of these two axioms, we find that with three 3-gun barbettes the *Dreadnought* broadside could be strengthened from eight to nine guns, and two barbettes and one gun less be carried. The extra weight the third gun would necessitate in each barbette would not equal a quarter of the weight removed. Again, the *Duke of Edinburgh* class, instead of carrying six 9.2's each in its own barbette, could, with three barbettes only, have saved 30 per cent of the weight carried and increased its broadside – i.e., its effective fire, by no less than 50 per cent.

Similarly, had these axioms been accepted, no *Lord Nelson* or *Agamemnon* would ever have been designed, with their burden of five useless 9.2's in three barbettes on the unengaged broadside; and, were these axioms accepted now, all haste would be made in getting a 13in. or a 13.5 gun developed, so that the ship of the future could carry six, that, at the ranges of the future, could penetrate or destroy the most powerful armour afloat.

Thus A.C., if successful, at once leads to certain tactical reforms – i.e., the known efficiency of fire at different ranges, and known flexibility of fire direction, would proscribe certain fleet formations at certain ranges, and, by deciding the tactics of engaging, indicate the tactical idea a ship should embody.

History is full of tales of victory following on the first development of a new means of fulfilling the fundamental tactical idea of securing superior force at the crucial point. Don John of Austria, with smaller ships and fewer men, defeated the Turks at Lepanto largely by fencing off a part abaft the bows of his galleys. In galley warfare the decisive element was hand to hand fighting. Don John knew it was hopeless trying to board the larger and better manned Turks; but, by limiting the numbers that the Turks could throw on his ships, he secured superiority of force where it was wanted. Similarly Frederick the Great counted wisely on the superiority of his infantry fire, because, with iron ram-rods, he knew they could load more quickly than Maria Theresa's troops could with wooden rods. Thus Nelson, too, had his gun crews trained so that in action he could always rely on a more rapid fire than he expected from the French. Man for man, and gun for gun, he knew he could count on superior fire; and thus inferior numbers constantly meant superior force. Similar cases can be multiplied for ever – Agincourt, Majuba, Sir G. Pakenham's defeat at New Orleans; victory has more often followed inevitably when superior technique in the use of arms has given an overwhelming tactical advantage.

Well, to-day the same result is to be got by getting the fire zone pushed out – i.e., increase in range; by securing a high frequency of hits, that is, hits in rapid succession — i.e., change of range certain; and by greater smashing power — i.e., larger and heavier guns. These things can only be secured with A.C., or some equally complete system of getting the data for hits before action — not during it — and once got numerical superiority in ships will be of very secondary importance, until of course other navies have the same, or as good, a system.

Of course if we, and as long as we alone, possess and practice such a system, we should be just as invincible as Kitchener's troops were at Khartoum, or Dewey was at Manilla. No enemy could approach his hitting range without passing through a fire zone of ours which should be impassable.

III*

ON THE FUTILITY OF SPOTTING

H.M.S. *Jupiter*,
Channel Fleet,
Glengarriff,
January, 1906.

DEAR ADMIRAL ——
It is rather difficult, without more elaborate mathematical work than I can do, to give you an exact statement of the full case against spotting as a method of long range shooting. But if you will allow of a certain amount of looseness of statement, it should not be difficult to state the case generally.

The requisite data for hitting continuously at long range are as follows:

1 Range ⎫
2 Speed ⎪
3 Course ⎬ of enemy
4 Bearing ⎭
5 Speed ⎫
6 Course ⎬ of own ship
7 Time of flight of projectile
8 Direction ⎫
9 Speed ⎬ of wind
10 Temperature ⎫
11 Density ⎬ of atmosphere
12 Temperature of ammunition

The first seven make up the elements that determine what I call geometric

---

* This letter has been greatly amplified.

range and geometric deflection. I call them geometric, because, assuming that there is no wind, and that the velocity of the projectile is normal, we do not have to consider any other elements to get hits.

Essentially spotting consists in judging, by the impact of shots on the water, how far the range they were fired at was accurate. Suppose in a given case it were perfectly done – i.e., that the exact shortage were read, the range being approximately 7,000 yards. All that can be said with certainty is that a shot fired 12 seconds ago was fired at a range 100 yards too short. Now it is no use raising the sights 100 yards and firing again. The range may well have changed a great many yards in the 12 seconds – or perhaps 15 – before the next shot could be fired. If the 100 yards correction, got by spotting, is to be of value, it must be in addition to the change that the shifting positions of gun and target have brought about, so that, in itself, the correction tells us nothing. But by keeping on firing, and still assuming that the corrections are rightly made, undoubtedly we shall be able to construct a kind of mental chart of what the enemy is doing. We can guess his speed, and guess his course, and feel for his range. But imagine at what a cost in ammunition it is done! Imagine, too, the cost in brain work! The difficulty must be enormous, if not insurmountable – scarcely realised to-day because no one has ever tried to do it. Remember that each mistake of 25 yards by the spotters will involve a hopeless error in the estimate of the enemy's speed and course, and at 7,000 yards, half the danger space – the required accuracy in the range – does not leave much margin for mistakes; so that anything like continuous successive hitting will be out of the question.

I have worked out roughly four test cases of a 5,500 yards range, in which the change of range for the first three minutes is 500 yards – say 83 yards per half minute. In each case the speeds of the two ships are slightly different – the bearing of the enemy altered, his angle of course changed. The rates for the next three minutes are respectively 84, 92, 75, and 147 yards per 30 seconds, and for the ensuing three minutes, 92, 100, 92, and 183 yards per 30 seconds. How can one suppose that in action, even assuming that the rate of change for the first three minutes were accurately got, that the mathematic formula for calculating the ensuing rates would be accurately found? It makes one's head ache to think of it!

Am I guilty of *lèse majesté* in suggesting that there is not a little amateurishness in the kind of way 'engaging under battle conditions' is sometimes discussed? When you think of it, if we went to war to-morrow every gunnery lieutenant would be liable to be called upon to engage a fast moving target at a moment's notice, and this a thing of which not a single one of them, nor any Captain, nor any Admiral has had any practical experience whatever.

It took hours the other day to convince —— and Admiral —— that the problem was going to be so very different from battle practice.

'Oh!' they said, 'with very few shots we should certainly get the enemy's speed within a knot or so, his course within a point or so; we can always get his bearing approximately.' That is the kind of talk that I call amateurish. They do not know that they can do what they say, and, if they could do it, it would not take them very far.

It is easy to construct a case where half a point's error in the enemy's course, and a knot error in his speed, and a trifling error in his bearing, would make the most deplorable difference. Using very exact instruments in the *Jupiter*, we have found that, without using gyroscopic correction, we could not, if there was any sea to speak of, count upon getting a bearing within 3 degrees. In anything like a considerable sea, a ship will yaw as much as 6 degrees in two or three seconds. Perhaps this is an extreme case, but the rapidity of the yaw is extraordinary. At present no ship in the service has any means whatever of taking bearings correctly, and, if we remember that we are dependent on 'spotting' for all the rest, the confusion is likely to be portentous.

I append a diagram that shows what happens when speed, bearing, and course of enemy are incorrectly judged. Line A shows the correct change of range; B the line when bearing is 5 degrees and the speed 1 knot out. C the line when bearing, speed are equally wrong, and his course is misjudged by half a point only. You will note that in the first case there is a steady increment in range. In the second, first a decrease, then an increment. In the last a steady decrease.

This shows how slight an error in any data may cause the most extraordinary errors in change of range, and seems to me to be a perfectly final argument in favour of tackling the problem the other way round, and trying to get data for hits before trying to hit, instead of trying first to hit and then to get the data by missing.

. . . has just been discussing some of the weaknesses of our present fire control. He maintains that the case of the *Bulwark* is crucial. She made no 12in. hits, not because her fire control and aiming were bad, but because both were too good. Had both been bad, every gun would have fired at different ranges, and hits must have been made. He says this is the explanation of such hits being made as were made by the Japanese at Tsushima. Had there been a well organised system by which all guns fired together, any mistake made by the system would have been repeated by every gun. There being no system, there were, of course, an awful lot of misses, but so many shots were fired and with such a fine freedom in selecting range, that there was inevitably a very serviceable proportion of hits. So he maintains that with our system it is, of course, just possible that now and then a ship would get in a broadside – but the chances are that we should get no hits at all on a fast-moving target. His complaint is that an unscientific system carried out with the mechanical precision only to be found in the British Navy is the most perfect means yet devised

Time - minutes

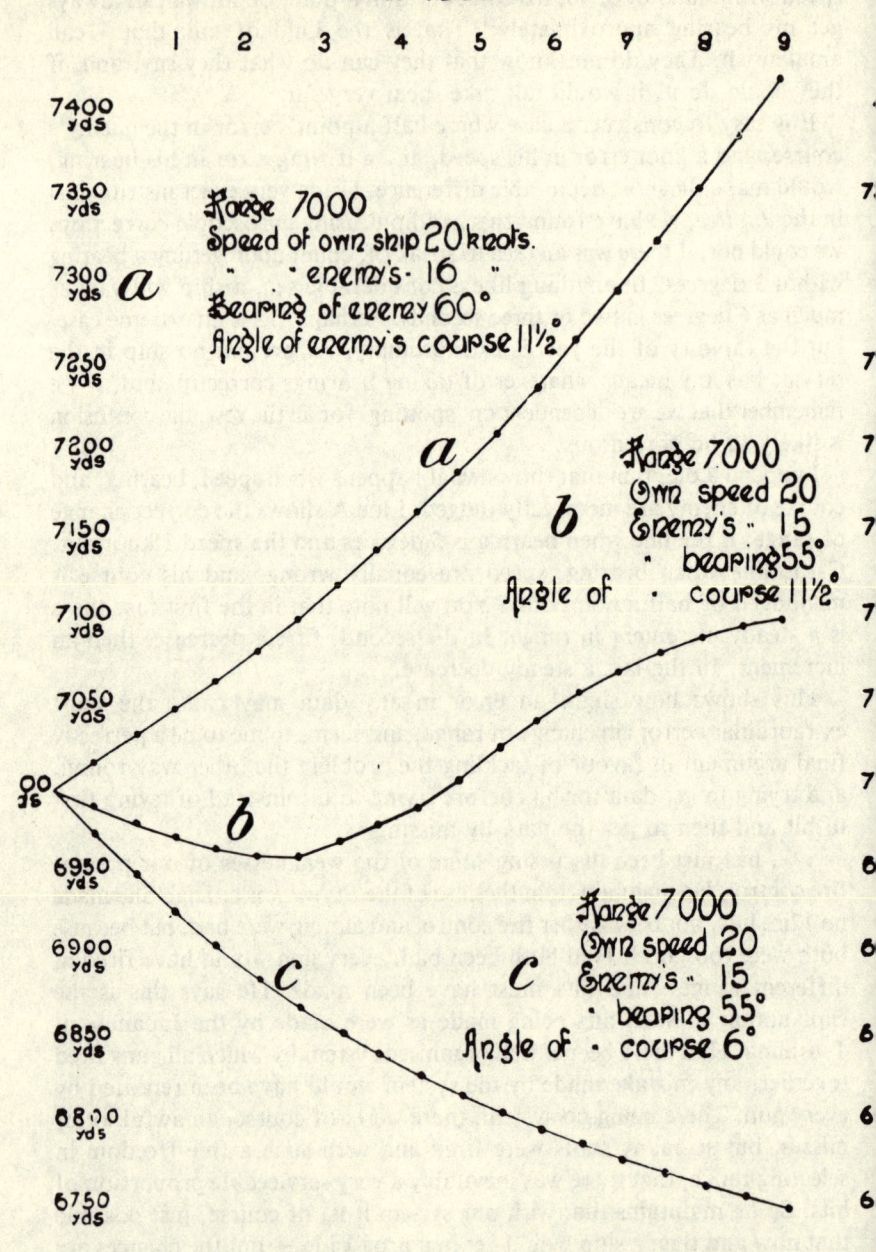

1 2 3 4 5 6 7 8 9

a
Range 7000
Speed of own ship 20 knots.
 "   " enemy's · 16   "
Bearing of enemy 60°
Angle of enemy's course 11½°

a

b
Range 7000
Own speed 20
Enemy's " 15
 " bearing 55°
Angle of " course 11½°

b

c   *c*
Range 7000
Own speed 20
Enemy's " 15
 " bearing 55°
Angle of " course 6°

for ensuring a minimum of hits. This sounds paradoxical, but it is well reasoned. Make the working of the system faultless, and you ensure that its errors dominate the result.

If you look back at the list of data necessary for getting successive hits, you will see that there are four to be got by observation of the enemy. Which seems the more reasonable system – to have instruments and appliances specially designed to get these data, or to fire 12in. guns at him, at a cost of some hundreds of pounds a minute, in the hope that out of the multitude of misses wisdom will emerge? The only parallel to our present procedure is Charles Lamb's story of the Chinese burning down their houses on the chance of a pig getting roasted in the process. It took them some years to discover a better means of getting pork and crackling.[23] So with our method of trying to get hits. The cost is prohibitive, the possibility of accurate observation exceedingly small and strictly limited to a comparatively short range – even if correctly read, the mathematical problems involved in calculating future ranges are almost insurmountable on guess work – and, above all, the entire personnel, instruments, and communications are necessarily exposed in the process. The Chinese analogy is pretty close here too! Their pig-roasting must have brought about some startling surprises when the wind got up.

The grounds, then, for condemning spotting as a method of gunnery may be summed up as follows:

1 It does not set out to get the information that is really necessary. Instead of seeking the range and bearing of the enemy and plotting it to get his actual course, from which his speed could be calculated, it only pretends to give, many seconds late, a range correction for any particular shot. Dealing only with the problem of a *continuous* supply of the geometric range, I point out that spotting does not even pretend to get it. But, of course, the allowances that must be made for abnormal conditions make the difficulties of using such data as may be obtained much greater. Spotting gives the range and deflection in terms of the sights – i.e., gun range. This must be explained by reference to data 8, 9, 10, 11, and 12, in the list given. There can be no attempt to solve the geometric problem apart from the ballistic problem. This is fatal to accuracy.

2 Next, I point out that to get such efficiency at 6,000 yards as is possible at 5,000 yards, the mast-head position must be raised by 22ft., and so on for each 1,000 yards. As this is impossible, it means that the difficulties of observation increase in exact proportion to the increase in importance of exactitude of observation.

3 As it is inconceivable that two, or more ships could spot together at the same target, and as the spotting system is dependent upon continuous spotting to supply continuous range, concentration of fire

is impossible. By this, of course, I mean concentration of long range fire at high frequency.

4 Whatever success it meets with is due to the possession of special gifts and good luck. ——— and ——— are miracles at spotting – their ships' performances show it. But are there enough of these genii to go round? Hence my feeble joke that A.C. substitutes a common but impregnable pianola for the rare and precarious Paderewski.[24] You can never have a reliable standard of fleet or ship efficiency, in other words, a groundwork for your tactics – as I argued in my first letter – if you are liable to be thrown out by the chance mistakes or misjudgment of your spotter. Each ship will have a different figure of merit in the line, and that a varying one. If you look at the 'battle practice' returns you will get crushing evidence of this.

5 But the unreliability of your spotter is not worse than his precarious tenure of office. You cannot protect him. His position is the most exposed and conspicuous mark there is for the enemy's shells. After August 10th,[25] Togo had to take all his fighting tops down – they were untenable in action. So that even your Paderewski may not last you five minutes, and your flagship, which stood first in your list of efficient ships, might by one shot be reduced to the uselessness of a third-class cruiser.

These opinions are not popular. They do not sort with the kind of gush which brilliant journalists in and out of the Service supply to the technical and lay press about naval gunnery and gunnery progress. But I doubt if there are half-a-dozen Admirals or Captains who do not, in their heart of hearts, share them. Have you ever met one who did not hold that our proper tactics in action would be to close to decisive range as quickly as possible? I have heard the phrase a dozen times. It tells you a good deal. It means that we have a well-founded belief that at ranges where range finding and change of range are not problems – i.e., 4,000 yards or under – our superior fire discipline and the excellence of our gun layers' training should give us a crushing advantage. But what a confession of impotence to solve the problem of using our artillery! It virtually admits that all our fire control is but an elaborate attempt to improve the chances for a few first hits only. It is not a *method* of gunnery complete and final in itself. Yet I have seen a 12in. gun placing shot after shot in a 40 yards circle at 8,000 yards. Why are we not to use them at these ranges, and at these ranges *only*? Merely because we have no means of making the hit – not want of skill in aiming – not want of brains in the officers – not want of zeal, keenness, discipline – solely want of right method of getting the data for the proper elevation and deflection of the gun.

IV

EXTRACTS FROM A LETTER TO ADMIRAL PARR

188, Fleet Street,
London, E.C.,
2nd February, 1906.

To VICE- ADMIRAL ARTHUR CHASE PARR
President of the Committee on Aim Correction.

Sir,
In accordance with your request at Monday's meeting of the Committee on Aim Correction, I am submitting to you herewith the observations I have to make on the experiments carried out with my apparatus on H.M.S. *Jupiter*, during the months of December, 1905, and January, 1906. I propose to cast my remarks in the form of a general review of what we have succeeded in demonstrating, and then to state how far we have fallen short in realising the system of Aim Correction set out in my pamphlet on 'Fire Control and Long-Range Firing'; to indicate the difficulties that prevented our fulfilling the requirements of the system; to show how these difficulties can be overcome; and, finally, to submit for the consideration of the Committee a simpler method of obtaining a large part of the advantages of my Aim Correction system without the use of two-observer range-finding.

At the time when the designs of the instruments tried in the *Jupiter* were put in hand, there was no certain evidence that it was possible to construct instruments that could be used on shipboard under any conditions for obtaining ranges by the use of two observers. The points of uncertainty were: (1) Whether in Azimuth telescopes the necessary accuracy could be obtained and maintained for measuring and recording angles with the required fineness; (2) whether the synchronism of observation could be made certain; and (3) whether the transmission of the angles observed, their computation and a charting of the target's course from the ranges and bearings obtained, could be carried on in a ship.

The experiments have shown without any question that we have succeeded in building Azimuth telescopes capable of reading and transmitting with an accuracy of two seconds of arc; and that these instruments maintain their accuracy after two months' usage on ship board without any special provision being made for protecting them from damp, and in positions very unfavourably exposed to the blast of the heaviest guns using full charges.

Under conditions in which the method of sub-dividing the arc and of training were applicable we have succeeded in getting ranges of hitherto unattained length and with unexampled accuracy.

We have demonstrated that if ranges can be read with approximate accuracy and the bearings of the target with regard to the centre line of a ship read with perfect accuracy, and corrected with gyroscopes, the chart can be made of a moving target's course with so close an approximation to truth that the exact data for obtaining the speed and course of the target are available.

The trials have, however, shown that with instruments as designed it was not possible to get reliable ranges continuously when there was any considerable movement of the ship. [It is to be noted that in the same stage of movements it would have been equally impossible to have maintained anything like continuous aim with guns, or to have done more in the way of shooting than taking occasional snap shots.] The conditions under which these trials were held, were, I think it must be acknowledged, exceedingly unfavourable to any system. The wind was fresh, the sea short and violent, and the speed of the ship exceedingly slow. It was noticeable, for instance, that in the course of the speed trial on Sunday, the 28th ult., when the ship was going sixteen knots, the movement of the ship was perfectly smooth, and that the conditions would have been favourable for observation even with the training mechanism as at present on the telescope.

It has, however, to be stated without any reservation that the instruments have failed to carry out the requirements of the system under weigh, because the arc of secondary training of the telescope was insufficient, the means of training were neither sufficiently rapid nor sufficiently sensitive, and the indicator, while securing that no message should be sent from either telescope without synchronism, was not able to prevent that message being confused by subsequent messages from one or both of the telescopes, and, consequently showing a set of figures which, for the purposes of accurate range finding, were misleading.

These defects were due to a fundamental misapprehension of the extent, the speed and the variety of the ship's movements in heavy weather.

[After discussing various possible improvements in the instruments, the writer proceeds.]

To sum up, the trials have demonstrated the capacity of the two-observer system to get exact range at very great distances under favourable conditions, and have shown us exactly what changes must be made to make all conditions favourable. But it is of more immediate moment to have proved that the charting system can be worked successfully and accurately when the gyroscope corrections are made. In my opinion the importance of getting long ranges accurately is in reality not so great as the plotting of all ranges and bearings, and so long as the latter are true the accuracy of the former is largely negligible. The reason is that range errors can be corrected by a minimum number of ranging shots, while change of range errors must cause the ranging shots to be almost indefinite in number. Now the charting system automatically averages out a greater proportion of range errors, and however erroneous the assumed range got, the value of the chart in giving the data for change of range will hardly be affected by it; certainly in no case in proportion to the error.

This leads me to suggest to the Committee that as the final development of the two-observer system must take some little time, the Aim Corrector System should be tried with one-observer range finders in the meantime. I have devised means for transmitting from such instruments the bearing of the line of sight relatively to the ship, concurrently with the range, and this would enable the instrument to be put into a barbette or gun shield – so that the most undesirable feature of the present fire control organisation, namely the exposure of the entire *personnel*, the instruments and communication in unarmoured tops, would be at once done away with. Doubtless even with the new long base instruments now being submitted by the makers, no such accuracy as is attainable with the two-observer system could be got; but I can see clearly that if range finders giving 8 per cent accuracy at 10,000 yards can be made, the charting of ranges and bearings should give results that would quite revolutionise modern gunnery.

The results of the recent battle practice show two things conclusively: The first is that the personal equation is far too prominent, the next is that if these are the best that can be got with a sitting target, we are to expect much less satisfactory results when the target is travelling at an unknown speed, and at an unknown angle. Charting gives speed and course independently of range accuracy, and so makes the closing of a bracket as simple as if both guns and target were stationary; and the bracket once closed, all errors are corrected. It secures this without stipulating any genius for judging distances, speed, or angle of course. In other words, it reduces the personal equation and inevitably raises the average of shooting. No doubt the best men will always get much better results than the average. It is sincerely to be hoped that they will. But no system and no training will produce men of genius at will; the few there are will not remain gunnery officers all their lives, and be they few or many, the present system puts them at the mercy of the first lucky shot.

I annex hereto a diagram that illustrates the value of charting to be irrespective of range accuracy; and shows how bracketting, when speed and course of enemy, and consequently change of range are known, differs not in degree but in kind from 'spotting' for hits. With spotting, the first shot must be fired with three essentials unknown: (1) range; (2) speed of enemy; (3) course of enemy. The diagram shows that if (2) and (3) are got

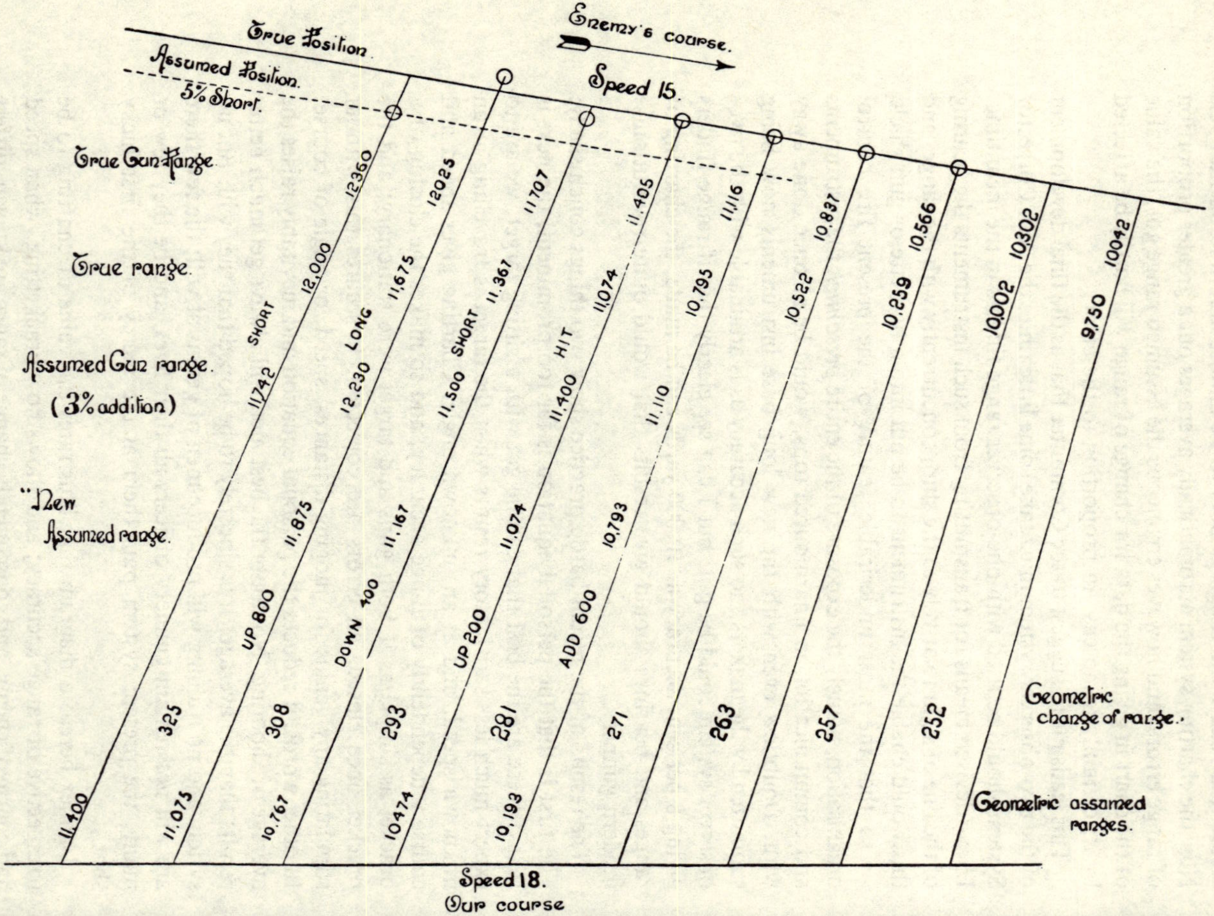

True Position.
Enemy's course.
Assumed Position.
5% Short.
Speed 15.
True Gun Range.
True range.
Assumed Gun range.
(3% addition)
"New
Assumed range.
12360
12025
11707
11.405
11.116
10.837
10.566
10302
10042
12,000
11,675
11.367
11.074
10.795
10.522
10.259
10,002
9750
SHORT
LONG
SHORT
HIT
11742
12.230
11.500
11.400
11.110
11.875
11.167
11.074
10,793
UP 800
DOWN 400
UP 200
ADD 600
325
308
293
281
271
263
257
252
Geometric change of range.
11.400
11.075
10.767
10474
10,193
Geometric assumed ranges.
Speed 18.
Our course

by charting, (1) can be bracketted for exactly as if the target were still and the gun at rest.

Our ship is steaming 18 knots. The figures written nearest its line show the range the enemy is assumed to be, on the supposition that his course as shown by our chart is correct. This course is indicated by a line approximately parallel to his true position. The lines connecting our course with the enemy's represent the changed ranges at intervals of 30 seconds; the figures between the lines in the second row from our course indicate the change of range for each interval.

Taking the first line, the geometric assumed range is shown 11,400; 3 per cent is to be added for the ballistic coefficient to give the gun range. The sight is set to 11,742 yards. But the true range is 12,000 yards (e.g., we are 5 per cent out in our chart), and the true gun range is 12,360 yards. The first shot is therefore short. We now fire a bracket shot 800 yards up. Passing on to the second line, we see that this 800 yards is added to the assumed range – now reduced by 325 yards – and with necessary corrections fired at 12,230 yards; the true gun range being 12,025 yards, it is long. The third line shows the bisecting of the bracket – i.e., 'down 400,' in reality the assumed range plus 400 – which is short. The fourth line the bisecting of the further half of the bracket – i.e., 'up 200,' in reality the assumed range plus 600. This, with corrections, gives a hit, and is within five yards of the true gun range. Note that if we treat this 600 as a geometric error only, and add it to our assumed geometric range, we practically get the true range, and this with the ballistic coefficient corrections gives us a succession of hits. In other words, though our range is wrong, our change of range, which means the rate and direction of the enemy's movement – vital for getting the deflection necessary for the time of flight, and for securing hits with successive shots, are almost automatically obtained with an accuracy that is quite adequate. As it is unlikely that it would ever be necessary to open a bracket exceeding 16 times the width of danger space, the maximum number of shots to close should never exceed six. There should be no difficulty in judging when a shot is short or over from the barbette or conning tower, as it will be quite unnecessary for any estimate to be made of the extent of the shortness.

I have not touched upon the question of ranging the splash of the shot for three reasons. The first is that our experience at Glengarriff was too limited for any useful deductions to be made as to the capacity of a two-observer system to do this. The next is that since writing 'Fire Control and Long-Range Firing', I have learnt that the formula for correcting variations in temperature and atmospheric density are now known and can be embodied mechanically in a change of range machine. Finally, it seems as if bracketting, as set out above, would never require more ranging shots than splash ranging, and would always be more certain in its results.

In conclusion, I know very well that to express my personal thanks to Captain Warren, M.V.O., Commander Bertram Smith, Commander Carr, and the other Officers of H.M.S. *Jupiter* for their energetic and enthusiastic assistance during the trials, often at the cost of great inconvenience to themselves, would be the merest impertinence on my part. No one could spend two months in the ship without being fully and pleasantly aware that anything that promised an advantage to the service would be certain of their devoted attention. At the same time, it is impossible to close these remarks without recording my deep appreciation of the fact that nothing I asked that could at all contribute to the success of the trials was ever refused in the ship, or denied me by the Admiralty.

I am, Sir,
Your obedient servant,
A. H. POLLEN

V

RANDOM NOTES ON SHIP DESIGN, ETC.

188, Fleet Street, E.C.,
February, 1906.

DEAR CAPTAIN ———

I had noted the passages you mark about the *Dreadnought*, apropos of her launch; and I am also a diligent reader of the so-called Service papers, and quite agree that you could not well have better instances of confusion of thought. The encyclicals of the so-called naval experts and the pronouncements of the Navy League are all in the same strain. The creed that seems to be professed is something like this:

1. The tactical value of a ship is the sum of her speed in knots, her armour protection in inches of Krupp steel, and aggregate of the foot ton energy of her broadside at the muzzle, multiplied by the number of rounds she can fire per minute, plus aggregate foot ton energy of her straight-ahead and straight-astern fire calculated on the same principle. The result is a hybrid formula that represents her figure of merit.
2. The strength of a fleet is the average of this figure of merit of all the ships in such fleet.
3. But if it desired to compare the British Navy with foreign Navies, these formulae are ignored, and a new classification of battleships, cruisers, etc, is adopted, each kind being divided into 1st, 2nd, and 3rd class, and numbers of such units only are counted. In these comparisons the

Worth and *Kaiser* class figure as 1st class battleships, as if they were the equals of the *King Edward VII*.
4   Guns are rated by the speed of loading and the striking energy of each shot at 3,000 yards, or the muzzle.

Imagine a ship armed with 9.4's engaging the *King Edward VII* at 6,000 yards! Why, she would not be a match for the *Duke of Edinburgh*!

These doctrines are a little contradictory, but as far as they betoken a frame of mind, it would appear that it could be illustrated thus by a comparison in a different field.

Six men with double-barrelled 12 bore guns is the normal complement for shooting certain covers. But if we had eight we should get more birds, and if they had 10 bore instead of 12 bore guns they would do better still. It does not matter that the original six men were six Earl de Grey's, and the eight 10 bore men were eight duffers. They are better armed and more numerous – they must get a bigger bag.

Of course, all this appallingly foolish talk is generated by an incapacity to realise what the unit of naval efficiency really is. The unit is, of course, the hit, and the hit only. Let it be far, frequent, formidable – but above all a hit. Who was it, Nelson? Moltke? Napoleon? who said 'only numbers can annihilate'? Translated into naval warfare this does not mean number of ships, or number of tons, nor inches of armour, nor knots of speed, nor number nor weight of nor muzzle energy of guns. But just simply number of hits. I wonder what 'The Globe'[26] would say if it were announced that owing to the British Fleet being able to secure 50 per cent more hits per gun than the navies of France, Germany, Russia, and Italy, therefore it was unnecessary to maintain numerical superiority? I am far from saying that that would be sound policy; but I am certain the superior number of ships will mean nothing if it does not mean superior numbers of hits also.

I suppose there has never been a similar case to that of modern navies in all countries. The men that use the machines with no certain data to go on to formulate a method for using them, and the men that build the machines having, until quite recently, no scheme to build to except to out-design their predecessors by increasing the speed, armour, and gun power. Should a genuine hitting system be developed we may see wide changes. I suppose a 13.5 gun would roughly have nearly double the smashing power of a 12in. gun at 10,000 yards if both had the same muzzle velocity. Of course, if 'decisive range' is to be 3,000 or 4,000 yards, there would be no object in having heavier guns. No armour could resist a 10in. gun at that range. More guns that are heavy enough would do as well or better. But with a hitting system, the decisive range would be the range at which penetrating hits could be counted on as a certainty. We can easily imagine a 13,000 or 14,000 ton battleship armoured as to vitals with 12in. of Krupp steel, capable of 18 or 20 knots, and carrying four 13.5 guns. Now suppose she could hit the *Dreadnought* at 10,000 yards and the *Dreadnought* hit her. It is true the *Dreadnought* would land twice as many hits as our hypothetical ship, but a 12in. projectile would not smash up 12in. of Krupp steel at 10,000 yards – it would hardly mark it. But the 13.5 would smash up the *Dreadnought's* armour.

It is a disturbing possibility, but one well within the pale – it only requires a development in gun making that is nothing compared to the development actually made in the last 25 years – and, I mustn't forget, a hitting system!

What becomes of the epoch-making changes of the last few years if this happens? Believe me, there can be no epoch-making change until the unit on which all tactical truths must be built is a certainty.

* * * * * *

Of course, I quite understand that the *Dreadnought* is a great advance on anything yet built, not only in result, but in origin. I have not denied that she is built to a tactical idea, and is thus in a new class. What I was saying amounted to this, that until we are technically sure of our weapons – i.e., know how far and how often we can hit – there must always be a quicksand in the foundation of our tactical ideas. I am not such a fool as to suppose that deciding a new design is easy. Cost, the chances of mines and torpedoes, the thousand possibilities that arise in thick weather – there are a heap of considerations that only experience and very full information can weigh properly. I am concerned with one consideration only – viz., the influence on tactical developments of certainty in hitting.

I suppose the *Dreadnought's* origin was largely the determination to build a ship that no foreign ship could destroy at some critical range, and to arm her so that at and beyond that range her armament would pierce any foreign ship afloat, and engine her so that she could keep that range. This is a great advance. But it would be folly to deny that the same process of reasoning would enable us to out-*Dreadnought* the *Dreadnought*, if, and as soon as, we can rely on any fixed percentage of hits beyond her critical range. The new 12in. guns can pierce 9in. of steel, say, at 6,500 yards, while her 12in. of armour are proof against 11in. shells at this range. But a larger gun would pierce 12in. and much beyond. A smaller ship could be built with equal or superior speed and equal armour, and carrying some of these – and there you are. Meanwhile – and this is the real point to insist on – you cannot be sure of the *Dreadnought* making any percentage of hits at all at her critical range. At a fast target – and the *Deutschland's* are supposed to do 19 knots – it is not likely that until something better in fire control is produced that she will do better than other ships. No skill in building, at any rate, can secure it.

* * * * * *

I don't see even now why so many guns in so many barbettes are to be put in. Aimed fire at long range, with a moving target, can never be so very quick. Would not six guns give you practically as many *aimed* rounds per minute as eight? There are a lot of things to think of. I admit the practice of to-day is not a guide, but let us take ideal conditions – such as you get in gun layers' competitions. It would be worth having a rapidity test, the *Exmouth* against the *Dreadnought* – a four-gun broadside against an eight-gun broadside. Eliminate fire control, and see how many more *aimed* rounds one ship can do than the other. I would bet heavily against the *Dreadnought* doing anything like double the number of rounds or making anything like double the number of hits.

If she can't justify the number of guns – i.e., eight, the construction of ships would be both simplified and cheapened by limiting guns to six – three barbettes in the centre line. The idea grows on me more strongly, ever since I first broached it to you last June. If she can get double the number of rounds and hits, why then try the three-gun barbette – a broadside of nine. It is rash to prophesy, but I am quite sure in my own mind that the ultimate ship will never have more than three barbettes, never more than six guns, and probably less.

The battle of the future will not be a mêlée but a stalking match – it will always be too touchy a business getting hits to justify endless salvoes. We must not allow the analogy of the magazine rifle and machine gun let rapidity and volume of fire obsess our minds. Frequency of hits is more likely to be got by quick succession of hits, than by simultaneous hits.

Pursuit is, of course, what is always thrown at one's head. But a simple case clears the air. Pursuit argues superior speed. Let the respective speeds be 15 knots in the pursuing, 12 in the pursued. The ships are end on say 6,000 yards apart. The pursuing ship wishes to destroy the pursued – does not wish to catch up for any other reason. She has a choice of gaining three knots per hour – i.e., 900 yards in nine minutes – by maintaining her present course, with offensives equal, and all the time exposed to the appalling risks of plunging fire, or losing 2,400 yards in nine minutes by turning 60 degrees to port or starboard. If she turns, say, 60 degrees, all her guns bear, and the target she presents is reduced to the minimum in size and is at its maximum in power of resistance. The question is this: Which is the better policy, to double her armament and halve her vulnerability, the enemy remaining at half armament and with engines, etc., exposed – i.e., alter the odds from equality to I don't know what in her favour, all with the range increasing – or remain on equal terms, with the range decreasing? If we assume that we can hit up to 8,000 yards, I can't understand there being a moment's hesitation.

The diagram shows the increase in range in the three cases of (1) equal speeds, (2) our ship having a superiority, and (3) the enemy having superiority.

Time - minutes.

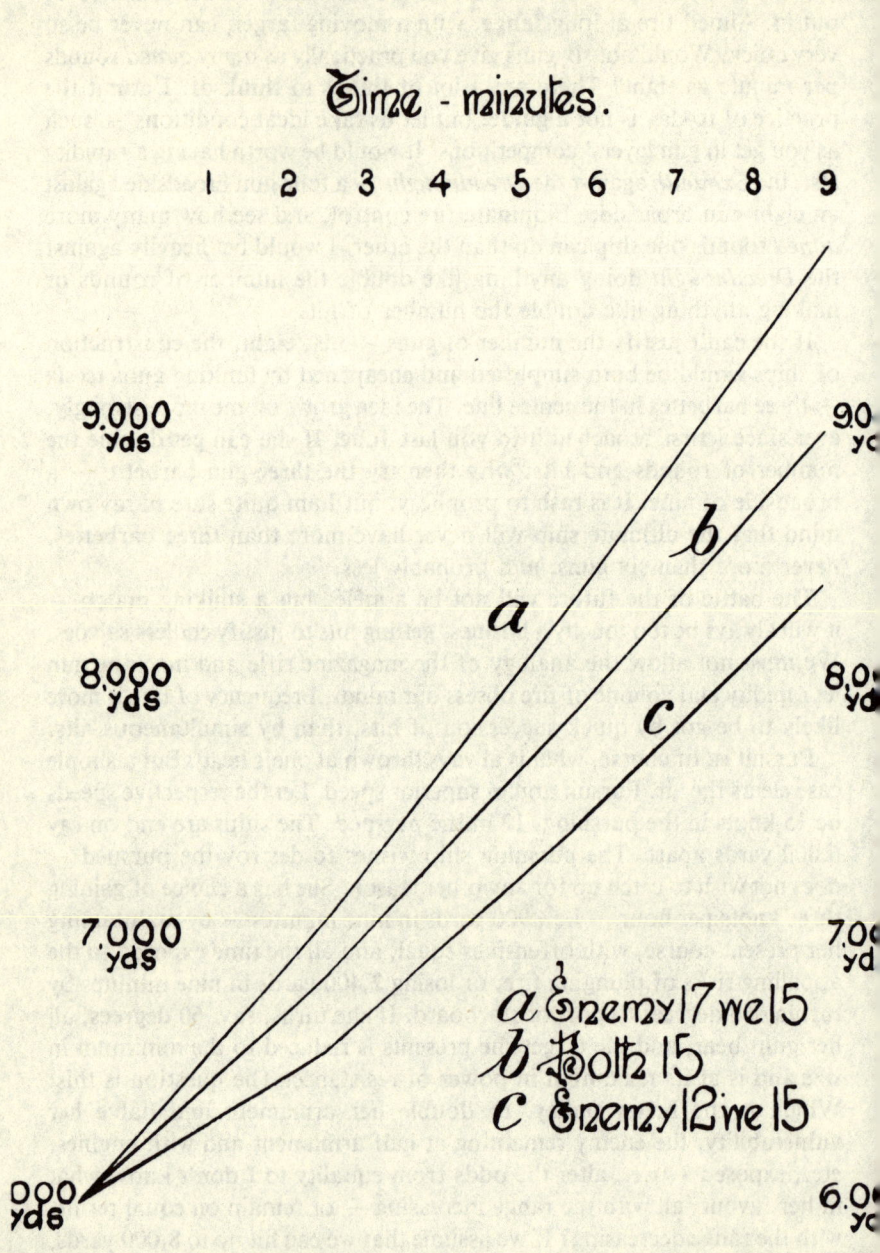

1 2 3 4 5 6 7 8 9

9,000 yds

8,000 yds

7,000 yds

6,000 yds

a Enemy 17 we 15
b Both 15
c Enemy 12 we 15

EXPLANATION OF DIAGRAMS OF THE ANGULAR SIZE OF A BATTLESHIP AT DIFFERENT RANGES

For the purposes of these diagrams, a battleship is assumed to be 400ft. long and 30ft. high and 75ft. wide. The three targets used in the Navy – viz., the gun layers' test target, the Whale Island 6 by 8 training target, and the battle practice target, are superimposed at their characteristic ranges, to afford a comparison. It will thus be seen that the battle practice target at 6,000 yards is the same angular height, and is, roughly speaking, a fifth of the length of a battleship broadside at 12,000 yards. It is little more than a third wider and is about one-third of the height of a battleship end on at 12,000 yards.

As far as aiming goes, there is no reason why a man who can hit the 6 by 8 target several times a minute should not hit a battleship at 12,000 yards broadside on several times a minute. End on the standard of accuracy both in getting the range, and aiming, would be materially reduced by the target being three times as favourable vertically.

The angular width of a battleship broadside on would be six times longer than the diagram at 6,000 yards, and at 12,000 yards three times longer.

VI

LONG RANGES AND NEW SHIPS
SPEED – ECONOMY

London, May, 1906.

DEAR ADMIRAL ——
I am delighted with 'Maga's' mysterious writer's history and provoked by his perverseness in applying it.[27] His and Captain Mahan's[28] views on speed seem to be very clearly erroneous. To an ignorant layman like me the most astonishing thing about naval controversies generally is the confusion that comes of mixing up the vital with the non-vital. Not till hitting is recognised as practically the only test of effectiveness will clearer thinking become the rule.

If we assume that guns can hit at long range – as, of course, they ought – such pestilent rubbish, for instance, as was talked about the *Triumph* and *Swiftsure* when they were purchased, would appear rubbish even to a naval expert in a halfpenny paper. 'Brassey's Annual', after dwelling

Angular target presented by battleship end on at different ranges
Size gun-layers target at 1600 yards is super-imposed.

1200 yards
1100 "
1000 "
9000 "
8000 · 7000 "
6000 yards

Angular size of Prize Firing Target
Gun-layers test 1600 yards

Angular size of 6×8 feet target 1600 yards (Whale Island Target)

Angular size of 30×50 battle practice target 6000 yds.

1905

Angular target presented by battleship broadside on at different ranges

6,000 yds
7000 "
12,000 "

lovingly on the rapidity of fire and high velocity of the 10in. and 7.5in. guns in the new ships, went so far as to say that the couple of them 'could fight the whole five ships of the *Royal Sovereign* class now in the Home Fleet, with a fair chance of success.' Much the same style of argument was used in Parliament to commend the purchase. It is the supreme illustration of the absurdities that ignoring tactical conditions involves. The comparison is, of course, a possible one if the range in the supposed engagement were 3,000 yards only. But change to the conditions at 10,000 yards, and the 13.5's of the *Royal Sovereign*, supposing both ships can hit, are overwhelmingly superior to the 10in. guns of the purchased ships.[29]

You cannot get away from this, that as soon as we know that our guns can hit at long ranges regularly, we shall have to base our tactics and our shipbuilding on the inevitable corollaries. The new factors are the increased angle of the hitting projectile's descent, and its decreased velocity. The increased angle will limit the positions that could safely be taken relative to the enemy − i.e., the end on position becomes barred − and make new distribution of guns and armour necessary. Guns all available for both broadsides, armour less vertical, more horizontal.

But beyond everything, you must have speed − for this is the determining factor both in choosing an initial range, and maintaining it with the minimum of change. Tactically, it is second in importance only to technical perfection in the use of guns. Protection − i.e., armour − is a bad third.

The old huntsman of the Llangibby pack was the hardest rider in any field in the island. But his contempt for horses was only equalled by the splendid use he made of them. To him a horse was just an expensive, stupid, ineffective and dangerous way of getting about the country. But the horse was also the only way of keeping with the hounds, and so indispensable. It didn't much matter to him what kind of horse he had; it mattered a good deal to the hunt whether he was with the hounds or no. Any horse that kept him with the hounds was good enough.

After all, a ship is only an expensive, dangerous, costly, and clumsy machine for carrying 12in. guns − or the biggest guns you can get. Its first duty is to carry the guns somehow − next to be quick − next to be safe. But when all is done, she is only a platform. It is the gun that counts.

Hence, I suppose, the argument that it is better to pay £1,750,000 to take 10 12in. guns into action than £1,000,000 to take only four.[30] In the first case, your guns cost us £175,000 apiece; in the second £250,000 apiece. But here I come to blows with the designers. We don't know that you can use anything like so large a number on so small a platform. There can be no excuse for not finding out without delay.

I hate the nonsense that is talked by some, as if it were a kind of advantage that the new ships should be so costly – a sort of 'Thank God, you're an Englishman, and can afford it,' as Du Maurier's truly British father says to his son when giving him champagne at a French restaurant.[31] Should the ideas on which these very costly ships are built prove to be faulty, the public indignation at the apparent extravagance will be natural. And cost would never deter other nations from building the like if the ideas prove right.

Meantime one thing is unquestionable. We have to-day built or building 'capital' ships fit to meet the best ships of foreign navies, carrying nearly 200 guns of the largest nature. These guns cost approximately £250,000 each to take into battle, and nearly £25,000 per year per gun to keep them fit for battle. As long as the fighting value of guns is as disproportionate as the differences between ships in the battle practices indicate, and until we have some assurance that any can be made to hit under battle conditions, it surely is better economy to devote more money to fathoming the mystery of hitting and less to adding new chapters in the history of missing.

When will it be finally realised that it is hits, and hits only, that count?

It would pay a hundred times over to cut the new building programme in half, and devote a couple of millions a year to nothing but practice, experiments, instruments – anything to raise the average hitting capacity of the 200 big guns we have already, rather than swell the list of guns that cannot hit at all.

Suppose you spend seven or eight millions a year to get five more *Dreadnought's*. Well, you have 40 more 12in. guns available for your next battle. But if you spend a fifth of this and can treble the efficiency of all the guns you have, you gain the equivalent of 600 new guns.

That will be a piece of new construction worth having; a secret worth working for, and finding; and once found, better worth guarding than any secrets of design that the womb of the pregnant future may conceal.

VI

NOTE ON THE POSSIBILITY OF DEMONSTRATING THE PRINCIPLE OF AIM CORRECTION WITHOUT THE USE OF INSTRUMENTS DESIGNED FOR THE PURPOSE

(Sent to the Director of Naval Ordnance, July 1906)

(PRESUMABLY JULY 1906)

In May and June 1906, Pollen provided the Admiralty with the plans to his reconceived system. The plans were accompanied by a report commissioned by Pollen and written by Charles Vernon Boys, one of England's leading applied physicists, that endorsed the design of Isherwood's gyroscope.[1] A three-man committee headed by Captain John Jellicoe, the Director of Naval Ordnance, evaluated the submissions.[2] In addition, Jellicoe sought the advice of the Inspector of Target Practice (Percy Scott) and his staff, the captain and staff of the HMS *Excellent*, the Whale Island Experimental Staff, the Director of Naval Intelligence, and the captain and staff of the War College.[3]

Following these consultations, Jellicoe advised Rear-Admiral Sir Henry Jackson, the Controller and thus his immediate superior, that it would be worth acquiring the secrecy rights to the system and funding the costs of prototypes for trial. But Jackson, who had been involved with the development of gyroscopes for torpedoes, where difficulties had been overcome only after great effort and expense, was highly sceptical of the claim that the Pollen-Isherwood gyroscope − upon which the success of the entire Pollen fire control system depended − could be made to work without prolonged and prohibitively expensive development. He refused, therefore, to agree to the proposal. Jellicoe then asked Pollen whether or not the feasibility of plotting could be proved using a single-observer range finder, a magnetic compass,[4] and a manual plotting table, in the hope that a success with cheap extemporised equipment would persuade Jackson to support a more ambitious experiment.[5]

Pollen argued in a memorandum sent to Jellicoe in July 1906, however, that the *Jupiter* trials had clearly shown that no successful demonstration could be made without the proper instruments, to which Jellicoe replied in agreement. Pollen subsequently printed the exchange as a 'Note on the Possibility of Demonstrating the Principle of Aim Correction Without the Use of Instruments Designed for the Purpose'. There is no record in the Pollen Papers of the recipients of this paper.

Aim correction consists essentially in making a chart of the successive positions of the enemy. These positions are arrived at by the observation of his bearing at successive moments of time, which may be regarded as lines drawn from the observing ship to the observed ship, and reading the range simultaneously, i.e., the exact point on that line at which the target is. The terminals of these lines are thus the raw materials of the system.

The observing ship being a swaying body, it is obvious that the correction of the bearing for such sway or yaw is part of the operation of taking the bearing.

It is quite clear, then, that the chart must be based upon the range, the bearing, and the yaw correction being made as accurately as possible and by one operation, or, what is the same thing, by three operations, two of which are automatic, and all of which are absolutely synchronous. To achieve this, the following instruments were devised:

(1) To get the range itself as accurately as possible, it was proposed to use the longest based one-observer rangefinder yet made.
(2) Means of taking the bearing automatically with the rangefinder by the same operation that takes the range.
(3) Automatic means of correcting the bearing so taken for the yaw of the ship, so that the correction is made simultaneously with the bearing observation.
(4) Means of transmitting the range and the corrected bearing simultaneously with the taking of the observation.

It is clear that if the bearings are accurately observed and corrected for yaw that the record of the successive observations of the range will enable an average to be made in the range errors, that is, assuming that the number of observations made of the range are sufficiently frequent to give a more or less continuous pattern on the chart.

It is equally clear that if the bearings are taken neither simultaneously, nor with synchronous correction for yaw, that the successive positions placed down on the chart will not be correctly located for sequence, and consequently the chart will be valueless for all the purposes for which it exists, viz., the averaged range and the map of the course of the enemy, from which the speed and the angle of his line of advance relatively to ours are to be deduced.

Finally, if the chart is to be of practical value, it must not only be made from ranges and corrected bearings taken simultaneously, but must be made with the smallest possible interval of time between the making of such observations of range and bearing and the recording of them. The interval between the taking of such observations, and the recording of such observations, can only be negligible as a source of error if it is exactly equal for all successive operations. If the interval could be reduced to an interval of time in itself negligible, say, for instance, something less than a second, and always the same, the ideal conditions for making a chart from a moving platform of the positions of a moving object would have been attained.

A consideration of the foregoing quite elementary and obvious truths will show that it is not, and cannot be, practicable to make a reliable chart, or, indeed, to prove that a reliable chart can be made, if means for taking observations of ranges, and bearings, and yaw corrections, are employed which do not ensure that all of these three operations are made and their results transmitted both synchronously and with the highest possible accuracy.

In the experiments carried out in the *Jupiter*, it was shown that such a chart could be made with gyroscopic correction, and, indeed, was made; but without a frequent supply of ranges and bearings, without sufficient accuracy and simultaneous accuracy in the incorporation of the yaw correction, it was shown to be impossible to ensure the making of a chart when desired. The principle, however, of making the chart was shown to be both feasible and correct, provided that ranges, bearings, and yaw correction could be made simultaneously.

In the instruments proposed to be built for a further trial of the A.C. system, it is proposed to get the greatest possible accuracy in range-taking by the employment of the longest possible base for a one-observer instrument; and, next, mechanical means of transmitting the range attained, which means would permit of more accurate readings than can be made visually by noting the position of a pointer on a scale.

Secondly, in the mounting of the rangefinder, it is proposed to arrange that the actual taking of the bearing, with regard to the ship, shall be in every case automatic and simultaneous with the taking of the range. This will ensure every bearing being taken correctly to something less than a quarter of a degree.

Thirdly, a device for the gyroscopic correction of such bearing, which will correct the record of the bearing transmitted, so as to ensure the true bearing from the ship's normal line, and not the bearing from the ship the moment the observation is taken, being used in the making of the chart.

It is essential in this yaw correction that the absolute accuracy of its results should be secured, not only for brief intervals of time, viz., those for which a single gyroscope can be trusted to run truly, but continuously, for the reason that at sea it is impossible to secure constant conditions for taking observations, and nothing should be introduced into the mechanism of the observing system to hamper the freedom of action of the observer.

It would be perfectly futile to attempt to experiment with this system if the observer were tied down to taking his observations in any given period of two or three minutes of time. The mechanism should be continuously at his disposal.

Finally, it will be proposed that the operation of charting should be made on a system very greatly improved on that used in the *Jupiter*, so that the time interval between the taking of the observation, and its recording on paper, shall be reduced to a minimum, say a fraction of a second, and that minimum a constant quantity.

The real point to bear in mind is that no charting can be reliable, or give results of the faintest value unless

(A) Range taking, bearing taking, and yaw correction are attained first by one operation; next
(B) Are obtainable at any moment; and
(C) Are transmitted synchronously and instantaneously to the chart.

For these reasons, the writer can see no possible good object being served by an experimental trial of Aim Correction by means of extemporised instruments, for the reason that no test from which the synchronous getting of the data and their instantaneous transmission are absent could throw light on a system which depended entirely upon them.

———

To this memorandum the Director of Naval Ordnance replied:

'Many thanks for your letter and memorandum. I agree with you.'

VII

SOME ASPECTS OF THE TACTICAL VALUE OF SPEED IN CAPITAL SHIPS
(NOVEMBER 1906)

By the mid-summer of 1906, Rear-Admiral Sir Henry Jackson, the Third Sea Lord and Controller, had agreed to support a compromise that had been worked out between Captain John Jellicoe, the Director of Naval Ordnance, and Pollen as a means of reducing initial costs, which would provide for the development of the change of range machine only after the successful trial of the gyroscope correction and automatic plotting instruments.[1] On 7 August 1906, the Board of Admiralty thus voted to open negotiations with Pollen for the rights to his system,[2] but three weeks of discussion and correspondence failed to produce an agreement.[3] On 10 September, however, Admiral Sir John Fisher, the First Sea Lord, insisted that the acquisition of the rights to Pollen's system was a matter of vital importance, after he had been briefed by Jellicoe, who, in turn, appears to have relied upon a report written by his assistant, Captain Edward Harding, Royal Marine Artillery.[4] The Admiralty thus agreed to meet Pollen's terms, and a tentative agreement was embodied in an Admiralty letter of 21 September 1906.[5]

Having secured the development of the better part of his fire control system, Pollen then began putting forward proposals for what he called a 'tactical machine'. This device, when set with the target range, bearing, speed and course obtained from Pollen's fire control system, and the known speed and course of one's own ship, would display the relative positions of ship and target and which, with manipulation, could then be made to display the relative positions of one's own ship and the target for any given time in the future for so long as the two vessels did not change their courses and speeds. By substituting the single ship tokens with those that represented several ships, the relative positions of two fleets could be displayed as well. The tactical machine, Pollen believed, would not only facilitate the study of naval tactics, but in battle enable a commanding officer to determine courses and speeds for his ship or fleet that would maximise the effect of gunnery.

The mechanism of the tactical machine had been developed from the proposals for a change of range machine that Pollen had presented to the special Admiralty Committee in December 1904.[6] A prototype had been delivered to the Admiralty on 14 May 1906[7] in fulfilment of Pollen's obligation to provide such a device as part of the agreement embodied in the Admiralty letter of 3 May 1905[8] and an application for a patent was made to the Patent Office on 6 June 1906.[9]

In October 1906, *Blackwood's Magazine* and *The United Service Magazine* published articles that discounted the tactical value of speed.[10] Pollen was convinced that a contrary argument could be made through the use of his tactical machine. He thus borrowed the prototype from the Ordnance Department in early November,[11] and presented his views that month in 'Some Aspects of the Tactical

Value of Speed in Capital Ships',[12] which, he explained, was intended as 'introductory to an exposition' of his tactical machine.[13] The Pollen Papers contain no record of the distribution of this paper.

The following notes on speed in battleships are somewhat hastily written as introductory to an exposition of my tactical machine. They were put together à propos of the articles in *Blackwood's Magazine* and in the *United Service Magazine*. No attempt has been made to reply in detail to the various statements in these articles, but rather to indicate the broad principles on which an answer might be written. It is intended later to illustrate the principles set out by diagrams taken from the machine.[14]

A.H.P.

1 In discussing the value of speed in battleships we must start with a few premises. The first is that tactics is the art of using your power of offence to the greatest advantage and your power of defence to the greatest advantage. Every tactical act is good or bad, according to its relation to the tactical act of the enemy. Hence, in land warfare, where the formation of ground has such an enormous influence on the value of arms and of armed men, Napoleon was able to say with truth that war was a matter of positions.

2 On the sea there are no chosen positions. There is no one geographical portion of the sea that it is tactically advantageous to occupy, rather than another. There are no hills, or dales, or commanding places to be secured before your enemy can secure them. There are advantages to be got in the matter of wind and sun which sometimes may be of importance, but hardly any of great importance. For instance, it was advantageous to Togo to have the wind blowing from him to the Russian Fleet, for several reasons. The wind carried away his smoke, drenched the Russians with spray from shots falling short, and drove the waves into whatever shot holes were made in the hulls.[15] As far as the conditions of weather are concerned in the selection of the advantageous position, speed must obviously be of help.

3 In the majority of prevailing conditions of weather, there will be practically no advantage, or only a very slight advantage, to be got in this way, and naval tactics must be looked at strictly from the point of view of naval conditions.

4 The unit of a fleet is the ship, and the unit of a ship is the gun. A naval battle is therefore a struggle between the opposing guns.

5 Excluding the *Dreadnought* the prevailing type of ship in the world to-day is represented by the enclosed diagram.[16] It is assumed that no commander would willingly allow the enemy to come closer than 5,500 yards, on the ground that fighting at such short ranges must

largely be a matter of accident and fluke, the range being so short that little practical advantage is to be got by superior skill or technical acquirement. It is also assumed that, outside 8,000 yards, the present gunnery technique makes it exceedingly unlikely that a serviceable number of hits will be made in proportion to the rounds fired. The diagram of the single ship has therefore been made showing its fire zone like one of Saturn's rings, the two sections of the ring opposite the ends of the ship being the part of the zone under fire from two guns, and the section of the ring opposite the broadside of the ship being under the fire of four guns.

6 Going back to the axiom that tactics are a matter of position, if we take two such diagrams as this and place the end of one ship in the four gun zone of the other, we get the end-on ship under the fire of four guns and the enemy under the fire of only two. This represents the position of greatest disadvantage for the end-on ship, whereas, if we put the second ship in the four gun zone, the first ship being also in the four gun zone, we have perfect equality.

7 It so happens that, with the distribution of armour now in vogue, the broadside position, which is the strongest in offence, is also the strongest in defence, for the reason that the end-on position exposes a long, largely-unarmoured target to a plunging fire.

8 The parallel position, then, for two ships, is necessarily a position of the greatest equality in offence and defence, whereas the right-angled position, or end-on position, is the one of maximum advantage or disadvantage.

9 Supposing we turn from the single ship diagram to the line ahead diagram, we shall find that parallel to the line ahead ships there is continuous fire zone, which, beginning with the four guns of the first ship, becomes 8, 12, 16, etc., as we approach the centre of the line. If we place another line of ships parallel to this with the end ships opposite to each other, we have again perfect equality, such as we had in the single ship diagram. But, if we place them at right angles to each other so that the end of one line is in the maximum fire zone of the other line, we have the position of maximum disadvantage shown in the single ship diagram multiplied prodigiously, because here the ships in the line that is broadside-on to the other are exposed to a perfectly insignificant fire, whereas the leading ships of the end-on line are under what should be an annihilating fire.

10 Starting with these two diagrams, the parallel and right-angled diagram, we can therefore see that the whole of naval tactics consist in getting the closest approximation we can to the right-angled diagram in action.

11 It is contended that if two fleets are approaching each other in any formation, say line abreast or converging on line ahead, that the

degree to which one can get into the more favourable position must be largely determined by speed.

12 But the selection of the right moment in the approach for making a decided move for a selected point must largely turn upon exact knowledge of the speed and course the enemy is steering. The freedom of movement of a line of ships steaming in line is strictly limited. It can continue on its existing course, it can wheel from the leader, or turn to line abreast on either side, but it cannot, without considerable loss of time (during which its guns must be silent) reverse its course, and become line ahead in the opposite direction. This limitation is one which affords great scope for tactical experiment.

13 A study of the writings on naval tactics that have appeared in the *Journal of the United Service Institution* during the last thirty years, will show that it has been regularly assumed by all writers that every gunnery action will resolve itself into an engagement on parallel lines. The recent articles by the writer in *Blackwood's Magazine,* and by 'Black Joke' in the *United Service Magazine,* show that this theory of parallel lines still holds the field, and as far as these two writers are concerned, is the foundation of their case that speed is of small if any tactical value.

14 But it is obvious that, with superior speed, one fleet could compel the other to form a line parallel to its own end under great disadvantages – that is to say, by straightway taking a course which would bring the rear ship in the enemy's line within its fire zone for a good number of minutes, and would compel the enemy's line to reverse course and form up parallel under fire. Ships turning one by one would present end-on targets, and be at a great disadvantage, and the broadsides could only be brought into bearing in succession. Bearing in mind that the victorious Admiral of Tsushima Straits, in his report on the battle,[17] says that within twenty minutes the issue was decided, and bearing in mind also that neither side had any tactical advantage over the other in that engagement, it can be readily seen that, with greatly improved gunnery technique, which means the certainty of making a considerable number of hits at a great range with great rapidity, the compulsion the enemy would be under to bring his broadside to bear in sections of one ship at a time would in all probability signify a disadvantage which would mean instantaneous defeat.

15 Without attempting to work out such diagrams as these in detail, it will quite clearly be seen that such diagrams could be worked out, and be of a perfectly convincing nature, and, what is more to the point at the moment, it will become obvious that the capacity of an enemy to thwart any movement of this kind will exist in inverse ratio to this inferiority in speed.

16   As far, then, as the approach to action is concerned, speed will certainly be advantageous, and, if it exists in sufficient superiority, might fairly confer an overwhelming advantage.

17   Hitherto it has been assumed in the argument that we are dealing with a homogeneous fleet of greater speed against a homogeneous fleet of inferior speed. Supposing, however, we assume that we have in a fleet a body of ships partly of equal speed with the enemy, and a small section of very greatly superior speed. In this case the tactics indicated as the proper ones for approach with superior speed in the instance we have taken become proper as a second stage in the action after the coming together of the fleets has been accomplished and parallel lines are established. A small squadron of swift ships, say, 2, or 3, or 4, could be located at either end of the enemy's line, and so bring about the effect of having a containing force and a superior attacking force, which is the true analysis of the Nelson touch at Trafalgar.

18   It is sometimes said that superior speed gives the choice of range. It would seem preferable to use the expression 'gives the choice of formation', which, as the diagrams show, is equivalent to saying that it confers superior concentration of fire. Given a certain number of ships of superior speed to the enemy's, the number of combinations which can be made by small squadrons to bring the enemy's fleet in succession under a superior fire is very great. These diagrams should be worked out in detail so as to ascertain the limit of usefulness in the size of each unit squadron.

19   A somewhat cursory view of the subject would seem to show that squadrons of three or four ships, approaching an enemy *en échelon,* the leading squadron being the nearest to the enemy, or the furthest, according as the enemy's line is approaching or retreating. But, without much detailed study, it would be impossible to be sure of this point.

To sum up, superior speed gives the choice of initial formation, because it enables the Admiral possessing it to determine his method of approach.

A small fast squadron, co-operating with a squadron of equal speed with the enemy's opens up the possibility of using a divided squadron, which use, *mutatis mutandis*, would give, under modern conditions, the effect that Nelson got at Trafalgar.

After the first decisive result has been got in action by the general demoralisation of the enemy's line, superior speed will of course be a determining factor in deciding the extent of completeness that can be got from the victory.

For instance, after Togo had lost the Russian Fleet in the fog and thick weather, he divided his squadron into two parts, sent his fast cruisers round on a circular course to the North, and, with his slower battleships, took the direct chord to the point of meeting. He thus made good a large amount of ground, and arrived with squadrons ready to unite where he judged the enemy must be, and where he successfully found them. In this case the superior speed existed in the case of ships of inferior armament. Cruisers, for instance, of even the *Natal* and *Black Prince* type, could not hope to engage successfully a few very slightly damaged battleships, which might have survived the first brunt of the action and be retreating. On the other hand, fast ships, armed with 12in. guns, with an extra knot or two of speed, would prove absolutely invaluable for this purpose.

Some very interesting diagrams can be made of the fire zone of such ships as the *Dreadnought,* and the new 25 knot capital ships,[18] owing to their arc of fire being so much greater than in the case of battleships up to the date of the *Agamemnon* and the *Lord Nelson*.

Looking at these diagrams, it must also be borne in mind that three *Dreadnought's* would have this further tactical advantage – viz., that their zone of higher concentration – viz., 24 12in. guns would be tactically far more flexibly applied than the smaller zone of the six *Duncan's*.

If we come back to the old saying that only numbers can annihilate, we have to apply it to present naval warfare by remembering that it is the number of hits that can be made on a given target in a given time at the minimum of risk. The truth of this statement has only to be realised for it to be seen that to make these numbers which alone can annihilate effective, we must be able to apply them as rapidly as possible to as many vulnerable points as possible in rapid succession and thus it may be said that, always premising the technical capacity to use guns, i.e., to hit, the tactical requirements of a ship are, first, the largest possible zone of fire, i.e., the biggest training arc; and, secondly, the greatest possible mobility, in order to bring successive portions of the enemy's fleet within these zones of concentration, and to remain for as short a time as possible within the enemy's corresponding zone.

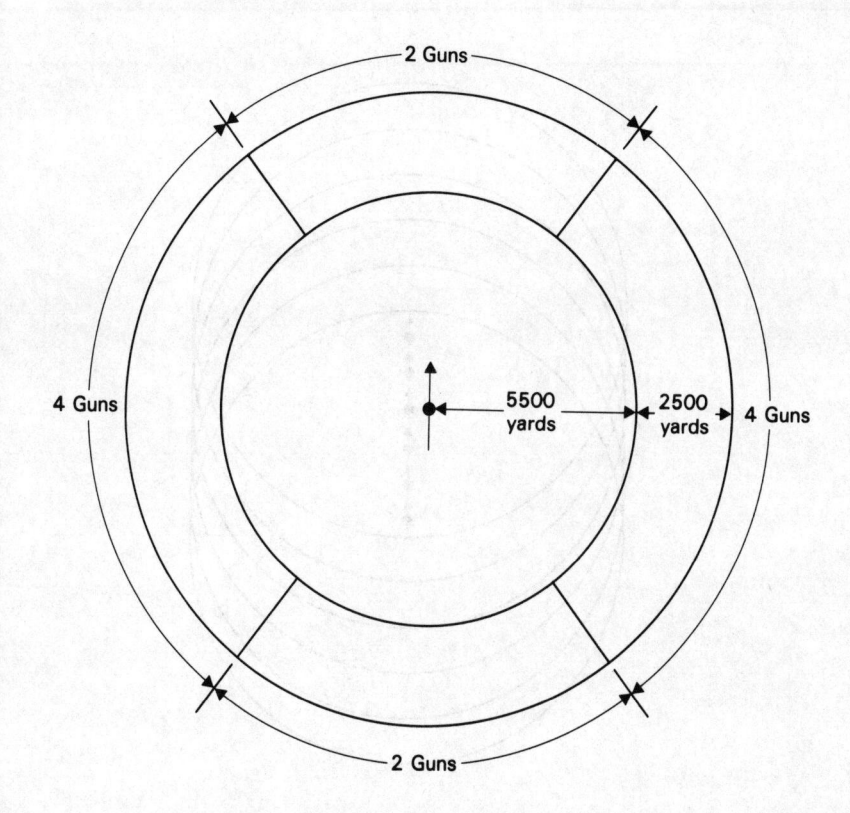

Diagram 1 *Fire Zones of the Main Armament of a Typical Predreadnought Battleship*

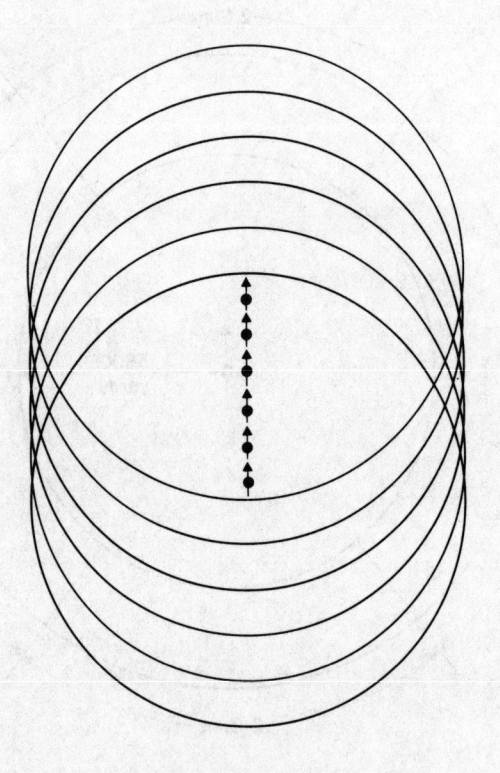

Diagram 2 *Fire Zones of the Main Armaments of Six Typical Predreadnought Battleships in Line Ahead Formation (For details, which have been omitted for the sake of clarity, see Diagram 1).*

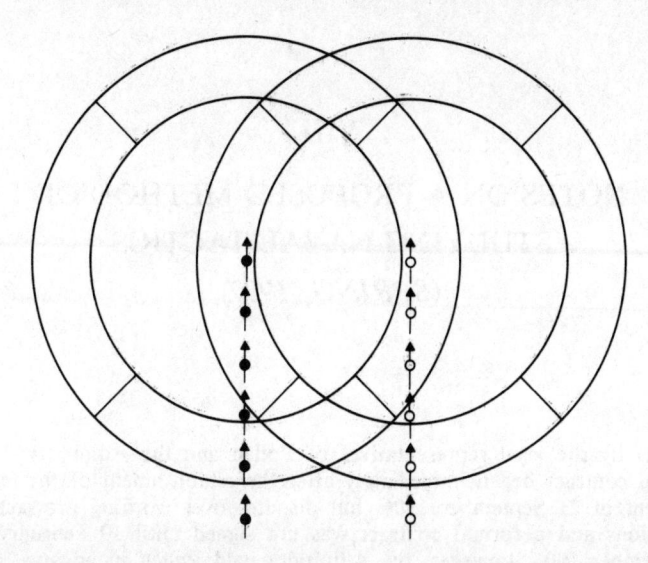

THE TACTICAL VALUE OF SPEED IN CAPITAL SHIPS   113

Diagram 3 *Two Opposing Fleets of Typical Predreadnought Battleships in Line Ahead Formation steaming on Parallel Course, with Fire Zones of the Main Armament shown for the Leading Ships only, in the interest of Clarity (For Fire Zone Legends and the Effect of Overlappings, see Diagrams 1 and 2).*

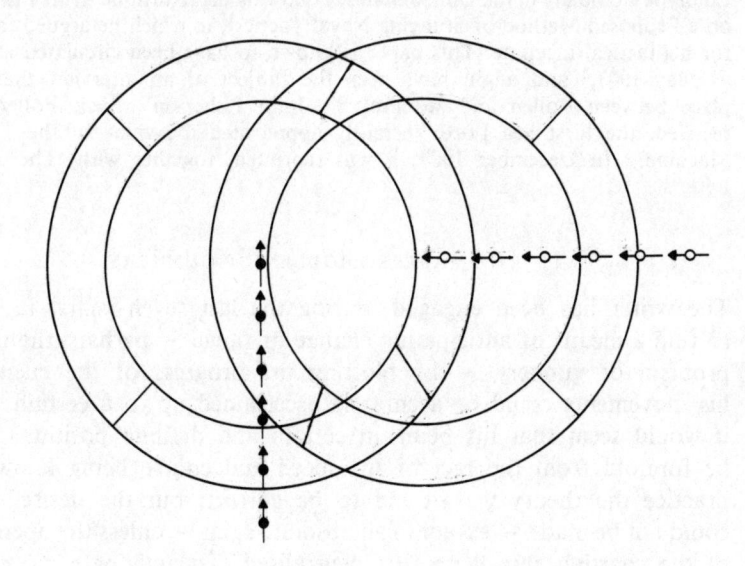

Diagram 4 *Two Opposing Fleets of Typical Predreadnought Battleships in Line Ahead Formation Approaching at Right Angles to Each Other so that the End of One Line is in the Maximum Fire Zone of the Other Line, with Fire Zones of the Main Armament shown for the Leading Ships only, in the interest of Clarity (For Fire Zone Legends and the Effect of Overlapping, see Diagrams 1 and 2).*

VIII

NOTES ON A PROPOSED METHOD OF STUDYING NAVAL TACTICS
(*SPRING, 1907*)

Attempts by the legal representatives of Pollen and the Admiralty to draft a formal contract began immediately after the establishment of the tentative agreement of 21 September 1906, but disputes over wording protracted the negotiations and a formal contract was not signed until 19 February 1908. In November 1906, however, the Admiralty paid Pollen in advance for the development of his instruments for trial – which had been stipulated by the tentative agreement – and he thus proceeded with the final design and manufacture of prototypes in spite of the lack of a formal contract.[1]

From January to mid-March 1907 Pollen was engaged with business for the Linotype Company in the United States. Following his return, he wrote his 'Notes on a Proposed Method of Studying Naval Tactics', in which he argued the case for his tactical machine. This paper is known to have been circulated as early as May 1907,[2] and might have been the subject of an interview that took place between Pollen and Admiral Sir John Fisher in which, Pollen later recalled, the First Sea Lord 'instantly appreciated the value of the Tactical Machine'.[3] In December 1907, it was reprinted, together with 'The *Jupiter* Letters'.

I – THE MASTER PROBLEM OF GUNNERY

The writer has been engaged, during the last seven years, in trying to find a means of anticipating change of range – perhaps the master problem of gunnery – by plotting the progress of the enemy. If his movements could be accurately ascertained up to a certain point, it would seem that his being in certain and definite positions could be foretold from the fact of his speed and course being known. In practice the theory was found to be correct, but the desired chart could not be made – except in an absolute calm – unless the operations of the charting ship were first neutralised. To anticipate movement, motion had first to be counteracted. It remains to be seen whether the means proposed for this are adequate; and, if they are, there would seem to be a possibility of approximating the problem of engaging a moving target to that of firing at a stationary target with a stationary gun.

II – ORIGIN OF THE TACTICAL MACHINE

In the course of his investigations into this problem, the writer made a series of charts of ships on divergent courses, and at different speeds, with a view to determining the law of change of range.

Taking a firing ship on a certain course and going at a certain speed, and a target ship on a different course and at a different speed, he divided their courses into lengths representing the distances each would travel in one minute, and then connected the two courses by lines at each one minute interval. The differences in the lengths of these lines – the plan being drawn to scale – represented the change of range from minute to minute. It then occurred to him to regard himself as being on board the firing ship, and to observe the bearing of the target ship to change, as if his ship were stationary and the target ship moving ahead or astern, according as the angles of the courses or the differences in the speeds dictated. In this situation it seemed he would wish to have a plan of what was happening. To bring the range lines from the second, third, and subsequent positions back to the first, and to draw these from the firing ship at the same angle they were drawn from in the extended chart, was an obvious way of getting the desired plan. If range circles were described round the firing ship, the apparent traverse of the enemy across these circles would show at a glance the changes in distances from minute to minute. From a plan so made the idea of the tactical machine grew. It was found that the mathematical formulæ embodying the functions of the bearing, speed, and course of the enemy, the speed of our own ship and the initial range could be expressed in mechanical linkages – and thus a ready means be formed for constructing a diagram such as is described above, that might be made both instantaneous in its action and universal in its scope.

The machine thus had its origin in the study of change of range, and although not of a kind to be immediately serviceable as an addition to a warship's organisation for gunnery, there remains the possibility of its being of use in another field – viz., as an assistance in the study of naval tactics and as a guide in naval engagements.

III – COMBINED MOVEMENTS THE ROOT PROBLEM OF TACTICS

If ships, and the targets they have to engage, were steady and stationary instead of oscillating and mobile, there would be no difficulties to distinguish naval from shore gunnery. If hostile squadrons always occupied definite or predeterminate positions, there would be no such art as naval tactics. It is then fundamentally true of naval warfare to say that movement is the chief cause of its complexities. If, in the field of tactics, these could be removed by the power of representing in diagrams the changes of

116 THE POLLEN PAPERS

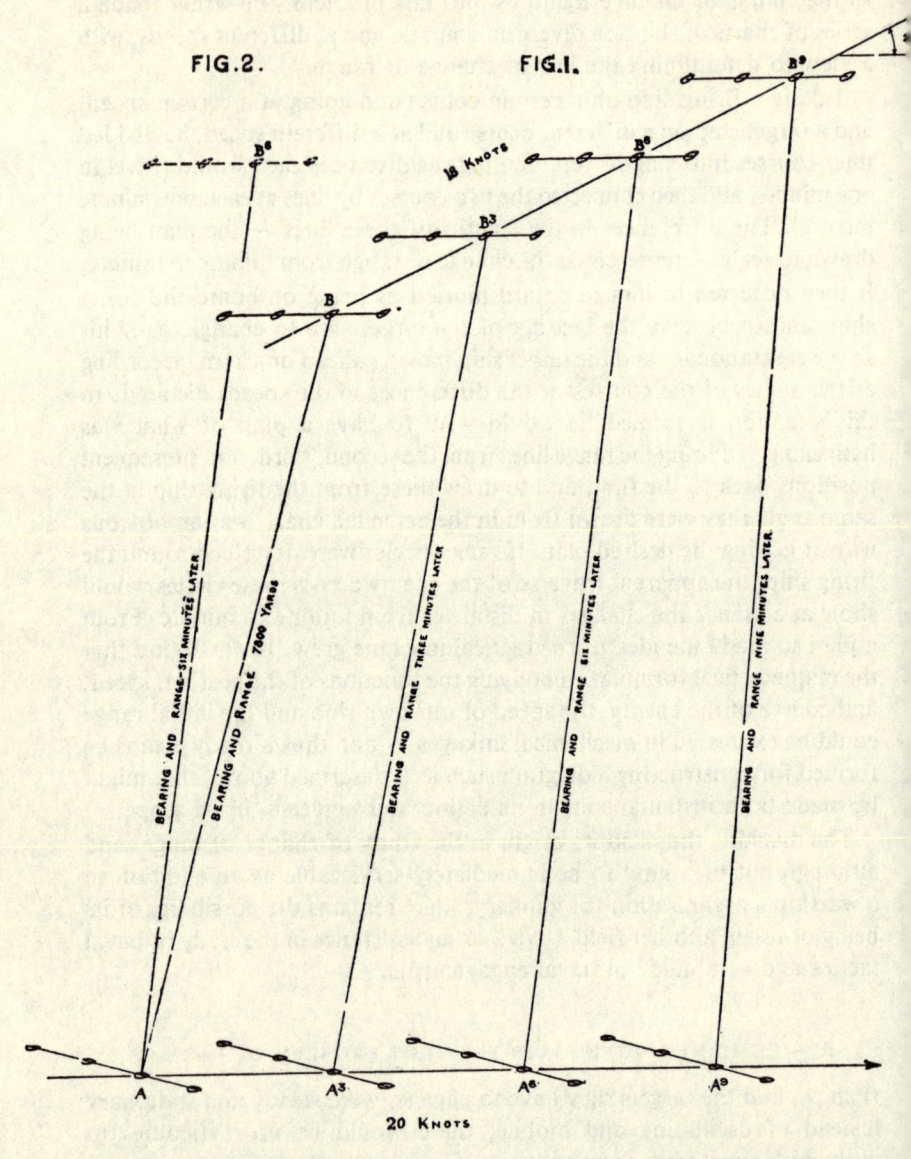

FIG. 2. **FIG. 1.**

20 KNOTS

The two figures show two fleets on lines of bearing. A to A³, A³ to A⁶, B to B³, B³ to B⁶, e
are the new positions every three minutes. By drawing A—B⁶, from A, parallel to A⁶—B⁶ and
equal length, we get a new line B—B⁶ in a line from B to B⁶, which represents the appar
course of B, assuming A to be stationary. If we draw similar parallel and equal lines, li
A—B³, A—B⁶, A—B⁹, we should get the tactical figure referred to in the text, and illustrated
an enlarged scale overleaf.

FIG. 3.

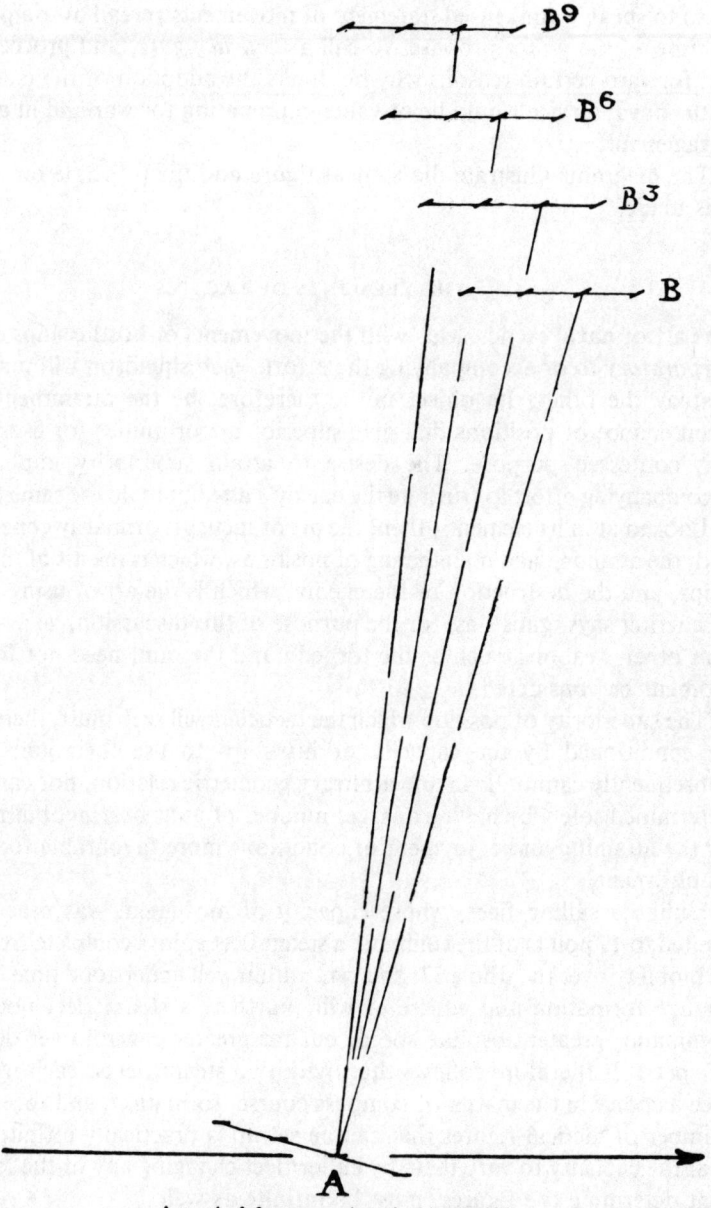

A tactical figure as given by the machine.

relation that will take place over a period of some minutes, it would seem as if some at least of the difficulties of tactics would be lessened. And this the tactical machine achieves, for, once set to the speed and course of both fleets, their bearing and distance apart, the machine gives the desired diagram instantly, thus showing in a stationary form successive phases of combined movements. The diagram thus given by the machine − so to speak a condensed statement of movements spread over a period of time − the writer proposes to call a *tactical figure*, and proceeds to put forward certain reasons why he thinks the adoption of the machine in the naval service would be of value in preparing for war and in actual engagement.

The diagrams illustrate the tactical figure and the principle on which it is made.

IV − THE ELEMENTS OF TACTICS

The art of naval tactics deals with the movements of hostile ships *either preparatory to* or accompanying the efforts each squadron will make to destroy the other. Its object must, therefore, be the attainment and maintenance of positions that give superior opportunities for executing this confessed purpose. The desire to attain superiority implies an accompanying effort to frustrate the enemy's attempt to do the same thing.

Looked at in its elements, then, the art of tactics is primarily concerned with the attaining and maintaining of positions, which is the art of moving ships, and the destruction of the enemy, which is the art of using guns. The writer says 'guns', as, for the purpose of this discussion, he assumes that other weapons, such as the torpedo and the ram, need not for the moment be considered.

The superiority of position which the tactician will seek must, therefore, be conditioned by the capacity of his ships to use their guns, and consequently cannot lie in any arbitrary geometric relation, nor can it be determined solely by having a larger number of guns bearing, but rather by the attaining and enjoyment of conditions more favourable for their employment.

Unlike a sailing fleet, whose capacity of movement was practically limited to 19 points of the compass, a steam fleet enjoys complete freedom of mobility over the whole 32; and can, within well understood time limits, change formation and course at will. Further, a steam fleet not only commands greater absolute speed, but far greater control over degrees of speed. It therefore follows that, with two steam fleets, each with so free a choice in the matter of compass course, formation, and speed, the number of tactical figures that can be set up is practically infinite, and that the capacity to vary them by either fleet changing any of the factors that determine the figures, must be infinite as well.

V – FIGHTING FIGURES

But while this is so, the number of figures in which the sustained employment of naval artillery with effect is possible is very limited indeed. This can be rendered impossible by three principal classes of conditions:

(1) The distance apart of the fleets may be so great that, even if both guns and target were stationary, continued hitting would be out of the question.
(2) A ship may have so much motion on her, due either to the weather or to the rapidity with which she is changing course, that the mere keeping of the guns trained and aimed would be impossible.
(3) Finally, the courses and speeds which the two ships are maintaining might bring about a change of range so rapid and so great that, though the guns might be quite correctly aimed, the supply of change of range and deflection necessary for continuous hitting would be out of the question.

For practical purposes, it is on the last set of conditions that we must concentrate attention in considering the tactical employment of guns, and that it is change of range and not mere range, that is the determining factor, will become apparent if we look at the tactical figures set up in the following cases, as illustrated in the following diagram.

In Figure 1, the 'A' fleet is steaming in line ahead at 23 knots. 'B' fleet bears 89 degrees from the bow, and is steering a parallel course in the same direction at 20 knots. The initial range is 8,000 yards. It will be seen that the change of range is infinitesimal, and that the conditions are not hostile to decisive engagement.

In Figure 2, 'A' has reversed course at the same speed, but 'B' bears 25 degrees from the bow, his course is at an opening angle of 60 degrees from that of 'A', the speed is still 20 knots, and the new range only 4,500 yards. Here it will be observed that the change of range is enormous in extent, and the rate of change varies from minute to minute. In the first figure, therefore, decisive engagement is possible; in the second it is probably impossible. It thus becomes natural to classify tactical figures into fighting figures and non-fighting figures; and the non-fighting figures into those that lead up to or are preparatory to fighting figures, and are hence approach figures; and those which lead up to nothing of tactical value.

This being so, the tactician's first business is to know exactly the conditions under which he can use his guns, that is, he must know the difference between a fighting and a non-fighting figure.

Secondly, he must know exactly what speed and manœuvring power he can count on in his fleet, so that he can have an equally precise

FIG. 2.

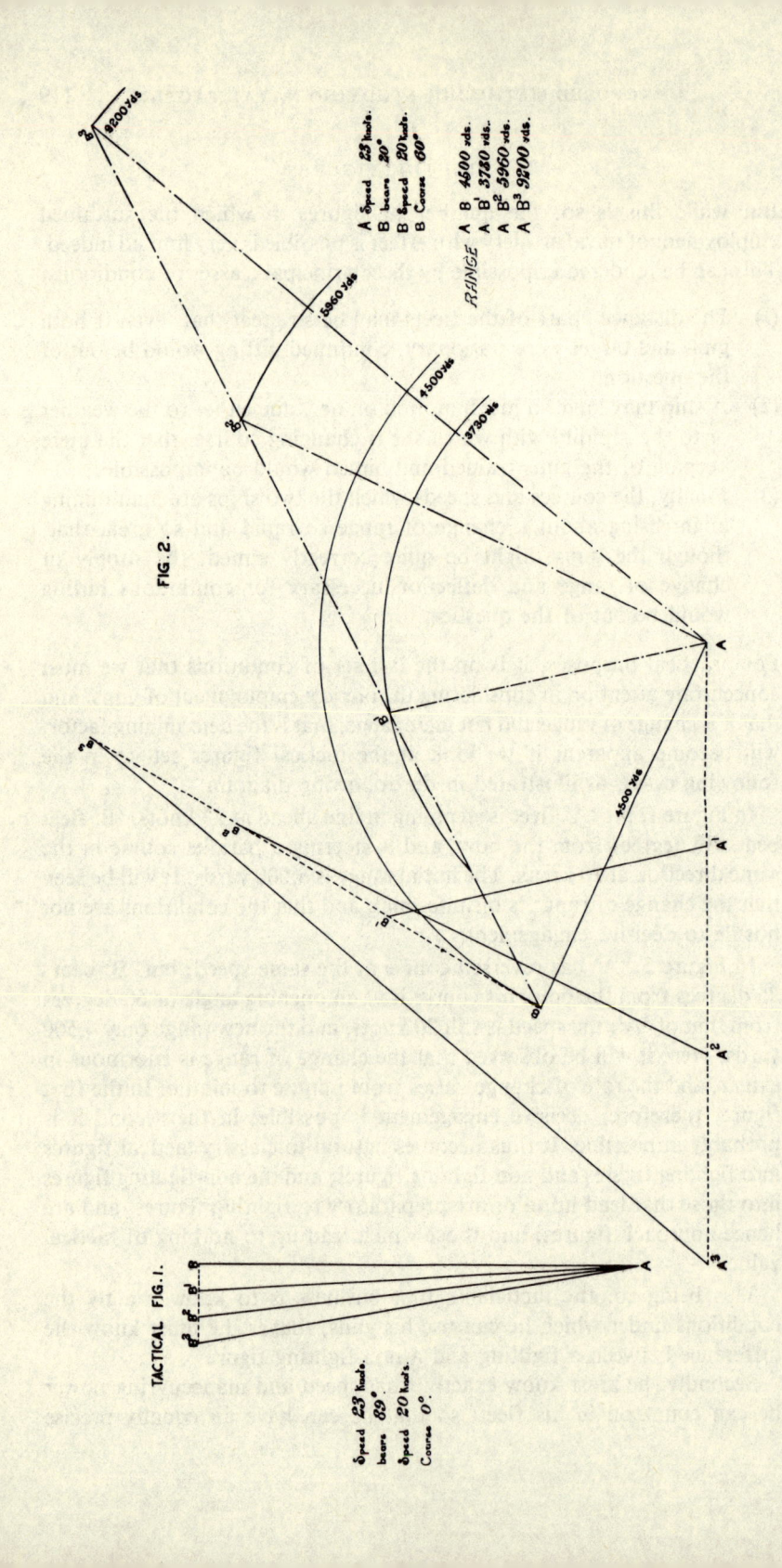

A Speed 23 knots
B bears 20°
B Speed 20 knots
B Course 60°

RANGE
A B 4500 yds.
A B¹ 3780 yds.
A B² 5960 yds.
A B³ 9200 yds.

TACTICAL FIG. 1.

Speed 23 knots
bears 69°
Speed 20 knots
Course 0°

knowledge of the approach figures at his disposal for getting at the fighting figures.

Thirdly, he should know by practice and experiments what are the best fighting figures and best approach figures in any circumstances or situation in which he may find himself.

Whether or not a particular figure is a fighting figure depends on the presence or absence of the prohibitive conditions set out above. But these conditions are not absolute; they vary with the state of artillery technique, and as technique progresses so must the prohibitive conditions diminish.

VI – GUNNERY PROGRESS

This progress has manifested itself in everything to do with the *materiel* – viz., more accurate guns of higher velocity and greater striking power, improved training gear, improved sights, improved methods of communication, etc.

These improvements have been partly the cause and partly the result of a far more striking advance in individual and organised skill in the handling of guns under different conditions. For instance, if we take the training and loading and aiming of the guns, it is probably true to say that perfection has been reached, and that the only progress to be looked for is to bring a larger number of individual gun layers and gun crews up to the standard of the best.

Again, the use of guns at great ranges under favourable conditions has enormously advanced, but there probably remains much to do in arriving at the best collective use of the artillery of a ship as a unit. The same is true of the use of the artillery of a fleet as a whole, the art of combining the gun power of several ships to the best advantage being only in its infancy.

VII – THE LIMITS OF ARTILLERY TECHNIQUE

But, while much has been done in all of the above directions, and the lines of progress in each direction give promise of further improvements, it still remains that the limiting problem of naval gunnery is ascertaining change of range when the target ship and firing ship are on different courses at different speeds, and range may vary rapidly.

The tactician having to manœuvre his ships with the sole object of using his guns must therefore come to his task equipped with a very exact knowledge of the state of technique at the moment he is to rely on it. This point is worth insisting upon, because there is sometimes a tendency to expect more of modern gunnery 'under battle conditions' than the known facts would warrant. The writer has called attention to this in his '*Jupiter* Letters' (pp. 82–3), where he gave an instance of unwillingness to push the analysis of modern gunnery technique to its logical and somewhat disillusioning limits.

But it does not need argument to prove that the tactician whose plans are founded in the quicksands of an over-estimate of the fighting efficiency of his guns must expect disaster. He only creates a fool's paradise for himself by supposing that performances possible under show conditions against a stationary target, will be reproduced under action conditions, against a target on an unknown course at perhaps a high speed. He must therefore know the exact circumstances when his fleet can get change of range, and know how to manœuvre to keep the range constant or changing only within the limits of his fleet's capacity to ascertain it. The tactical machine shows how sometimes a change in course of a point will make all the difference between a constant distance and a baffling rate of change.

VIII – APPROACH FIGURES

Just as his selection of fighting figures must be made infallible by a bedrock and ruthless realisation of the limits of his gunnery, so too must he have an equally uncompromising acquaintance with the speed and mobility of the squadrons he commands. The advantages to be gained by position may be decisive in a very few minutes. There must therefore be precision in prescribing formations and directions of course, and exactitude in following them out – above all, certainty that what he asks for can and will be given. If his selection of the correct fighting figure will be conditioned by gunnery technique, so will his approach figure be conditioned by the technique of the companion naval arts. He must know the degree of flexibility possessed by his units – i.e., their manœuvring skill and capacity; and, next, the time they will take to assume the formations and positions he designs for them – i.e., their speed. For on these depends his ultimate goal – i.e., the fighting figure that is advantageous.

IX – IMPORTANCE OF EXACT KNOWLEDGE OF TECHNIQUE

If the writer has laboured these points with some detail and repetition, it is because the confusion that follows on inability to get at the bedrock facts of technique – whether of gunnery or of manœuvring ships – is a constant menace to ordered progress. The long obsession of the naval mind by the idea of ramming would have been impossible if gunnery technique had been at all vigorously experimented with. Nor would ship after ship have been built with their big guns bearing only 30 degrees before and abaft the beam – a flaw in so many of our battleships that handicaps their tactical employment deplorably – if consideration had been given to the somewhat obvious fact that battleships are units in a squadron, and that, other things being equal, the tactical efficiency of a squadron varies in proportion to the area it can bring under fire without altering course.

That there is still considerable vagueness in thought about technique – and consequently confused thinking about tactics, was very apparent in the recent discussions to which the building of the *Dreadnought* and *Invincible* gave rise. And this confusion is conspicuous, even amongst the most erudite of the controversionalists, both in this country and abroad; so much so that while much has been contributed during the discussion of the highest scientific and historical importance, the value of the whole is greatly lessened by appeals to maxims and phrases that indicate that the writers have not pursued their analysis of technique to its logical conclusion.

X – INCONSEQUENT ARGUMENTS AND MISLEADING PHRASES

Now, without quoting textually from the gallant and learned protagonists in the recent controversy, the following hardly misrepresents the doctrines they lay down.

(a) Speed is not a weapon; therefore to spend money on getting higher speed in fighting ships is waste, or implies – what is the same thing – a sacrifice of fighting power to speed: as an example, the design of the *Dreadnought* is inferior to that of the *Lord Nelson*, because 2 per cent of *weight* has been transferred from armament to hull and machinery.

(b) Superior speed gives no fighting advantage when fleets are drawn up opposite each other in parallel lines on the same course at fighting ranges – and therefore is of no tactical value at all, or at most of the smallest value.

(c) The chief aim of tactics is to bring more guns to bear on the enemy than he can bring to bear on you.

(d) It is an error – due to matériélistic megalomania – to abolish the secondary (6in. or 9.2in.) armament and substitute for it more guns of the greatest calibre, because at decisive ranges, namely 5,000 yards or under, the smaller guns can be used with equal or superior effect.

XI – THE LOGICAL RESULT OF THE FOREGOING

There is, of course, an element of truth in each of the above statements – but they do not amount to general truths or doctrines because they are not universally applicable. Thus, while speed is undoubtedly not a weapon, it has still to be proved that it is only on weapons that money must be spent. Are Krupp steel armour, anchors, even Admirals weapons?

How does the transference of 2 per cent of *weight* from armament to machinery prove a sacrifice of fighting power? Are weight of armament and fighting power synonymous? The answer can only be 'Yes' – and this is surely the very apotheosis of materialistic ([sic] ed.) unorthodoxy. The argument for speed, apart from its strategical value, is not that it is a weapon, but that it may facilitate the use of weapons to advantage. It can be granted that with fleets maintaining and *wishing* to maintain parallel courses, as shown in the published plans, superior speed is of negligible fighting value, without admitting that it is of no tactical advantage. How do fleets get into fighting formation? Are there no tactics of approach? Is superior speed of no value in approach? Would it be scientific to expound the tactical principles underlying the Nelson touch, and ignore all that took place *before* the hostile ships were actually engaged? Is it untrue to say of the sailing era that, if we compare the relative importance of the tactical disposition of fleets before and after the opening of fire, the tactics of the approach were everything, and the tactics of the purely fighting phase, nothing? Is it too sweeping an exaggeration even if we remember such brilliant exploits as the raking of the *Bucentaur*,[4] to say that the battles of the golden age were fought out – i.e., that the decisive fighting took place – between what were virtually stationary fleets? In other words, in considering action tactics, the approach is just as integral a part of the action, as the fighting phases. Can then the value of speed be discussed solely with reference to only part of the operations? The obvious answers to these questions must at least show that the doctrines laid down are hopelessly misleading if they are to be accepted as of universal application. Again, while it is clearly an advantage to secure numerical superiority of effective guns, this is not the same thing as numerical superiority of bearing guns, and to imply that the latter should be the chief aim of tactics, is to ignore the fact that unless guns can hit, they may as well not bear at all. Once more, what can decisive range mean, except the range that is always known, and therefore one at which hits are made? It is shown in paragraph v. that 8,000 yards may be a decisive range in one set of circumstances, and 5,000 yards an impossible range in another. The argument for the small gun is that 5,000 is always, and 8,000 never, a decisive range.

XII – THE TECHNICAL SCHOOL OF THOUGHT

We are asked to believe that the Navy is divided into an historical school and a matériel school. There may be two such schools, but they do not exhaust the categories. The writer suggests that the technical school holds the field to-day, and that any hope there may be for a sound school of tactics must be grounded on the outcome of their work.

It is just as great an error to suppose that the whole of tactics can be learnt from history without scrupulous reference to modern technique, as to suppose that latter day developments have made the teachings of history valueless. As a fact, however, it was not ignorance of history, but ignorance of the true use of artillery that created the gulf separating the sailing from the steam era. The dark ages of the Lissa[5] and Whitehead[6] domination do not owe their origin to infidelity to history, but to the idolatry of false weapons. Nor in the changed conditions of speed, long range, and torpedoes can more be learnt from history than sound principles; the crowning resource of our ancestors − the interposition of their own ships to break up the enemy's line, is for obvious reasons denied to us, a fact which shows that to apply the lessons of their tactics to-day with correctness, it is necessary to have mastered the technique of to-day in every particular. We can learn their principles, but we cannot copy their dispositions. A modern captain, in doubt as to his admiral's wishes, would go very far wrong, in laying himself alongside the nearest ship of the enemy!

XIII − INEQUALITY IN TECHNIQUE MAKES TACTICS OBVIOUS OR UNNECESSARY

Perhaps the most curious instance of the frame of mind the writer has already commented on is to be found in Barfleur's 'Naval Policy', pp. 181–182.[7] Speaking of Tsushima, the author says: 'This study of the battle supplies strong evidence that the Japanese victory was due to superior skill in tactics and not to superior speed. The formation of the Russian Fleet was faulty in the extreme. Admiral Togo took full advantage of this . . . *No doubt something was due to the better shooting of the veteran Japanese seamen*. But this was largely a consequence arising from superior tactics, in that, during the critical period of the battle the Japanese showed a straight course, and kept their guns bearing steadily on the enemy; whereas the Russians, at that time, did not do so.' The sentence italicized explains victory without dragging in tactics. The whole passage ignores the fact − already exemplified by Dewey's victory at Manilla − that if one side possesses any very marked superiority of technical skill in the use of guns, the tactical skill required for using such superiority is of a very ordinary kind, so much so that it is almost true to say that such superiority makes the fleet possessing it independent of tactical skill altogether, if the phrase is used in the usual sense. In the case of Dewey this simply meant keeping outside the range of the inferior Spanish guns; in the case of Togo it meant attacking the Russians while half their fleet masked the guns of the other half, and hammering away at easy range and taking full advantage of the admittedly better gunnery of his crews. Togo's tactics would have been anything but good had the Russian gunnery technique been equal or superior to the Japanese. For that matter the fact that the *Mikasa* was hit so frequently in the first phases of the action would point to no great initial inequality in mere skill, but it is obvious that whatever skill the Russians possessed was unequal to the strain of battle. Superior morale made the Japanese gunnery keep its quality in spite of casualties. Looked at in this light, the moral of Tsushima is neither the importance of tactical superiority nor of technical superiority – viewed as a peace acquirement – but of better morale. And if we regard morale – as perhaps we should – as an integral factor in *enduring* technique, why then Tsushima, like Manilla, is simply a case of better technique – i.e., gunnery skill – making superiority in tactics superfluous. The art of tactics only becomes of really vital importance, when the technique of both sides is equal.

XIV – ASSUMED RESTRICTIONS ON MODERN TACTICS

If we assume the battle formation set out by some of the writers referred to – viz., two parallel lines ahead on the same course – as the inevitable and only formation possible, for this is implied, we must be driven to the conclusion that the resources of the modern tactician are woefully restricted. That they are restricted is obvious until the problem of change of range is solved. But every day the art of fire control improves – and with each improvement a new addition is made to the number of fighting figures. Each extension of speed throws more approach figures open. And so the enfranchisement of the tactician will be found in every aspect of technical advance.

Meanwhile the writer does not admit that the resources of tactics are so narrow as these writers imply; and he believes that the explanation of their being so held, lies largely in the very limited opportunity that exists to-day for tactical experiment. It cannot be insisted on too strongly that the three requisites for the right understanding of tactics are:

First: Knowledge of the technique of weapons.
Second: Knowledge of the technique of mobility.
Third: Experimental knowledge of the technique of combining the first and the second.

Before any study of tactics can be usefully begun, the student must possess the first and second of these requirements. Such knowledge can only be acquired in the school of actual work with guns and ships; and in these two fields experiment is not only the school of study, but an indispensable condition of progress.

But if he possesses a sound knowledge of what the fighting value of the gunnery of the day really is, and of the manoeuvring power and speed of his ships, he is equipped for experimenting in the art that includes all other naval arts. The question is which is the best method of study?

XV – THE CONDITIONS OF A SOUND METHOD OF STUDY

Looking back once more at the deplorable results that have before followed from ignoring technique, it would seem that a condition of any method being sound is that, first, it must be one in which the adoption of each advance or modification of the use of weapons or manœuvring power should be easy and immediate; next which should continually show the possibilities that any particular advance would throw open. The danger in all great services is the stereotyping of thought; and the only way to counteract this is to maintain some organised system of experiment at work that will refuse to take things for granted until they have been put to the test. Just as the lessons of Lissa were misapplied a generation ago, so is there a danger of the teachings of Tsushima being misread to-day.

To test each conclusion is one thing; to show what is wanted is even of greater value, for this is to breed the frame of mind that, finding existing limitations intolerable, will impatiently drive research into the right line of escape. If ingenuity, invention, and resourcefulness are to be made the most of, they must be guided into the most useful channels. The ideal method of tactical study would therefore seem to be one in which improvements and advances in technique can be adopted without encountering the conservative objections of stereotyped minds; which is of a kind that will indicate the direction in which technical progress is required; and, lastly, lends itself to the exhaustive trial of every combination that tactical ingenuity can suggest.

XVI – INADEQUACY OF MANŒUVRES

If, then, a wide and exhaustive test of tactical possibilities is a necessary preliminary to a right grasp of how to combine all the contributory arts in one grand art of naval war, the writer would respectfully suggest that the actual manœuvring of fleets is neither the only, nor indeed the best method of experiment. Manœuvres consume much coal, occupy an immense number of persons in each experiment, and, consequently, would make the strain upon both men and ships intolerable before a tithe of the problems had been exhausted. It would appear, then, that the ground to be harvested is too large for so slow and costly a system of husbandry.

But, further, in manœuvres, the effect of fire is eliminated, and the tendency is for them to become utterly unreal, because purely conventional. Even where, as in the case of P.Z. exercises,[8] and the war game, an attempt is made to allow for the effect of fire, the variation in such effect due to the variation in the favourableness, or otherwise, of gunnery conditions, is not always, or fully reproduced; and, consequently, the motives for the different tactical decisions are either entirely absent or exist only in a weakened and unreal form. While the manœuvring of ships has a real value in teaching and practising the art of handling ships in groups, it has, for those reasons, only a modified value in teaching tactics.

Again, work with a mooring board, with rulers, compasses and protractors, affords no sufficient alternative, for the making of accurate plans is an exercise at which few excel and of which all soon weary.

XVII – A SUGGESTED METHOD OF STUDY

The writer suggests with some confidence that his tactical machine affords a convenient means for the scientific study of tactics, that is free from the objections applying to other methods. For any given set of conditions it supplies the tactical figure that results as soon as the factors are recorded on the separate dials and scales. It supplies a bird's-eye view of what will happen over a period of several minutes, and shows at a glance the consequences of any move. It thus enables experiments to be made with the utmost rapidity, and assuming that the student is familiar with the limitations that exist in changing the formation, course and speed of ships, and the circumstances when gun fire may be expected to be effective, he will be able to judge at a glance whether the figure presented is a fighting figure or not; and what modification is desirable to obtain and keep superiority of position. The tactics of approach will be apparent when the best initial position is found, and the value or influence of speed in making the correct approach to secure it. Similarly the feasibility of fitting Nelson's conceptions to modern conditions can be experimented with by splitting a fleet into two squadrons, and the nice and complex problems involved in reuniting the divided squadrons, or bringing the flying section to some desired point for future co-operation, successfully solved, while if two machines be used side by side, the correct answering or counterstroke to the attack developed in each fighting figure can be simply and expeditiously arrived at.

XVIII – A NEW FORM OF WAR GAME

Using one machine against another should also make a new form of war game possible. In actions decisions will have to be come to with very great rapidity, and as the machine gives the result of each movement instantaneously, it ought to make the reproduction of battle conditions very real and useful. A code of rules would have to be made to meet the case. Thus the efficiency of gun fire should be held to be proportioned to the range and change of range, helm, and any other elements that should be considered. Changes of course and speed must be given their approximate time allowance, and might be assumed to become known to the enemy within a certain number of minutes, and be announced to the opponent at the end of the suitable period. But, of course, the rules should be made to correspond as closely as possible with what might be expected would be the actual facts of action.

XIX – THE VALUE OF THE MACHINE IN ACTION

If systematic study is possible by these means, it has only to be supposed that the means are open to all admirals and captains in important ships, and at the chief centres of naval study and research, to foresee the speedy development of a school of tactical thought, based not on the presumed but inconclusive lessons of peace manoeuvres, but on the critical and scientific examination of accurately propounded problems. It could not be long before a stage were reached when the ideal figure to aim for in any given conditions would be as obvious to each captain as it would be to his commander-in-chief. How to approach, when to increase and when to decrease speed, how to seek and ensure superiority of position, the right counterstroke to each possible attack the enemy may develope – long familiarity with the solution of such problems would make the right decision instinctive. If we assume that peace study with the machine has taught admirals and captains to think tactically in the figures the machine supplies, it will be obvious that its importance in action as a guide to what to do in every circumstance will be very great. The speed, course and formation of both fleets, the initial range and bearings will, as far as they are ascertainable, have been put on the machine in every ship, and thus each officer in command will have an actual plan in front of him of what is on foot. With this guide, the significance of every order of the admiral will be diagrammatically plain, and the function of each ship in the general scheme apparent. Signals in action would be simplified when they still remained necessary – and should command have to pass from flagship to flagship, its passage would cause no confusion in a fleet where the chief's plans were necessarily the subjects of his captains' divination. Indeed, the point seems hardly to be worth labouring, that if the use of the machine is admitted to be of value for the serious study of tactics, it must be of indispensable assistance when the actual carrying out of the operations, long studied by its means, has in grim earnest to be done.

XX – TRUE PREPAREDNESS FOR WAR

If we assume that all the technical arts of war – that is, the navigation and control of ships, technique of gunnery, and so forth, are brought to the highest perfection by diligent practice, by open-minded willingness to experiment with new principles and new gear, and by the concerted enthusiasm and self-devotion so characteristic in the British Navy, there still remains a master task to be performed before the Navy is truly and perfectly prepared for war.

This task is the development within the Navy of a highly organised knowledge as to how all these arts not only are to be, but will actually be combined together in the art of war. The possession of great genius, either in strategy or tactics, by individual men is a thing entirely independent of anything that administrative effort can bring about, and while we may hope for the production of such genius when the time of crisis comes, nothing can consciously be done to produce it.

But the value of such a man of genius can be almost indefinitely increased, and his absence may best be compensated for, by the growth within the Service, widely diffused and deeply rooted, of an exact knowledge of all that can be gathered from scientific study and close acquaintance with the art of war as the embodiment of all the naval arts.

It would seem as if the method of study that the writer has proposed should go far towards making the creation of the desired form of knowledge possible and diffusing it amongst the greatest number, and that such diffusion would necessarily result in that unity of effort without which neither sound technical progress is to be expected, nor perfect co-operation in war secured.

IX

AN APOLOGY FOR THE A.C. BATTLE SYSTEM: BEING NOTES FOR A LECTURE TO THE WAR COURSE COLLEGE, PORTSMOUTH
(AUGUST–DECEMBER 1907)

On 11 June 1907, a party of naval officers that included Rear-Admiral Sir Henry Jackson, the Controller; Rear-Admiral John Jellicoe, the Director of Naval Ordnance; and Captain (RMA) Edward Harding, Jellicoe's assistant; inspected the Linotype Company works near Manchester at Broadheath.[1] There they were much impressed by demonstrations of the action of the automatic plotting mechanism and the gyroscopically stabilised range finder and bearing indicator, upon which work was well advanced.[2] In August 1907, however, Jellicoe completed his term as head of the Ordnance Department, and was succeeded by Captain Reginald Bacon, who was opposed to the mechanisation of fire control on the grounds that machines were more likely to fail under service conditions than men.[3] And while Bacon was bound by the actions of his predecessor to give the Pollen system a trial, there can be little doubt that he was determined from the beginning to prevent its being adopted for service.

Shortly after taking over from Jellicoe, Bacon replaced Captain (RMA) Edward Harding, who had been a strong supporter of the Pollen system, with Lieutenant Frederic Dreyer as his assistant on matters pertaining to fire control. Dreyer, like Bacon, was opposed to the mechanisation of fire control and had even proposed that manual fire control methods of his own be adopted for service.[4] Dreyer's advice on gunnery matters, moreover, was trusted by Admiral of the Fleet Sir Arthur Wilson, who in November 1907 was directed by the Admiralty – undoubtedly at Bacon's suggestion – to supervise the trials of Pollen's instruments.[5] In 1905, Wilson, then Commander-in-Chief of the Channel Fleet, had apparently been angered by the detachment of the battleship *Jupiter* from his force for trials of Pollen's fire control gear,[6] and his disposition towards the inventor could not have been improved by the opinions of Dreyer, who in December informed Pollen that he would do his best 'to crab' his efforts.[7] 'If your gear can be broken down,' Pollen was warned by Rear-Admiral Sir Henry Jackson, 'Wilson will break it – so look out.'[8]

On 20 November 1907, Pollen accompanied Wilson on an inspection of the trials instruments, which were then being installed in the protected cruiser HMS *Ariadne*. Apparently ignorant of the compromise that had been agreed to during the summer of 1906,[9] Wilson, at the end of the inspection, complained that the Pollen system lacked any means of getting data to the guns – that is, that the system was incomplete.[10] A few weeks later, on 8 December 1907, Wilson wrote to Pollen and asked for 'a written statement showing exactly what are the advantages you claim for your system as fitted in the *Ariadne* and for which the Admiralty are asked to pay £100,000.'[11] On 9 December, Pollen thus wrote to Wilson, informing him that the first copy of a recently completed paper would be sent to him 'within 36 hours post'.[12]

The text of the paper, which was entitled 'An Apology for the A.C. Battle System: Being Notes for a Lecture to the War Course College, Portsmouth', had been written the previous August in response to an invitation from Captain Edmond Slade, the President of the War Course College at Portsmouth, to deliver a lecture to the officers attending the War Course Programme.[13] At Bacon's insistence the lecture had been postponed pending the outcome of the trials of Pollen's instruments,[14] but on 4 December 1907 Pollen had improvised a 'Foreword' for a planned printed version of his lecture for distribution 'to those who are concerned with the experiments about to be undertaken.'[15] The preparation of the text for publication must, therefore, have been well advanced when Pollen received Wilson's request for a written statement that would justify his system. In addition to Wilson, copies of the paper are known to have been sent to Bacon, Dreyer, Jellicoe and Vice-Admiral Sir Reginald Custance.[16]

In 'An Apology for the A.C. Battle System', Pollen presented the case for a completely mechanised system of fire control. In addition to the gyroscopically stabilised mounting, automatic plotter, change of range and deflection machines, and tactical machine, Pollen thus called for the elimination of manual sight setting and its replacement by gun sights that moved automatically in response to signals sent from the change of range and deflection machines. Pollen's argument was directed against those who believed that the fire control problem could be solved by little more than an extension of the methods of better organisation, intensive drill, and new practice procedures that had been introduced in the preceding few years with great success. Their opposition to the alternative of completely mechanising the fire control process, Pollen maintained, was that of progressives whose policies had been overtaken by new developments.

Pollen found an appreciative reader in Jellicoe. 'The great necessity,' Jellicoe wrote to Pollen in December, 'is to convince those who have the decision of the value or rather the need of a method of obtaining correct rate of change and your notes for the lecture are very convincing to those who know the difficulties now.' But Jellicoe then warned that 'to those who do not appreciate these difficulties, it will not be so easy to demonstrate it.'[17] Wilson, perhaps not unsurprisingly, was dissatisfied, and even angered, by Pollen's submission. On 11 December 1907, he wrote,

> I have read your lecture but it does not quite answer my question which was, what are exactly the advantages you claim for your system as fitted in the *Ariadne* and for which the Admiralty are asked to pay £100,000. As far as I can see the only paragraphs in the lecture which refer to the apparatus which is fitted in the *Ariadne* are 17 and 18 and possibly part of 23. The rest I understand has not even been designed and in any case is not included in the portion of the system for which the Admiralty would obtain for their £100,000. . . . Am I right in the above and is there anything further that you claim. If so what is it?[18]

FOREWORD

In the autumn of this year, the writer was asked to deliver, to the Officers attending the War Course College at Portsmouth, a lecture explanatory of his A.C. Battle System, and the following notes were therefore prepared. It was afterwards thought that any such general exposition of the system had better be postponed until after the preliminary trial had been completed in the Ariadne, so that the importance, or otherwise, of taking special steps to preserve secrecy could be established. The lecture, therefore, has not yet been given, but the notes have been printed for the convenience of those who are concerned with the experiments about to be undertaken.

The writer is conscious of their many literary shortcomings, and the inconsectivity of occasional statements; but he has not time to re-write with the deliberation proper for a treatise. He trusts the critical reader will remember that in notes for a viva voce lecture, many points would be left for further illustration, and certain arguments for clearer elucidation on delivery. It was also intended to exhibit lantern slide diagrams, which could not be prepared for reproduction in print.

H.M.S. Ariadne,
 Spithead, December 4th, 1907.

(1) THE LIMITATIONS OF THE LANDSMAN

There is no necessity to expatiate on the delicate position of a layman who addresses a professional audience on their own subject. Whatever the normal difficulties of a man so situated, they are aggravated when the case is that of a landsman addressing a seafaring audience. Nothing better illustrates the contrast that their respective lives and experience present than the fact that the popular expression which denotes the utmost confusion on shore summarises the normal condition of a sailor's life — viz., the expression 'at sea'. It is therefore necessary to crave the indulgence of the audience, and to ask them to remember that no one is more conscious than the lecturer that any lay presentations of the presumed facts of the sea, and every lay suggestion for dealing with them, must always be exceedingly incomplete, and that, whatever merit they may ultimately be found to possess in themselves, it is a merit that can only become effective when the ideas, or instruments, or whatever may be proposed, are taken by the naval service, divested of their amateur characteristics, and used in obedience, not to the *a priori* reasonings of the abstract thinker, but in the light of the hard-won knowledge of the practised expert. Any layman, then, who aspires to take a place amongst those who have assisted in advancing the naval art in any of its branches, can only do so successfully if he begins by recognising that the individuality of his work must be sunk and transmogrified in the great and mysterious organism that absorbs it. As Ariel sings:

> Full fathom five thy father lies;
>   Of his bones are coral made;
> Those are pearls that were his eyes:
>   Nothing of him that doth fade.
> But doth suffer a sea-change
>   Into something rich and strange.[19]

(2) HIS PROVINCE AND HIS DIFFICULTIES

While the lecturer is, then, conscious that the chasm between the layman and the sailor cannot be passed over by the landsman, however heroic his efforts, he does not wish to be understood to say that it is impossible for a layman to know anything about sea problems, or to contribute towards their solution. As a matter of historical fact, for the most part, every particular, which distinguishes the *Dreadnought* from Drake's flagship,[20] is the outcome of lay thought and progress and advance in the different arts of which landsmen have a monopoly. It illustrates the point still further to remember that though it needed the genius of Nelson to make it fructify, the victory of Trafalgar was in a measure the fruit of Clerk of Eldin's work.[21] So that there is truth in the old saying that lookers-on sometimes see most of the game; but all will agree that it has, perhaps, been the misfortune of the modern Navy that the different steps in its material progress have been made, and are now being made, with such startling rapidity, that there has been some real difficulty in getting the Naval Service to assimilate them, and make them their own; and simply because, on the one hand, in these strenuous days, it is the business of the Navy to be every day ready for war, and, to this end, each individual in it has his daily task, which is far too exacting to permit anyone either to dream of the future or theorise in the present; and, on the other, the Service has been without the kind of outside help that would make the process of assimilation easier.

The lecturer has explained, he hopes frankly, the layman's limitations; may he suggest a possible danger from the professional side? If lay suggestions are only to be useful when used in light of hard-won professional knowledge, there is manifestly a danger that this knowledge will prove an obstacle. The harder it is won, and the more firmly it is held, the greater the danger.

The tragedy of progress is when those who have earned the title of leader, find their enemies behind them instead of in front; when the pioneer has to use the barriers he is cutting down to fortify his own position; when he is only saved from being trampled down by the on-rushing mass taking a new and shorter road to victory. Nor is there a more pathetic spectacle than the warnings of the adventurous against any adventure but their own. Hence there is no conservatism like that of the progressive. It was not the duffers, but the greatest astronomers in the world, who found Galileo's discovery hard to swallow.[22]

(3) THE TOUCHSTONE OF SEA-CHANGE IS TECHNIQUE

Be this as it may, it has been a mark of recent material progress that the gulf between lay work and sailors' work has not yet been bridged, and there is even to-day an undue domination of lay counsels, and perhaps of lay standards, in naval affairs, because these counsels are apt to be limited only to such aspects of naval affairs as landsmen, *qua* landsmen, understand. Engineers, electricians, armour-plate makers, manufacturers of ordnance, engines, and ammunition, ship designers and constructors, are masters of the development of their respective crafts. They invent, develop, and produce prodigies of progress in their respective lines – and it is impossible for the Navy to refuse their sometimes unwelcome gifts – but they contribute nothing when it comes to saying how their productions are to be used.

Those who say that there is a school whose only thought is of the *matériel*, and that this school dominates naval counsel, are witnesses of this undoubted phenomenon, though we may hesitate to accept their condemnation as too sweeping. Certainly no man can complain that in the last few years a most formidable effort has not been made to get to the bottom of the best way of using the marvellously improved weapons and instruments of war that the advance of science has placed in the sailor's hands, or that this effort has not modified and set its mark upon these instruments in turn; and, consequently, it seems as if it would be more in consonance with the apparent facts to say that the domination of purely material ideals is already over, and that the Navy is to-day in the hands of a technical rather than a *matériel* school. And this is no more than to say that the sea-change of the poet is at work.

(4) TECHNIQUE THE CRITERION OF PROGRESS AND WAR-FITNESS

The lecturer will attempt to define the word technique. It is often a source of confusion and misunderstanding that words are used by a speaker in one sense, and understood by his audience in another: the danger is greater where the audience may, for all the speaker knows to the contrary, have a limited meaning of their own. So, without further apology, except to ask his hearers to accept his definition for the purpose of his argument, he proceeds. Technique is the method by which we attain any particular end by the use of tools, instruments and material. The technique of painting is the method of using brushes and oil paint – of playing, the art of using fingers and notes. Then must there be many choices of technical methods in every art or craft. The unit of basic activity in the art of war is the use of weapons: and as the efficiency of any weapon must depend on the way it is used, the technique of weapons is necessarily the foundation stone on which the whole structure of the art of war is reared; and, as technique varies, the best art of war can only be built up on the best technique. If this is true of all warfare, how crucially true is it of modern sea warfare, where the ratio of weapons to men is in such extraordinary contrast to what it is on land. Japan had for many months, many hundreds of thousands of men fighting in Manchuria, handling almost as many weapons as they numbered themselves; but at Tsushima the fate and destiny of 50,000,000 people were settled in forty minutes by Togo's 200 guns alone.

(5) THE PROPERTIES ON WHICH TECHNIQUE OPERATES – THE VITAL PROPERTIES

We can say of weapons that their qualities are of two kinds – first there are their accidental qualities, as of a gun for instance, its size, the power of its propellant, its projectile, its accuracy, etc., and next its inherent qualities – i.e., those that constitute its handiness and so lend themselves to a more perfect method of employment, such as its superior training, elevating, sighting, and loading facilities. Technique is then the art of developing these inherent qualities to their utmost; it is the only condition which can make material additions of value; it is, in point of fact, the vital, converting principle of life, that enables an art to make a foreign body part of its own organism – just as a tree embraces and assimilates a grafted branch; and consequently, it is obvious that in the evolution of any weapons or other tool of war, improvements in the inherent qualities are vastly more important than mere improvements in their accidental qualities. This will be clearly seen if we compare, for instance, the extremest examples extant of difference in both qualities, namely, the *Victory*[23] and the *Dreadnought*. Anyone can prepare a list of the accidental differences, size, strength, and impenetrability of structure, weight, and power of armament and speed. But over and above these are the crowning differences that the *Victory* has not constant command of its prime motive power, nor ever command of motive power except on 19 points of the compass, nor has the power of handling her guns in the sense of doing anything towards, for instance, maintaining continuous aim or rapid loading. The *Dreadnought*, on the other hand, has great command of speed at all times, and can almost at will proceed instantly towards any selected spot. Her guns, moreover, can also at will be pointed, and be kept pointed, at any object.

That these two inherent qualities, the dirigibility, so to speak, of the ship itself and the possession of dirigible guns, are the real landmarks of advance and consequently the root distinction between the two ships, can be demonstrated by imagining a *Victory* having at no time a greater speed than the real *Victory*, except that her speed should be available over all courses, and having no heavier or more powerful guns than her prototype, but those as handy to aim and load as modern guns of the weight. What chance would the old *Victory* stand against the new? Or, again, suppose a *Dreadnought*, in every point as the actual ship, but robbed of these inherent powers of command over her mobility and her armament, and of what earthly use would she be?

(6) THE EPOCH MARKS OF PROGRESS

If this analysis is a true one, it follows that the master significance of steam, as applied to ships of war, lies not in the conferring of greater speed, but in making the ship answer at a constant speed to any helm – and that the master steps in the progress of artillery are not those that have marked its growth in size and power, but those that have permitted an advance in the capacity to use it.

Note that the dirigibility of the ship preceded the dirigibility of the gun. In this may perhaps be found at least one element that explains the long predominance of the ram. Seamen, who understood ships and could trust their eyes, may have preferred the weapon they could control to the one they could not, and thus for nearly forty years put the ship's ram to the forefront, not only as beyond question their most formidable, but in all probability their most reliable weapon. Those interested in historical parallels may find entertainment in the fact that the argument that killed the ram was the impossibility of judging the range, speed, and course of the enemy, with sufficient accuracy to use the ram with precision. Half a length error, and the rammer became the rammed.

(7) THE RECENT TECHNICAL ADVANCE

There is, of course, nothing new in emphasising the paramount importance of technique. Between 1890 and 1905, while there was little obvious advance in the design of battleships, there was a marked addition to their inherent qualities; first, the speed advanced from 13 to 19 knots; and, next, there was a notable modification in armament, due to the technical preference for more easily manipulated ordnance. The 16.25 gave way to the 13.5, and this in turn to the 12 inch, and even to the 10 inch, not because power was not greatly desired, but because handiness was desired more greatly still. The Q.F. armament again was advocated (and is still advocated) for the reason that its inherent qualities were held to ensure superior hitting capacity. The fact, of course, is that the period named was one of marked technical advance, because it was a period of most vigorous experiment and study. Very significant was a powerful essay on Naval Strategy in the *Brassey* of 1901, written, one supposes, in 1900, which was a trumpet call to discard conservative pre-possessions in view of the 'unknown possibilities' of modern weapons.[24] The efforts to plumb these unknown depths were ingenious, continuous, universal, and, in a marked degree, successful. The era of manoeuvres, antedated what is unsympathetically called the craze for gunnery, but the upshot of development, experiment, and practice in the handling of ships and guns is that no Admiral to-day would take a fleet, say, of *Majestics*, into action as he would have when they were first commissioned. The technique of gunnery has changed, and with it the technique of fleet handling. Fighting ranges have been pushed out, hits can be counted on in the ordinarily expected conditions of action at 5,000, 6,000, or even 7,000 yards – whereas ten years ago the opening range could not, it was conceived by the late Admiral May,[25] a very high authority, exceed 4,000 yards. And yet the guns are literally the same guns. Their inherent qualities have not, except in one respect – i.e., their sights – been improved, and the explanation lies in the development of modern fire control.

(8) IS IT BASED ON FINAL PRINCIPLES?

It is, then, a palpable understatement of the case that the Navy has successfully developed a new technique in recent years; we are more concerned to enquire whether the lines of advance that have been followed so far, are lines that will serve us as well in the future as they have done in the past? Do they hold the seeds of finality in them? Or are they approaching the limit of their possible development? Or, put another way, are they such as to keep pace with the material progress that is making around us? Finally, will they enable us to meet the challenge of the essayist, and fathom the mysteries of the uncharted deep? The speaker frankly states that his apology for being here is that his answer to each question is a negative, and proceeds to call the Navy itself as a witness to, at any rate, the tenability of his position.

(9) EVIDENCE THAT IT IS NOT SO BASED

Quite the most interesting event in the development of modern navies has been the inauguration of what is called the *Dreadnought* type of ship. In itself, the fact of a ship having a ten per cent increase in length and weight, a two-knot speed advantage over her predecessors, and a broadside of eight 12in., against four 12in., and five 9.2's, should not, one would think, have created so great a stir. The stir arose because the new type brought to a head an uneasiness as to the trend of things, that had been growing for years. Many and eminent officers of different nationalities openly said that neither speed nor more distant striking power conferred any tactical advantage. We were told that speed was no weapon – was indeed only useful for running away, that the choice of range was a delusion, that the slower fleet could not be prevented from closing, that games at long bowls were so much opportunity and ammunition wasted, and that at decisive ranges – 6,000 yards or less – an armament of quicker firing guns was preferable, for the reason that ten hits that would shake up the personnel were worth more than one that would shake up the ship. And to complete the picture, the battles of the tenth of August[26] and Tsushima were contrasted; and the latter victory was clearly attributed by a very brilliant lay critic almost solely to the Japanese having a preponderance of lighter pieces. Those who took these views defended them ably, and they have not been as ably answered. To those that held them, it seemed as if an era of senseless megalomania was threatened. Back of the arguments was the consciousness that whatever the legend or accidental power of these ships, there was not known to exist a technique that could convert their paper power into actuality on the day of battle; and it would look as if the advocates and opponents of the new type might be supposed to have taken one or other of two views. The first, with a wistful eye on the future, said 'Give us the new powers, and just as we have made a new gun out of the Mark VIII. 12in. by our development of technique, so we will find a way of using these also to the utmost'; and the others, with a practical eye on present needs, said, 'Don't waste money on speed and guns we can't use, but give us more of the kind of ships and guns we can.' In other words, it was consciousness of the limitations of modern technique that was the backbone of opposition – hope of a better technique the justification of their advocacy. And if this view of the situation is not altogether wide of the mark, it is worth analysing present-day technique, and this will enable us, in reference to a concrete case to examine its title to finality or power of progress – and, if it is found wanting, we shall be led into asking what new functions or services, not at present rendered, are necessary to the technique that is wanted. And this enquiry is vital, because we shall surely never progress in the direction we need to go in, unless we very clearly understand what that direction can only be.

(10) THE REAL NATURE OF MODERN ARTILLERY TECHNIQUE

If we look at the gunnery technique of the day, the first thing that will strike us – apart altogether, of course, from the amazing character of its results – is that, in spite of a high differentiation of functions, it is, on the ultimate analysis, identical with the method used by the first gunner who ever tried to make a hit at sea. He may be supposed to have formed a rough idea of the direction in which his piece was pointing, to have waited until his ship's movement made that line, with the requisite allowance for trajectory (if he thought of that!) coincide with the target, to have fired the gun, and then watched for the shot to strike somewhere. When he missed, he saw the shot splash – and profited, let us hope, by what he learnt. From that day until we put a man into the tops to watch the shot there was no functional departure from this method. It is true there were first invented sights – then guns that could not only be trained but aimed – then telescopic sights, and better and better gear for handling, until continuous aim in even baddish weather was possible – but in spite of the gradual perfection in the inherent qualities of artillery the man behind the gun still did as his forefather did – he fired, and watched, and corrected.

But there came a day when the guns were known to be both powerful to strike and accurate, at far longer distances than the man behind could see. And so another man was sent aloft to do the seeing for him, and the man aloft needed another to transmit the message, and then a third and a fourth to help calculate the significance of each correction; then all of these needed their separate instruments, range finders, bearing keeping devices, rate of change calculators, rate of change clocks; and proper and prompt communications were required to get the results of all these activities to the gun, so that most ingenious and accurate transmitting and indicator gear was installed, which in time involved new adjuncts in the gun position – namely, sight setters – and then the gun layer was further helped by assistants, who may be called sub-trainers and sub-elevators; and now at last we hear of sub-training and sub-elevating of the range finder. All this implies a minute sub-division of the operation. If we suppose the whole organisation called into activity to fight a single gun, it would involve mobilising some dozen persons at least. The point is that, while you have multiplied your functionaries, you have not increased the functions. You improved the gun captain's sight, but at the cost of his communication, and it remains that, in the result, the combined activities of all these persons and devices are nothing more or less than the old gun captain, turned, like many another old business, into a public and almost unlimited company, and their method remains as it began, a hit or miss method.

Such efficiency as the gun captain had – such efficiency as the gun layer shows in his tests to-day – is partly a matter of manual skill, partly of that quick unconscious co-operation of eye, brain, and hand, that defies analysis. It is a combination of activities that cannot, for a different scale, be synthetically reconstructed. All the drill in the world will not make a dozen men act as one. In other words, the putting of the gun captaincy into a commission of twelve fails because you cannot syndicate instinct.

In the lecturer's opinion, this is the fundamental oversight of the present fire control; the thing has grown by degrees; it has increased marvellously in efficiency; small wonder then that it has been hard to realize that the moment it was attempted to make hits beyond the range at which the eye of a single man – and that man the pointer – could gather all that was necessary to control the gun, the problem became, not the old problem with a different degree of difficulty, but a new one altogether, differing not in degree, but in kind from the old. It must, then, be solved by means differing as the problem does – another way of saying that a new technique was imperative.

(11) THE *LACUNAE* IN THE PRESENT METHOD
(a) THE DISTANCE LIMIT

The trouble with the spotting method is that its unit of activity is watching the impact of misses in the water. Clearly the larger the angle at the spotter's eye, subtended by a given measurement of sea-surface, the more efficient the spotting will be. Experience shows it to be very efficient under certain conditions with a stationary target at 5,000, 6,000, and even 7,000 yards. But in a heavy sea its efficiency would be lessened by the size of the waves – and the only way to increase its efficiency under any circumstances would be to increase the height of the observer. The formula is 22 feet for every 1,000 yards, so that without any more fundamental objections to present-day technique we get a first-class objection right away – viz., that if, for instance, we take 6,000 yards as a standard range for good spotting results, we must raise our masts 44 feet to get equally good results at 8,000 yards, and so on – which is impossible.

Now it is commonly held that modern guns keep their accuracy at much greater ranges than this. If they do, it is well to remember that a battleship half as far away again – namely at 12,000 yards, is a much larger target than a battle practice target. She presents no aiming difficulty then. Is it conceivable that engagement could ever take place under circumstances of a changing range at such a distance with present methods? We see in point of fact that the limitations in range of spotting forbid it.

(12) *LACUNA* (b) IN ITS NATURE UNRELIABLE

Further, the service has had a good deal of experience now in firing at a stationary target, and the results, as shown by the battle practice returns, while indicating marked improvement, also indicate that the method of gunnery gives a very uneven figure of merit to the ships – a more uneven figure than the gunlayer's test, for instance. Indeed, it is a notoriously difficult, and consequently an exceedingly chancy, method. And all this is at the stationary target, where so long as the bearing of the target can be approximately taken, there is no difficulty in making a table of the change in range for considerable limits. What then are we to expect when the target is not stationary but moving, with speed and angle of course unknown, and spotting made almost impossible by smoke? One thing, at any rate, we can surely say, and that is that the results will not be better, under action conditions, than at battle practice — and if these are not good enough for us, the other will most assuredly be worse.

(13) *LACUNA* (c) DOES NOT ADDRESS ITSELF TO THE REQUIRED DATA

It is, of course, a commonplace now, in any analysis of the data for hits, to say that the governing factors in change of range, are the speed, the angle of course, and the bearing of the enemy. But while this is so, present-day technique does not address itself to the task of collecting them. Indeed, the nature of the operation, estimating the relative position of misses in the water, makes the getting of such data a virtual impossibility. So astonishing is the skill of latter-day officers that there would be nothing surprising in these data being often quite correctly guessed. But that may only be done while guns are firing, and in very few combined variations in the speed and course of the firing and target ships. Experiments made without firing are not very encouraging if rumour is to be believed; and the experiment with guns firing back has still to be tried.

In short, the spotting method is based on a clearly limited possibility of correction — viz., the height of the masthead makes *efficiency inconceivable* long *before* the *limits* of even existing *guns* have been *reached*; next, it is not a technique that has any of the advantages of *mechanical repetition* — and this means so long as it remains the only technique in use, it will not be possible to count on a *standard of hitting values* for every ship — the first desideratum of a fighting force; while, lastly, it cannot be at all effective in cases of extremely rapid changes of range, because, even within its other limitations, it is not directed towards ascertaining the *causes of change*.

(14) THE MISSING LINK

But now, to come back to our point of departure, we are to see what is wanted in an ideal technique that shall enable us to use our new and fast mastodons to the top of their inherent capacity. As we have seen, this inherent capacity is, first, the power of reaching any desired point, or the lecturer would prefer to say, maintaining any selected course, at a higher speed — and striking with a greater power at a greater range. [The distinction between 'reaching a point or position' and 'maintaining a selected course' is important, because naval tactics do not present static but mobile problems. There are not tactical positions at sea — the position relationship of two moving bodies is so seldom continuous without change – and to think of this relationship apart from the laws that in each particular case govern this change, can only lead to confusion of thought.]

These two inherent capacities are interdependent, viz., there can be no object in speeding on any particular course, except as preparatory to, or to accompany, the use of the said striking power. But further, there are no moves in tactics that are in the abstract the best – the merit or demerit of any particular move is conditioned by the disposition and movement of the enemy – and, again, there can be no such thing as using our more powerful and more distant striking power, unless, as we have just seen, we know the range and its change. And so, at the end of all, we find that we cannot use either our greater speed or our greater gun power unless we have certain knowledge – namely, distance, bearing, speed, and angle of course of the enemy. On that depends our choice of position relationship resulting from the course and speed we select; on that depends our power of hitting him while we maintain it. In other words, incapacity to get the data that are the clue to the enemy's movements, makes high speed of small tactical value, just as a faster polo pony would give a player, unable to judge of his opponents' movements, no advantage in the game – and incapacity to get the same data, which will give us a continuous knowledge of change of range, makes sustained hitting unattainable, and hence our long ranging artillery is as useless as closely choked guns in the hands of a man with no instinct for judging the speed of a high rocketter. The tactical value of speed and distant gun power are not proportional to knowledge of the enemy's movement. It simply doesn't exist without such knowledge.

(15) THE A.C. BATTLE SYSTEM DIRECTED TOWARDS
FILLING THE THREE *LACUNAE*

And so we arrive at last at the subject matter of the lecturer's address. Up to now, he has been setting out in a brief form the analysis of the problems of gunnery and tactics that has led him in the course of the last eight years to develop a system which he hopes to see made the starting point of the new technique that must ultimately be admitted to be inevitable.

The subject cannot be considered of other than the first importance, for the whole art of naval warfare stands to be reformed if such a technique proves possible, and if we ask what is wanting to-day to enable us to employ such a technique in the case of the fastest and most powerful battleships afloat, we find it is simply and solely, first, continuous knowledge of the movements, bearing, and distance of the enemy; and, secondly, means of using such knowledge in handling ships and their armaments.

The system he will now put before you was urged upon Whitehall primarily as a gunnery system, and was christened with the deliberately misleading name of the Aim Corrector System. But, as has been explained, the data for right gunnery and right tactics are identical, and the speaker has steadily kept this in view from the first, and has consequently attempted, not only to get these data and make them available for gunnery, but to make them available as a guide in tactics as well; and he calls the system the A.C. Battle System because he believes that as experience brings it to a useful stage of development, there will be found in it the germ of that technique in the use of ships and guns which will make a reality of their latent powers.

The task in hand is to explain this system. To many it will certainly appear to be so ambitious as to be Utopian, using that adjective as pertaining to a state of affairs unreasonably good for a workaday world. Accepting this challenge, the lecturer proposes to adopt the method of the writer of Utopia, and to ask you in fancy to accompany him on board one of the new *Invincibles* to see the system at work, and to receive an explanation of how the various operations are performed. He has found, from his own experience, that the most effective way of bringing a mechanical problem home to himself, especially if it is a complicated one, is to picture the problem actually solved, and to work back from the results got to the means inevitably necessary to attain them.

(16) A HYPOTHETICAL CASE

We are, then, to imagine ourselves on board the *Invincible*, steaming at 25 knots. Another ship bears about 30 degrees from our bow, and is apparently steaming at a speed as high as our own. Our guns are pointed towards her. A shot is fired from the forward barbette, which we observe falls short. The range is considerable − 7,000 yards to 8,000 yards at least. From our position in the conning tower, we have no idea how near to the target this particular shot may be. A second shot is fired, which we do not see at all. A third shot is again short, and the fourth shot, we notice, is a hit. For the next two minutes, independent firing is kept up by all our guns, the forward guns firing first and the rest in sequence, down to the aftermost gun, when the first begins again. There is a very powerful following wind, and so the smoke from each discharge is carried in front of the guns that are loading, and the loaded gun waiting to fire has a clear field. As far as we are able to judge, about two out of three shots hit, which is the limit of what would be expected, with perfect aiming and exact range and deflection, on a target of that size at that range.

We have looked in vain for fire control positions on the mast, and the only difference we observe, in the general appearance of the ship, is that the conning tower is in two storeys and of larger diameter, that there is a second conning tower aft, and that protruding from either side of each are the two ends of a large-sized range finder. There is not a soul in an exposed position on the deck, nor apparently any communications which can be reached by shot or shell.

At the end of the operation we ask to have the method explained to us, and this is the story that we hear:

(17) GROUP I – RANGE AND BEARING

We are first taken to the upper storey of the forward conning tower, and in it we see an ordinary type of range finder of rather greater length than is now used, so mounted that the operator is seated in a comfortable position. On one side is a cased-in apparatus, connected by a covered spindle with the base of the range finder. Instead of the knurled roller, or other ordinary means of adjusting the lenses, so as to measure the range, we notice something encased and clamped to the range finder, and from it protruding a handle, which the operator manipulates. Under his foot is a contact maker. From the first cased-in mechanism clamped to the tube goes an armoured cable. From another mechanism clamped to part of the pedestal, another cable. It is then explained to us that the mechanism on one side of the conning tower, connected to the pedestal by a spindle, is a gyroscopically-controlled electric drive, which keeps the range finder on the selected keel line of the ship, and consequently neutralises those rapid and minute involuntary changes of course due to the yawing of the ship. The knurled head the operator controls with his right hand is for setting the lenses to obtain a conjuncture of images. The hand wheel under his left hand enables him to train the range finder to keep the image central. The two cables connect transmitting gear of similar character, one of which transmits the range electrically, instead of it being necessary for the operator to read it. The other transmits the bearing, and both are operated simultaneously. Except that the operator does not read optically, or transmit vocally, the ranges that he gets, and that instead of standing he is sitting, and, instead of being compelled himself to counteract the movements of the ship, to obtain a continuous and steady view of his target, this is done for him automatically, the actual functions of the operator are not different from those of an operator using an ordinary range finder under the conditions we are all familiar with. We see, however, that the gyroscopic control has removed many obstacles to his keeping the target permanently in view, that being seated instead of standing he is more at his ease, and has his instrument under better control, and we observe that, as he does not have to cease the operation of keeping his images in conjunction, a great obstacle has been removed in the task of continuous range keeping.

(18) GROUP II – BEARING, SPEED, COURSE, OF ENEMY

We are now taken from this position to a station immediately under the conning tower and below the armoured deck. This is the A.C. Central, and in it we find a charting table, with a wide strip of paper fed from a roll at one end, and drawn across the table at a steady speed. A clock mechanism runs this, and, above the clock mechanism, we see a dial which is set for a speed of 25 knots, and an indicator from the engine room, which shows us that this is the speed at which our ship is going. We are told that the scale of the paper is so many inches to the thousand yards, and that the speed at which it is moving exactly corresponds with the speed at which the sea is going past the ship. The paper, then, may be regarded as a blank map or chart of the sea, such as is used in the hunting of the Snark.[27] Now we observe that the charting arm is moving radially, and that a small carriage on it, carrying a pen, is travelling along the arm, and that at certain points the pen is depressed, and makes a mark upon the paper, and that these marks are fairly continuous, until, in the course of a minute or two, we have a series upon the paper, all at slightly different distances from the edge of the paper from which the arm radiates. From this series of lines we have no difficulty in constructing a median line, which will average the ranges given. Hinged to the further edge of the table are a parallel rule and protractor. The officer in charge of the charting table, having marked the time that the observations began, and having marked off minute intervals in their progress, now brings the ruler down to the chart, and reads off on a protractor how much out of parallel to the further edge of the paper the new charted line happens to be, and, then, turning his ruler on to the minute intervals, informs us that the ship under observation is travelling at so many knots speed on a course so many degrees divergent from ours, and that at the moment her bearing is so many degrees from our bows.

(19) GROUP III – CONTINUOUS RANGE AND DEFLECTION

Alongside of the charting table is a large enclosed box, with a series of dials upon it. The dials are marked 'Our Speed', 'Their Range', 'Their Bearing', 'Their Speed', 'Their Angle of Course'; and upon these dials he proceeds to set the data that he has just gathered from the chart upon the charting table. There are, moreover, other dials upon the machine which have already been set, and, when we ask upon what principle these have been set, our attention is turned to the indicators showing what is the temperature and density of the air, what is the force and direction of the wind, what is the temperature and any other abnormal conditions of the ammunition, what is the nature of the projectile to be used in the guns, and, finally, the time of flight at the range.

AN APOLOGY FOR THE A.C. BATTLE SYSTEM

The officer in charge explains to us that, the range tables being only true for a certain set of conditions, we have each day practically to select a range table for the occasion, that these are the elements that have to be taken into consideration in making up such a range table, and that consequently the data, from which the machine on the wall is going to calculate the range, have to be varied to suit the conditions of the day.

A second machine is set to the same data, and this machine, we are told, is going to calculate the deflection. All the data used in the first machine are put into this, and we are informed that each of these machines will independently calculate the exact range and deflection that should go on to a normal gun – i.e., the range given by your chart in terms of the presumed performance of the gun under the conditions that the indicators have shown us to be the conditions of the day.

We are further informed that not only do these machines calculate this range and deflection, but that, from their clock-like character, they will continuously calculate them on the assumption that both ships are maintaining a steady course and a level speed. In other words, we perceive that we have got out of the region of assuming a rate of change, and waiting for experience to show us that the rate is wrong, and that we are dealing with the mathematical bedrock facts controlling the causes of change. At this point we ask what, as a matter of fact, was the actual curve of change in the piece of gunnery practice of which we have been witnesses. We are given the figures of a similar combination of speed, courses, and bearing angles, but of 2,000 yards less range, as the diagram shows.

(20) GROUP IV – CONTINUOUSLY SET SIGHTS

We should naturally have guessed that the object of these calculating clocks is to make the exact range from second to second available for the guns, and when we see the startling rapidity with which the range has changed in the case we have witnessed, and remember that the firing was independent and not in salvoes, we shall wish to know how the sights have been adjusted, or the gun has been made to respond, to all these changes as they happen. We shall probably guess that, the unit of sight adjustment being one movement for every 25 yards of change in range, it has hardly been feasible to transmit these unit changes to indicators, and to have even the best drilled sight setters maintaining an adjustment of the sight synchronous with each change in the range. When we are taken, therefore, to one of the barbettes, we shall be prepared to find that the range indicator and the sight setter have been abolished, and that, as a substitute for them, two mechanisms at right angles to each other have been attached to the sights themselves. On inspecting them, we shall find that, exactly as the pen on the charting table has been made to move in azimuth and for

Time

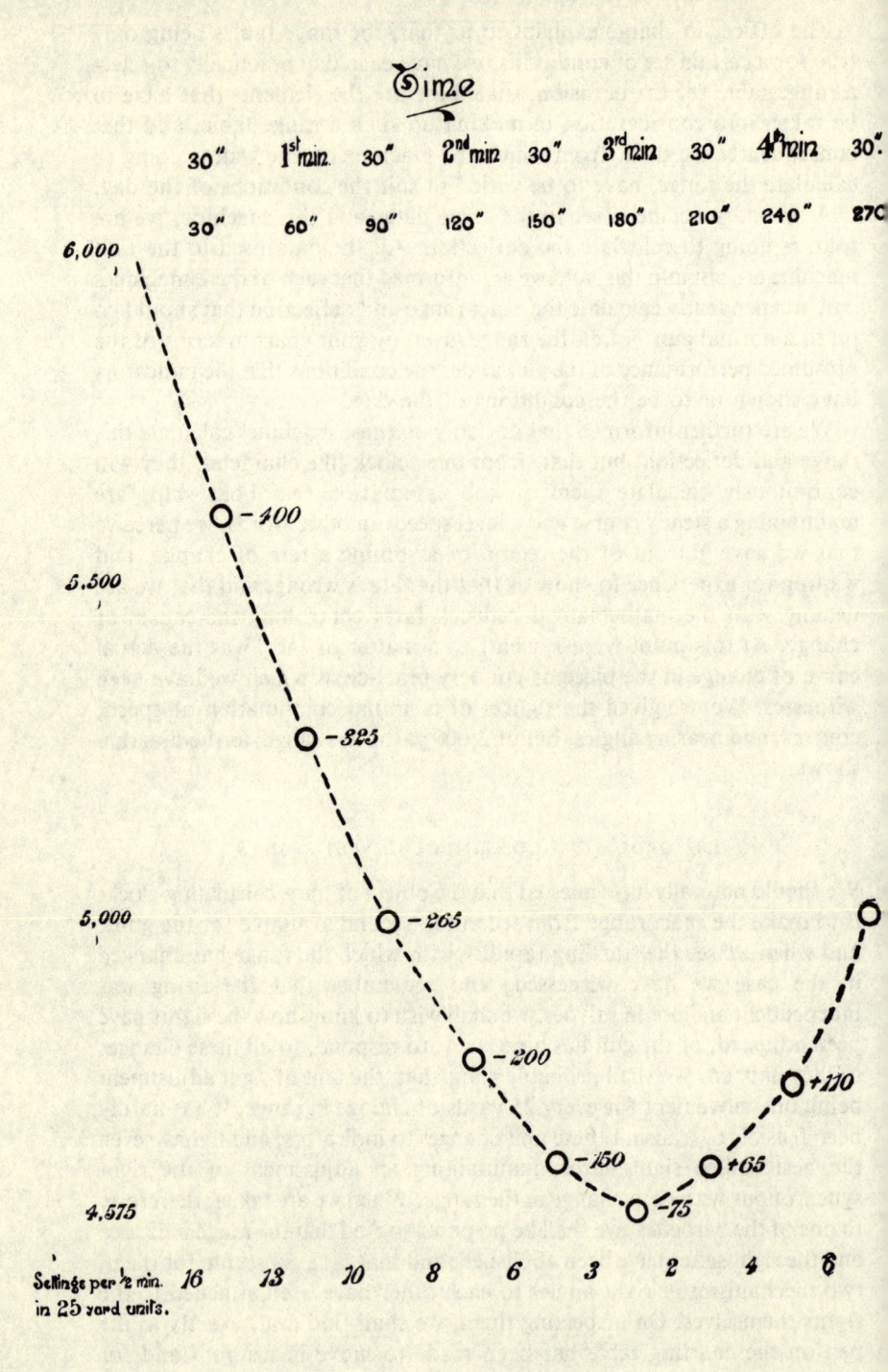

	30″	1st min	30″	2nd min	30″	3rd min	30″	4th min	30″
	30″	60″	90″	120″	150″	180″	210″	240″	270″

6,000 ○ −400

5,500 ○ −325

5,000 ○ −265

○ −200

○ −150 ○ +110

4,575 ○ −75 ○ +65

Settings per ½ min.   16   13   10   8   6   3   2   4   6
in 25 yard units.

distance in exact correspondence with the operator's movements of the range finder in azimuth or in range adjustment, so, too, have the sights of the gun been made to move for elevation and deflection in exact correspondence with the calculated results of the range and deflection machines.

But our attention will also be drawn to an intermediate link between the mechanism controlled by the change of range machine and the actual sight; and this is, the differential arrangement put on to each gun, so that the range, as finally used by the gun pointer, shall not necessarily be that calculated by the change of range machine, but a modification of it, suited to such variation in natural velocity which each separate gun has been found, by the most recent calibration, to possess.

It will now be a natural question for us to ask, assuming that all these operations have been successfully, and as accurately as possible, performed, how it was that the process of firing the preliminary shots that were fired was carried through successfully, without there being apparently any observers or spotters in a particular elevated position to make those corrections which were the presumed object of the preliminary shots.

We should then be told that experience has shown that, even with the utmost care, a certain residual error in the total result has to be taken into acount; but that, as the main disturbing factor − viz., the change of range due to the speed and angle of the course and bearing of the two ships, is accurately known, the process of converting the assumed range to hitting range can be as simply and expeditiously performed, under the circumstances of the most rapid movement, as if both ships were stationary. The initial range being 8,000 yards, and half the danger space being consequently 60 yards, it is quite simply demonstrated to us that 250 yards residual error can be successfully eliminated in four shots, and this with no further knowledge of the performance of any of the trial shots than that they were either short, or over, or on the target; in other words, that the only observation of fire necessary was such as would be necessary were a bracket closed on a stationary target with a stationary gun; and, finally, that such observation is feasible from the armoured positions.

(22)* CHECK ON ENEMY'S MOVEMENTS

We shall not, however, probably be satisfied that this is the only observation for error that will be needed, and we shall ask what is going to happen if the enemy changes his course or speed. We are therefore taken back to the central station below the armoured deck, and shown that the change of range machine has not only been calculating the range that the enemy is presumably at at every moment, but is also calculating his exact bearing, and we are told that, while, after engagement has been begun,

* [Editor's note: no section (21) in original.]

all hope has to be abandoned of obtaining continuous readings, nevertheless it is practicable to take occasional readings; that, while the operators do not, under the circumstances, attempt to obtain accurate ranges, they are under orders to send down every conceivable reading of the enemy's bearing, and that consequently a comparison between the bearings sent down from the range finder, and the bearing indicated by the change of range machine, is a continuous check upon any change of course or speed that the enemy may either voluntarily or by compulsion adopt.

(23) TACTICAL CONTROL OF SHIP OR FLEET

Finally, we are taken to the conning tower, below that part where the range finder is housed, and, here, in a suitable position, we see a mechanical linkage, fitted with dials, showing that it is supplied with the same data as those put upon the change of range machine below. They are marked, 'Range', 'Speed', 'Course', 'Bearing of Enemy', and 'Own Speed', and, as these are set to show us the working of the machine, we observe that the lazy tongs, connected with the point indicating the enemy, show us a series of future positions, which we are told represent the bearings and ranges of the enemy at those intervals of time, assuming that the factors remain constant. Above the linkage is a sheet of plate glass ruled in range circles. The point marked 'Our Ship' is stationary in space, but can revolve round its centre; and the principle of the machine, as explained to us, is that the lazy tongs show us the virtual course of the enemy, as we should see it in plan were we suspended exactly over our ship, so that, as long as we are aware of the speed of our ship, we can imagine ourselves motionless, and only aware of the changed position relationship from ship to ship by the alteration in the enemy's bearing and distance from us. If, therefore, we alter our own course on the machine, we see at a glance what is the new series of positions that the enemy gets into by our doing so; and consequently we are in a position to see in plan form exactly how we can modify the tactical relationship of one fleet to another, and what additional advantage or disadvantage results from our doing so; in other words, we find that this tactical machine, used in conjunction with the charting system, gives the officer in command just that same foreknowledge of the movements of the enemy's fleet that the gun sights get of the range of any particular target ship; and it is of course very clear to us that the main value of this machine, in the tactical command that it gives, is not for the purpose of altering course, or changing formation, during action, but in selecting the disposition, and establishing the initial position relationship, under which the action is to commence; because of course we shall realise that the capacity of a fleet of ships to alter their position, relative to an enemy's fleet, is far more limited than the capacity of a single ship, *vis-à-vis* to a single ship; and, if we experiment at all with the machine, the first thing that will come home to us will be that, whatever other tactical qualities a ship may possess, two are supreme – namely, the greatest possible training angles of the largest guns, and the completest control of the highest speed.

(24) VITAL IMPORTANCE OF THE TRUTH THAT SEA PROBLEMS ARE BASED ON THE CONTINUING MOVEMENTS OF TWO BODIES

Nobody is more aware than the lecturer that the foregoing must appear a highly fanciful and much overdrawn picture. But ideals are the things to work for, because the only way to get the best possible is to aim at perfection. Really, if the picture is examined, at no point does there appear to be a mechanical function which presents more than quite superficial difficulties. No doubt he pre-supposes an accurate knowledge of what may be called the biological properties of big guns. Although minute examination into this subject is not very old, what an enormous deal has already been accomplished in the last few years! The lecturer had the good fortune, and the privilege, to attend the first calibration experiments made not three years ago. There was no serious problem involved in the velocities of guns differing, until it was thought possible to use them at ranges where the difference of velocity became material. In the same way, the possible difficulties of ascertaining change of range were not really appreciated, and, in many ways, are not appreciated now, because it would seem as if very few people had contemplated the possibility of engaging under conditions in which accurate knowledge of change of range is vital. The speaker has made it his business to read all that he can find that has been written on modern sea tactics, both abroad and in this country, and has found so far no serious effort made to deal with the possibility of engaging with a target other than on an approximately parallel course. It is a commonplace that no navy in the world has any experience of firing at a fast moving target at long range. This, of course, is not to say that the problem of trying to ascertain change of range has not been tackled, because, as a matter of fact, in the established battle practice, change of range is far more rapid than in many cases where the target will be moving, and the courses diverging; but it is a characteristic of engaging a stationary target, under the usual conditions, that the range changes at an approximately uniform rate for a considerable time; and it is a curious fact that, if there were any means of taking the bearing of the target, the rate of change could be given off hand, so long as the speed of the firing ship is approximately known, and the range known within 500 yards. The fact that it is on an experience of this character, that most thinking on the subject of long range gunnery has been done, explains the predominance of the phrase 'Rate of Change', as if there were always a rate which was maintained for an appreciable or sufficient period. Hence,

DIAGRAM OF AIM CORRECTOR BATTLE SYSTEM.
as worked by one or more single observer instruments.

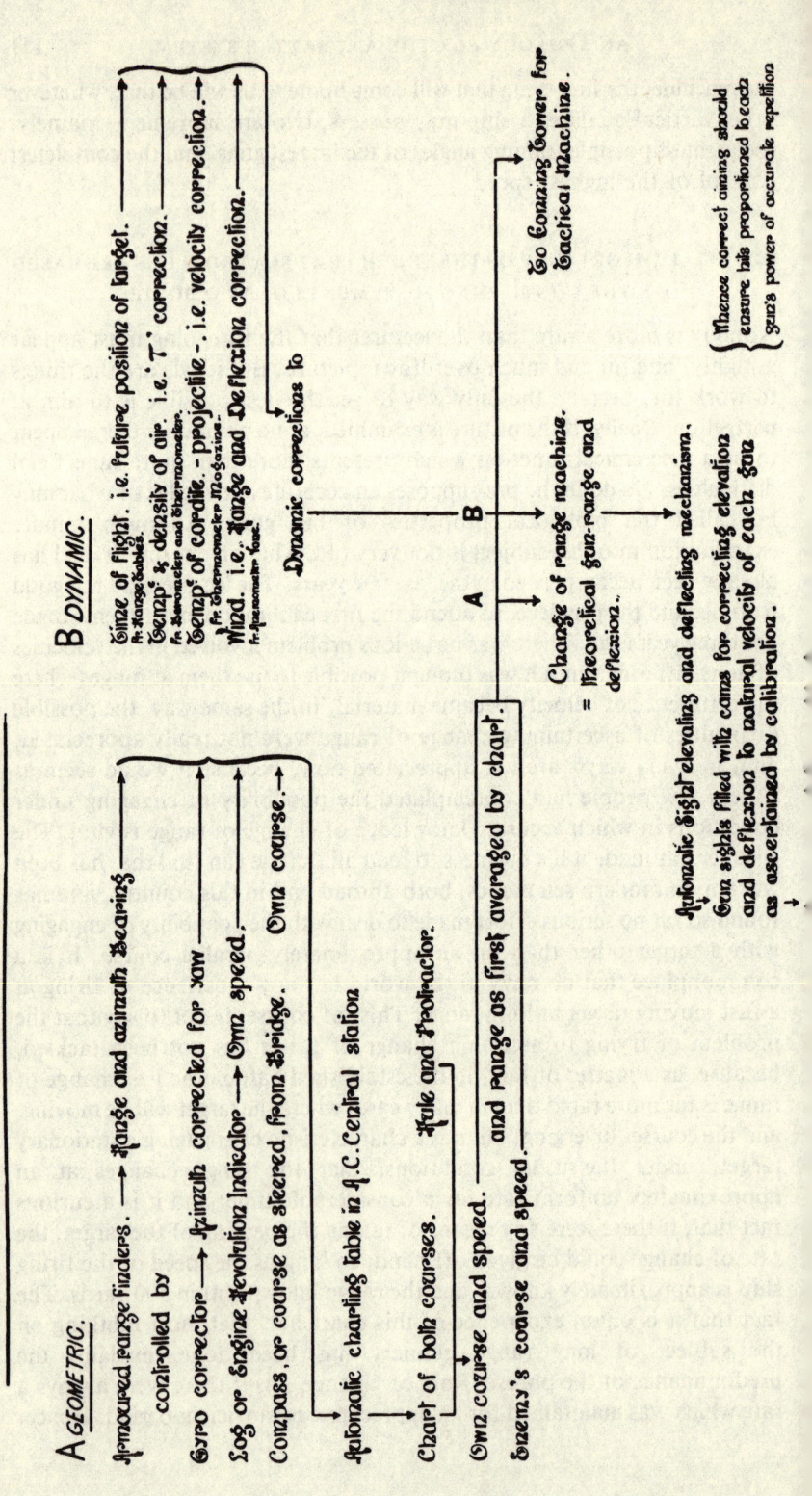

A GEOMETRIC.

Measured range finders. → Range and azimuth bearing
                              controlled by
Gyro corrector → Azimuth corrected for yaw
Log or Engine Revolution Indicator → Own speed.
Compass, or course as steered, from Bridge.   Own course. →

Automatic charting table in g.C. Central station
Chart of both courses.   Rule and Protractor.

Own course and speed.
Enemy's course and speed.   and range as first averaged by chart.

B DYNAMIC.

Time of flight. i.e. Future position of target. →
    fr. Range Tables.
Temp^r & density of air. i.e. 'T' correction. →
    fr. Barometer and thermometer.
Temp^r of cordite. — projectile. i.e. velocity correction. →
    fr. thermometer magazines.
Wind. i.e. Range and Deflexion correction. →
    fr. Anemometer vane.

Dynamic corrections to →

A → Change of range machine.
    = Theoretical gun-range and deflexion.

B → To Conning Tower for Tactical Machine.

Automatic sight-elevating and deflecting mechanism.
Gun sights fitted with cams for correcting elevation and deflexion to natural velocity of each gun as ascertained by calibration.

{ Whence correct aiming should ensure hits proportioned to each gun's power of accurate repetition too, the fire control instruments in use are based on rates, not on a continuing change. If the hypothetical case, as shown by the range diagram (p. 148), is consulted, it will be seen that the 'Rate' varies per 30 seconds, 16, 13, 10, 8, 6, 3 down – up 2, 4, 6, for unit settings of 25 yards each – and at long range the units cannot be larger. There is no sense in talking about a 'Rate' in a case like this. Practically, no doubt, in very many instances, there is a rate, and a steady rate; but the audience will probably agree that no system of getting ranges and changes of range can be considered perfectly satisfactory unless it deals with all cases that may arise, and the advantage of being able to deal with a rapidly altering range when your enemy cannot, is just one of those tactical advantages that all commanders dream about, but seldom possess.

(25) A.C. GIVES, NOT CHOICE OF RANGE, BUT CHOICE OF CHANGE OF RANGE

Perhaps the strongest point in favour of the A.C. System being a scientifically sound one is that there are no instances of combinations of speed and course where it is harder and more exacting to get the change of range than is the case with any of the remaining combinations; and consequently although a ship nowadays, by present methods, might get 50 per cent of hits with a stationary target at a range of from 7,500 to 5,500 yards, without this affording the least guarantee that she could do anything in a case of less rapid change when the speed and course of the enemy are unknown, it is demonstrable that, if such a series of hits are to be got by the A.C. System at a stationary target, at such ranges, those results would equally be got were the target moving at any speed and on any course within similar ranges. The significance of this is that the ship can always be turned to a course that will make the rate of change vary almost at will.

(26) CONCLUSION

The speaker fears that much of what he has said may sound, from one point of view, far too boastful and ambitious; and, from another point of view, at the best premature, in view of the fact that the initial problem that the system has to get over is not yet demonstrated as solved. He has no wish to fall short of propriety in either of these two ways, but a very long series of experiments has convinced him that the problem, if not solved, is most certainly soluble. In the light of previous experience, he expects the trials to show a substantial solution, though, quite possibly, not a sufficient solution to make the instruments in their present form suited for ship work. After all, his sea experience in the *Jupiter* was extremely short and inadequate. What he principally hopes is that enough will be proved to earn the co-operation of the Service in future work, instead of merely its criticism. In any event, the instruments as they stand now are only one part of the system, if we view it as a whole; and, assuming that it worked correctly, it would be a very marked advance on the present system of fire control, and would enable ships to engage under conditions in which they could either not engage at all or only engage under a grave disadvantage – viz., when there is a considerable change of range due to the target ship moving on a course and at a speed that is not ascertained by present methods. But the lecturer is of opinion that it can only be of its full value to the Service when the change of range and deflection machine, and the corresponding sight setting mechanisms, are satisfactorily tested.

'I have to thank you, gentlemen, for your patient and forbearing attention; and, through you, I have to thank a large number of friends and acquaintances in the Service, who have been most encouraging well-wishers in the eight years that I have been engaged upon this work. Like everyone else who has attempted to break new ground in a new way, I have had long and severe spells of discouragement. It has often been a source, both of disappointment and anxiety, that I have not succeeded in getting the official representatives of the Navy to recognise in me a person anxious to be one of the flock, instead of, an uninvited and unwelcome shepherd, whose only presumptive pastoral function was to fleece. But private individuals have encouraged me where official encouragement was perhaps impossible. My experience is long enough now to know how great the obstacles must necessarily be to an important department of State engaging in experimental work; and, in the circumstances, I ought to be less inclined – and am really less inclined – to complain than to be astonished that, the facts being what they are, I have succeeded in getting the large amount of official support that has been accorded me.'

'At a first presentation, the scheme may seem chimerical – but after a certain number of years' work on one subject, any man, otherwise rational, is entitled to be heard; and so, for what it is worth, I wish to tell you that never, since I put my first scheme forward – now nearly eight years ago – have I doubted, or wavered in maintaining, that the foundation of right gunnery could only be laid on a fore-knowledge of the enemy's movements; that this could only be built up on a past knowledge of his movements; and that this, again, could only be obtained by a surveying system, such as I have explained; and that, from the day such a system succeeded, not only gunnery, but tactics and strategy would have "unknown possibilities" thrown open to them. The first part of the system will throw the possibilities open – the tactical machine should go some way towards making their "unknown" character "knowable"; an automatic sight-setting system should complete the work. And so I still think that day will come, though possibly, as I began by saying, only on the condition of my own devices suffering the relentless sea-change of the singer.'

'Whether there is good reason for such faith, or not, is hardly worth discussing; but you will readily understand that it is quite an exciting and pleasant belief to hold, and that whether it materialises in whole or in part, or fails altogether, the time and work I have devoted to it have been well repaid to me, first, in the many pleasant friendships that I have come by on the way, next, in a very inspiriting, because intimate, knowledge of the Naval Service, a thing that can come to few laymen, except to those who touch it at some vital point of its work, and finally in the exhilaration that comes from a not dishonourable hope that the work, such as it is, may some day in more powerful hands than mine, prove of service to one's country in the hour of need.'

X

NOTES, CORRESPONDENCE, ETC., ON THE POLLEN A.C. SYSTEM, INSTALLED AND TRIED IN HMS *ARIADNE* (*DECEMBER 1907–JANUARY 1908*)

Preliminary trials of Pollen's instruments were conducted in the second week of December 1907, in the absence of Admiral of the Fleet Sir Arthur Wilson. They were conducted by tracking the armoured cruiser *Shannon* with the instruments on the *Ariadne*, while both ships steamed at speeds and on courses that were such as to pose the problem of moderately high and continuous changes in the change of range rate. The first trial runs were undertaken in relatively calm weather, but on 9 December a heavy storm struck, resulting in very rough seas for several days. In tests runs made on 10, 11 and 12 December, however, the gyroscopically stabilised range finder and bearing indicator proved to be capable of compensating for the very considerable yaw of the *Ariadne* resulting from the extreme wave conditions, and therefore enabled the range taker to make numerous and accurate range and bearing observations, which were transmitted without mishap to the automatic plotter, which in its turn produced a clear chart of the movements of the *Shannon* and the *Ariadne*. Accurate measurements of the *Shannon*'s course and speed were thus obtained although the courses and speeds of the two vessels were such that the change of range rates were as high as over 500 yards a minute and varying continuously, and the ranges were as great as 12,000 yards.[1] In other words, in these trials Pollen's instruments demonstrated their ability to more than satisfy the requirements for complete success that had been agreed upon in September 1906.[2]

But in the meanwhile, Wilson, with the assistance of Lieutenant Frederic Dreyer, formulated a manual fire control system that could be put up as an alternative to Pollen's mechanised approach. The Wilson-Dreyer system used a standard 9-foot range finder and mounting, which was operated by a range taker whose sole task was to take ranges, and a range finder layer who altered the train of the range finder as required by changes of bearing caused by yaw or by alterations in the relative positions of his ship and the target. Simultaneously observed bearings were taken with a telescope mounted on a torpedo director, which consisted of a flat plate with a bearing scale engraved on its face. The bearing observation was made by training the telescope on the target, and reading off the bearing from the scale as indicated by the train of the telescope in relation to the scale. After the range and bearing had been obtained, the data were transmitted from the mast to a plotting station below by voice pipe where it was plotted on a fixed chart to produce a virtual course diagram of the relative movement of the firing ship and target in which the firing ship was presumed to be stationary.[3] The plotting officer then used special rulers and protractors to measure from the plot either the target's virtual course and speed or the change of range rate and the change of bearing in the time of flight of the projectile.

The two sets of data offered a choice of methods for supplying the gun sights with ranges and bearings that had been corrected by time of flight of the projectile and ballistical data. A dumaresq could be set with the target's virtual course and speed established by the plot. The dumaresq settings for the firing ship's course and speed would thus be set at zero. The dumaresq would then indicate change of range rates and deflections as the bearing indicator was adjusted for changes in the target bearing. The change of range rates indicated by the dumaresq would be set on the Vickers Clock, which would generate ranges that would be combined with change of range in the time of flight of the projectile and known ballistical data on a special slide rule, and this would be manipulated to calculate a corrected target range. The deflection indicated by the dumaresq would be combined with bearing observations, known ballistical data, and change of bearing in the time of flight of the projectile data on a slide rule that would be manipulated to calculate a corrected target bearing. Alternatively, the change of range rates as measured directly off the plot could be set on the Vickers Clock, whose generated ranges would be corrected as described above to provide the required corrected ranges. The changes of bearing in the time of flight of the projectile, as measured directly off the plot, would be combined with bearing observations and known ballistical data on a special slide rule that would be manipulated to calculate the required corrected bearings. After the corrected ranges and bearings had been calculated by either method, they were then to be relayed by voice pipe to the sight setter at each gun, who would use the data to set the gun sights for elevation and angle of train.[4]

The Wilson-Dreyer manual system of fire control made no allowance for the effect of yaw because bearings were taken with respect to the keel-line rather than the course-line-movement of the ship from which the observations were made. The manual approach, moreover, required a minimum of sixteen men – five men to make and transmit the range and bearing observations, five men to plot the observations and to measure and transmit the results, five men to set the clock and calculate corrected ranges and bearings, and one man to transmit the corrected ranges and bearings to the gun positions.[5] In contrast, Pollen's mechanised system of observation, transmission of data and plotting, and proposed mechanised system of computing, transmission of results, and sight setting took account of yaw and could be operated by only two men – one man to make the observations and the other to measure the target's true course and speed off the plot and to set this data onto the clock. Although Pollen's mechanised system risked the possibility of mechanical failure where the Wilson-Dreyer manual system to all intents and purposes did not, the *Jupiter* trials had clearly demonstrated that yaw-corrected bearings were essential and that manual methods were too slow when the change of range rates were high and changing and unreliable due to human error.

Competitive trials between the Pollen mechanised range and bearing measurement and true-course plotting methods in the *Ariadne*, and the Wilson-Dreyer system's manual range and bearing measurement and virtual-course plotting methods in the pre-dreadnought battleship *Vengeance*, were held on 11, 13 and 15 January 1908 with Wilson umpiring the proceedings from on board the *Ariadne*. Although the trials were concerned with the accuracy of the observing and plotting methods only, the performances of the competing mechanised and manual approaches were evaluated on the basis of the accuracy of ranges produced by the dumaresq-Vickers Clock combination after being set with data taken from the plots. The Wilson-Dreyer system was manned by naval personnel led by Dreyer.

But while Pollen's instruments were worked by Lieutenant George Gipps and an assistant, who were both trained in their operation, the dumaresq and Vickers Clock were operated by Pollen and Isherwood, who were unpractised in the use of their instruments. They were thus unable to produce results with the same rapidity and accuracy as their experienced counterparts in the *Vengeance*. This inequity, which favoured the Wilson-Dreyer system, apparently suited Wilson, who, Pollen later recalled, had watched their clumsy efforts 'with great amusement'.[6]

The trials of 11 and 13 January were both conducted in Torbay, whose calm waters meant that yaw was practically nil. Visibility, in addition, was unusually clear, which greatly facilitated the taking of numerous range and bearing observations. Both of these factors favoured the Wilson-Dreyer manual method compared with the mechanised method of Pollen, because yaw would have introduced inconsistent errors in the bearings used by the manual method thus seriously undermining the integrity of the virtual course diagram. Added to this, whatever inaccuracies in range and bearing observation did occur for want of gyroscopic stabilisation could be meaned out given a large number of plotted observations. The two plotting methods were tested simultaneously by having them each plot the movement of the opposing test ship while both test ships steamed in and out of a rectangle marked by four moored destroyers. In these trials the range never exceeded 8,000 yards, and the courses and speeds of the two ships were such that the change of range was never greater than 500 yards per minute. Two runs were made on 11 January and two more on 13 January, each run lasting no more than five minutes. The accuracy of the ranges generated from the plotted data were checked against observations that had been made from the moored destroyers, and both methods were found to have produced valid plots, which was hardly surprising given the absence of yaw, clear visibility, the relatively low maximum ranges, and moderate change of range rates.

Preparations were then made for the competitive trial of the two methods of measurement and plotting on the open sea. Testing was to occur simultaneously as before, but without the presence of the moored destroyers. The course and speed of each test ship was determined by a lottery in which courses varying by half a degree and speeds from eight to fourteen knots were written on slips of paper, placed in a cap, and then drawn. The outcome of the drawing was such that the course and speed of the *Ariadne* was practically identical to that of the *Vengeance*. Under such conditions the change of range would be almost non-existent, and the virtual plot would thus be useless because the simultaneously observed ranges and bearings would simply be superimposed over one another given the little change in the relative positions of the two vessels. Dreyer suggested, therefore, that the virtual plot be replaced during the trial by a manual plot of ranges against time – sometimes referred to as a 'time curve' – from which a single change of range rate could be measured. This expedient suited the particular conditions of the trial as determined by the lottery because the alteration in the change of range rate would be negligible. Dreyer's method, moreover, did not require bearings, which obviated any difficulties otherwise arising from the fact that the telescope-torpedo director arrangement could not provide accurate bearings on the course-line movement of the ship from which the observations were made in the face of yaw, caused by the wave action of the open sea.

Wilson adopted Dreyer's time against range plot proposal, and the competitive plotting trials at sea were held on 15 January 1908. Gipps and an assistant operated the Pollen instruments in the *Ariadne* as in the earlier trials. The same naval personnel operated the range finders in the *Vengeance* as before, but the virtual plot was replaced by an improvised time and range plot operated by Dreyer and an assistant. This consisted of nothing more than a table on which was placed a chart showing a time and range graph with ranges marked in pencil, the sequence of ranges meaned with a ruled pencil line, while the slope of the line was measured with a protractor to obtain a change of range rate. Wilson again umpired from the *Ariadne*. The two test vessels steamed on roughly the same course and at the same speed – as had been determined by the lottery – in clear weather, in a moderate sea, and separated by a distance of 8,000 yards. Fifty minutes after the trial had begun, Wilson received a flag signal from the *Vengeance* that informed him of the success of the time and range plot. He then called a halt to all further testing, and told Pollen that his system had been superseded by one that was 'vastly superior'.[7] Wilson at this time gave Pollen permission to submit any comments that he might have on the trials in writing.[8]

On 21 January 1908, Pollen informed Wilson that his report would soon be ready and that it would be printed 'as it will reduce the bulk of what has to be sent to you, and make it a great deal easier to read and refer to.'[9] The paper, which was entitled 'Notes, Correspondence, Etc., on the Pollen A.C. System, installed and tried in HMS *Ariadne*', was sent to Wilson on 24 January. It consisted of copies of two letters from Pollen to Wilson dated 17 December 1907 and 24 January 1908, notes on the preliminary trials of December and official trials of January, a technical description of Pollen's instruments, and a number of charts and illustrations (the notes, technical description, charts, and illustrations have been omitted from this volume).[10] Wilson's views were not, however, changed by Pollen's printed remarks. In his report to Captain Reginald Bacon, the Director of Naval Ordnance, of 31 January 1908 he claimed that the gyroscope arrangement was unworkable and that the instruments in general were mechanically unreliable, and thus recommended that the Pollen system be rejected.[11]

STRICTLY PRIVATE AND CONFIDENTIAL
[COPY]

188, FLEET STREET,
LONDON, E.C.,
January 24th, 1908.

Dear SIR ARTHUR WILSON,

Herewith are sent some examples of the work done with the A.C. apparatus before Christmas, with notes as to weather conditions and other matters that explain the results. A further note on the actual making of the forecasts in the tests is included.

I also append a longer memorandum descriptive of the functions of the rangefinder mounting, the transmission gear, and the charting table, drawn up by my partner and co-inventor, Mr Isherwood, assisted by Lieut. G. Gipps, R.N. This explains the manner in which the rangefinder, when worked by a single observer only, is converted into a continuous, allweather, and automatic indicator of the enemy's position, speed, and direction.

At our last conversation, when I asked if you could give me a general indication of your views as to the experiments, I gathered, first, that you considered the principal credit for whatever success had been achieved was due to the rangefinder. Next, that any importance previously attaching to the invention had been obliterated by the subsequent development of a better system. You quoted to me the saying of Lafontaine: *'Le mieux est l'ennemi du bien.'*[12] And, finally, you gave me permission to send you any comments the position seemed to me to call for.

The matter always having been considered one of such grave importance to the King's Service, you would not wish me, in whatever remarks I have to make, to be less than straightforward and open. If, then, I seem to combat your views, you will not think me lacking in respect for one, to whose opinion the highest rank and unique experience give such weight and authority. It is just because your authority is, beyond question, so great, that the obligation to discharge this difficult and distasteful duty is the more stringent.

There is fortunately, at this time, no likelihood of my being thought to be influenced by pecuniary considerations in what I say. The gear has fulfilled the 'complete success' conditions. The sum the Lords Commissioners have agreed to pay for monopoly, if desired, can now neither be increased by my system being shown to be better than has already been proved, nor diminished, as partial failure might have caused it to be. The suggestion in the Admiralty letter to me of the third of this month, that a refusal of monopoly might affect the value of the system, is, frankly, to me unintelligible. The testimony of experts as to the vital character of its subject matter, and now the demonstration of its final success, are in this respect conclusive. If I had such attractive offers from foreign navies when nothing was proved, is it likely that they will not materialise when everything is established? The exercise, by the Lords Commissioners, of their option, can, it is true, *limit* the value to the agreed sum, and I should rejoice, the system having surpassed all hopes, at their getting a better bargain than was expected. But if, after bearing so large a share of the cost and affording all the opportunities for trial and experience, they set me free, their waiver can only make the prospect of commercial profit incalculable. Had I ever preferred financial security to the hope of distinguishing myself by working, at least in part, for the public service, the choice, to their Lordships' knowledge, was open to me, until I voluntarily closed it. Should this distinction, then, be put beyond my reach now, I can have no uneasiness as to the certainty of finding whatever consolation may be derived from a more material reward. Financial considerations, then, can be put on one side.

The burden of what I have to say is this. The policy of the Admiralty leading up to the trials, and the policy of the trials seem to be inconsistent and at cross purposes, and, if this is so, it can be explained only by a fundamental misunderstanding.

Will you allow me to show why the contrast appears on any other hypothesis to be almost incomprehensible? The following facts are, I believe, indisputable, and would be confirmed by those who took part in them.

In the autumn of 1906 the Naval Ordnance Department, acting in conjunction with the considered opinions of the gunnery experts at its disposal, advised that, could my inventions achieve what was claimed, they would revolutionise naval gunnery and might be justifiably monopolised for the agreed sum. I understand that, while this was the advice that the Board acted on, the Department did not commit itself to any sanguine hope of success being likely. On the contrary, if the view expressed to me was put before the Board, it was that, while success was ardently to be desired, the difficulties made a happy issue problematical. I know as a fact that the most acute scientific intellect in the Service considered the obstacles prohibitory, and was opposed to further expenditure, on the ground that the experiment was foredoomed to failure. Apparently the argument that carried the day was that the bare possibility of success made an option on monopoly imperative.

In the month of June last year, the Controller, the Director of Naval Ordnance, the Inspector of Target Practice, Commanders Smith and Craig, of H.M.S. *Excellent*, Lieut. (now Commander) Dreyer, and Capt. Harding, R.M.A., inspected the gear at Manchester while under construction.[13]

From what was said it was obvious that, although each mechanical principle was shown at work, the majority were still unshaken in their incredulity as to success being attained; but every member of the party wished me success, and complimented my partner on his chief share in devising such ingenious means for realising the scheme so long advocated. Certainly there was no evidence of change in the opinion that the importance of – perhaps unattainable – success was as great as ever.

By the end of the second week in December last, the preliminary trials had shown that the A.C. gear was doing not only all that had ever been claimed for it, but was plotting speed and course at greater distances and in more severe weather than had ever been dreamt of. The 'complete success' conditions that had been drawn up as a standard of almost visionary excellence (so that all questions of the agreed sum being excessive should be finally answered if complete success was achieved) now assumed an innocent and easy character, and those who had been most insistent in their doubts, became instantly the most generous in their congratulations.

These congratulations came from every single specialist who had been concerned with the previous discussion and experiments, and had been informed of the result of the preliminary tests.[14] Right up, then, to my being made aware of the limited character of the final tests by being informed of their termination, I had no reason for supposing that expert opinion had wavered on the point of the overwhelming importance of what

I was trying to do, and, consequently, on the portentous significance of my complete and, to the expert opinion, unexpected, success.

If the above facts set out correctly the former policy of the Admiralty, as shown by the history of my relations both with the Board and its expert advisers, you will, I think, be the first to admit the contrast it presents to the policy of the trials.

Excluding the abortive run, there were, to test the results for accuracy, four experiments with the rectangle; each was limited to five minutes' run; all were within 8,000 yards; and the maximum change of range in any one of them was not 500 yards a minute. These four runs occupied Saturday and Monday. Tuesday was a day off. On Wednesday the one experiment made extended over thirty minutes. The range in this experiment was again limited to about 8,000 yards, and in this, the only experiment made at sea, there was no change of range at all. And at the close, I heard that the trials were over, and the opinions set out above. It cannot be thought strange that I was dumbfounded by the irony of so bewildering a situation.

Instead of the production of a course and speed indicator being an achievement almost beyond human expectation, it was considered, when achieved, to prove the merit of the rangefinder, and nothing else. Instead of the production of means of always knowing the exact data for continuous hitting, being a triumphant advance, it seemed, when the trials had shown it actually to exist, a thing of such negligible importance, in view of another system having appeared, that it was unnecessary to test it at sea on any problem that the present Service gear could not solve.

I assume, in generously giving me so straightforward a warning of what was in your mind, while you desired in fairness to prepare me for whatever disappointment an adverse recommendation might involve, you did not wish me to consider your first impression as final; and I therefore take your permission to present my views to you, as indicating the two points I should discuss: First, is there more in the A.C. gear than the mere product of the rangefinder? and, next, to what extent does another system make the A.C. success of negligible importance? And in my answers to these questions, I shall try to find an explanation of that contrast in policies for which, at first sight, it is so difficult to account.

The product of the Barr and Stroud rangefinder, as it existed before my gear was constructed, consisted in occasional accurate ranges, their frequency and accuracy depending on the ship's steadiness. The product of the same instrument, with my mounting, bearing taking, and transmitting and plotting devices, is a continuous and graphic statement of the position, speed, and course of the enemy in all variations of the ship's movement. To say that the credit for this result is due to the rangefinder, and not to the gear, seems to me equivalent to saying that the credit of making hits with a 12-inch gun is due to the gun, and not to the elevating and training mechanism, and those who have invented improved methods of using them. If I may repeat what I have said in my proposed lecture, advance in the *technique* of gunnery has made a new weapon of the 12-inch gun. Ten years ago it was a 3,000 yards gun: today it is, in some circumstances, an 8,000 yards gun; and the credit for the change should be given – as in fact it is – to the authors of the new *technique*, and not to the makers of the ordnance.

I make a like claim for the *Ariadne* gear – viz., it has made a new instrument out of the rangefinder, and while the product of the instrument itself is just as essential a factor in getting our results, as the accuracy of the gun is in getting hits, it is of small or no value in getting data for gunnery, until it is combined with the other factor – i.e., bearing; and, by some system like mine, both are made continuously obtainable in all weather and under all conditions of speed and course of the firing ship and target, and recorded in such a form as to constitute an unquestionable guide, not to the distance only, but to the *distance, speed,* and *course* of the enemy.

The question of the A.C. system being superseded, I can only discuss on general principles, since the actual nature of the devices in the *Vengeance* is of course not known to me. But I could see four 9ft. rangefinders in the ship, some of which seemed to be manned by four (!) operators each. In what I say, then, I shall presume that the new system is certainly dependent on getting ranges, and, one must suppose, bearings.

As such, it would at least be a very flattering, if belated, recognition of my main contention since I first occupied myself with this question, viz., that no gunnery system not based on the *synchronous taking of ranges and corrected bearings*, could be of value with a moving target. It is worth enquiring, however, whether the large *personnel* of rangefinder operators, compass bearing takers, and those that transmit the readings, and the compass itself, the rangefinders and other instruments, can all be as *economically*, if at all, protected by armour, as the single operator and single instrument of the A.C. system; and, again, whether any adaption of the surveying system – of which I was for so long a lonely advocate – based on substituting a very multitude of persons for automatic machinery, can be proof against error and disorganisation in action, and not be vitiated in all difficult conditions – i.e., rough weather and rapid change of range and rate, by loss of synchronism and time lag in getting results. I merely ask these questions, because the feasibility of getting ranges and bearings with Service gear in conditions of small change, smooth water, and without provision either for protection, or the elimination of personal and co-ordination errors, was quite understood before the A.C. instruments were designed, and has been demonstrated on several occasions. I have been helping certain gunnery lieutenants – who are anxious about next year's battle practice and have paid me the compliment of asking my leave to adapt my system to the available gear – with suggestions as to the best way of doing it.

For the rest, it is to be supposed that the new scheme must successfully deal with some, or all, of the other operations that must be done before the geometric data become material for getting hits. What those operations are is common knowledge. Range and deflection must be kept, forecasted, corrected for ballistics, transmitted and set upon the sights. It is possible, then, that the *Vengeance* gear may have covered all the ground between (1) getting the geometric data by a somewhat reactionary attempt at a surveying system, and (2) getting them corrected and set upon the sights by a first-class automatic system, and covered it successfully. But, even if it does, any success in the second half of the process leaves the A.C. gear exactly where it was, and simply because the A.C. gear addresses itself to the first operation only.

Now it should be remembered that neither the Board of Admiralty nor the late D.N.O., nor any of those he acted with or who advised him in the policy of 1906, or accompanied him to Broadheath in June, 1907, or expressed their unbounded satisfaction at the success of the December experiments, were ever for a single moment under the illusion that the A.C. gear, authorised and tried, was a complete system in itself. It was always known that the experiment was limited to the production solely of a course and speed indicator, to be operated by one man, and lending itself to armour protection. If this could be done, everything else seemed easy.

It should be clear, then, that of the two main operations in procuring hits, the A.C. gear was limited to the first. If, then, it is in competition with the *Vengeance* gear, the two can be compared only in relation to their success in this one operation.

And it is exactly because of this that an entirely new, and, to me, quite unexpected, significance attaches to the trials having been limited to calm weather and inconsiderable change of range conditions. If it was only in respect of getting *data* that the systems competed, surely it was of primary importance to exhaust the capacity of each by the severest possible tests in the matter of range, change of range, and change of rate; to test them in the worst weather conditions in which guns could possibly be used; and not to limit the only sea experiment to a problem which presents no difficulties to the fire control instruments to be found in every one of the King's ships of the line.

I ventured to ask for one test (*a*) in getting the range at extreme distances; (*b*) under conditions of maximum change per minute, and (*c*) the most rapid variation in the rate. For example, had the *Ariadne* been run at 18 knots, the *Vengeance* at 14, had the *Vengeance* been stationed 20 degrees from the *Ariadne's* bow at 17,000 yards, and the ships made to approach, so that they should pass at about 3,500 yards, there would have been a change of range of well over 1,000 yards a minute, and in the last four minutes before they were abeam, the rate would have varied from some 800 yards to nil. A light is thrown on the refusal of this request by your telling me, after our only experiment made at sea, that you did not regret the accident by which there was no change of range at all, as this was the way in which fleets did in fact approach each other.

My whole contention for years has been that present tactical methods of approach are necessarily limited by the existing capacity of the guns to make hits; in other words, to combinations of speed and course in which change of range can somehow or other be got. I had fully expected by my gear, not only to make sure of getting this change more accurately and rapidly in all cases where it can now only be got approximately, but to get it with equal ease and certainty in cases where, by present methods, it cannot be got at all. I was, and am, confident that we should have got the change of range as correctly in the test proposed in rough weather, as we did in the experiment which was actually carried out in smooth; and I submit that if my gear can do this, it opens up entirely new tactical possibilities of employing fleets.

I have, from the first, understood that it was to ascertain whether so ambitious and revolutionary a claim could be substantiated, that the Admiralty entered into the unprecedented agreements of 1905 and 1906. Consequently, not to try this out, until this claim was either known to be true or proved to be false, was to ignore the one thing that made the gear worth building.

Is it possible to suggest any explanation of so radical a difference in the point of view of the Admiralty and its gunnery advisers, on the one hand, and what I may call the interim judgment given me at the conclusion of the trials, on the other? If I may, with the utmost respect, suggest an explanation of what seems a very important, if, in the circumstances, quite natural, misunderstanding, it is as follows.

You will remember that, at your first inspection of the *Ariadne* gear, you told me that you were — was it horrified? — at the absence of any means of dealing with the data and getting them to the guns. Not realising at the time either your point of view, or its fateful significance, my explanations, both verbal and in my letter of December 17th, 1907 (copy of which I annex for you convenience), were doubtless insufficient.

You will further remember the importance you attached, in your written and oral instructions, to the Service gear being worked during the experiments to give you the clock rate, etc. From the first, then, you wished the experiments worked *not* simply as a test of the gear, but as a test of skill in getting the data off the chart and on to the clock. Does not all this point to a probability of your having expected, and consequently directed the experiments to test, a complete system of gunnery, whereas what you found was what may well have appeared an exceedingly costly fragment, and of much overrated importance because it was a fragment?

I put this possibility to you simply because I cannot think that the entire body of expert opinion, that defined the limits of this gear, that regarded it as designed to settle the most momentous of the unsolved problems of naval war, that prescribed the tests, and has now welcomed its success as the deliverance from the master difficulty of sea-fighting, can be so extraordinarily mistaken. Had you been fully informed of the previous history of the negotiations, of the careful analysis on which the arguments for this experiment were based, and of the microscopic hope of its succeeding, I do not think you would have been led either to be astonished at its being directed to a portion of the gunnery problem only, or to suppose our achievement insignificant.

Had you supposed from the first that it was the getting of data alone that was on trial, is it possible you would have been content with *fifty minutes' test in a flat calm?* because the very essence of the matter is Battle, not Bisley, conditions, and, had this been explained, you could not have compared the systems until they had been *competitively* tested both for facility of use in the roughest weather in which guns can be handled, and at the extremest distances and most rapid changes of range and rate under which guns could be trained and aimed with any possible chance of hitting.

If I am right in supposing the character of the *Vengeance* gear to be as I have suggested, it would seem that the misunderstanding which has arisen is due to your having been led to suppose that they were rivals, not as getters of data, but as ultimate and complete gunnery systems. Is it not more true to say that while they cannot, in this respect, be competitive, they may be complementary?

Certainly, after coming successfully through the Torbay tests for exactness in results, seeing the A.C. gear at work in the roughest weather, and ranging so many moving ships up to 10,000, 11,000, and 12,000 yards, with uniform ease and constant precision, I find it difficult to suppose – until any competing system is tested against mine to the point where one or the other breaks down – that it is valueless to the British Fleet, or negligible in the hands of any other navy.

I have to end, as I began, by asking you to forgive my possibly too great freedom, and to attribute it, not to any improper motive, but to perhaps an over-zealous anxiety that whatever is done shall at least be done in full knowledge of the facts. If I offend, it is a poor return for your uniform candour and courtesy to me during the trials. You know that I think them inadequate – but now you know why, I wish also to add that, on the supposition that they were adequate for their purpose, they were conducted with the most scrupulous fairness to me, and with relentless determination to verify each result as far as verification was feasible. For all this I owe you my sincere thanks.

I owe much, too, to Capt. Lafone and the officers of the *Ariadne* –

and especially to Lieut. G. Gipps, of H.M.S. *Excellent*, who certainly went to the limit of zeal and hard work in carrying out his instructions to do the best he could for my instruments. If the request is not impertinent, I should be glad if you could convey to their Lordships, on my behalf, my gratitude for the characteristic thoroughness with which all concerned threw themselves into the task of testing the new system.

I am, Sir,
Your obedient Servant,
A. H. POLLEN.

To Admiral of the Fleet,
  SIR ARTHUR KNYVET WILSON, V.C., G.C.B., G.C.V.O.

COPY – STRICTLY PRIVATE AND CONFIDENTIAL

188, FLEET STREET,
LONDON, E.C.,
December 17th, 1907.

Dear SIR ARTHUR WILSON,
Your question is: 'Why should the Admiralty pay [the agreed sum] for the monopoly of the A.C. system if it succeeds?'

Expanded, the question is: 'Before I advise the Admiralty to buy, I must know exactly what it is they are buying – i.e., I must be sure it is something new and valuable that we do not possess already. Next, I must be sure that we can profit by its value, viz., that we are not buying something we cannot use until we buy something else not included in the agreement – because this is simply for us to put ourselves into your hands – supposing you to have the exclusive possession of this second something. Lastly, if it is a new and useful thing we are to buy, while I assume it is *ipso facto* worth buying, I must be satisfied that the price is not exorbitant.'

So there are three questions to answer:

1 Does A.C. do anything new?
2 If yes, does this new thing increase fighting efficiency without further gear?
3 If yes, is it worth [the agreed sum]?

Question 1.–Does A.C. do anything new?
Answer 1.–The Raw Material of Tactics and Gunnery are knowledge of the enemy's movements.

The A.C. system gives in very few minutes the position, present speed, and direction of movement of any ship met in the open sea, at any range at which the rangefinder can be used, in any weather in which guns could be used — but at distances beyond those at which guns could be used.

This raw material can be instantly put upon the tactical machine — so that, as a guide in tactical decisions, before and during action, the system supplies not only something new, but means for using it as well, and in this respect it is complete.

Do present methods give this raw material?

Spotting means watching misses pitch into the water, and, judging, from their positions relative to the enemy, the error in the range, and hence the probable course and speed of the enemy.

This operation is confined to the limits of vision, say 7,000 to 8,000 yards. A.C. gives results at the limit of the rangefinder — say 12,000 to 14,000 yards.

Spotting is only possible when the hull of the ship and the surrounding water are visible. These are only intermittently visible if the enemy fires back, or our shells are bursting on or near him.

But his masts are visible when his hull and the surrounding water are shrouded in smoke. Hence A.C. gives results irrespective of smoke.

Again, the course and speed of the enemy are guessed by the spotting method — not deduced from accurate data.

Hence A.C. gives a new thing, viz., speed and course of the enemy under all conditions. Whether it can be used for gunnery or not, at least the provision of raw material is new, so that the answer to Question 1 is 'Yes.'

Question 2.–Does this new thing — i.e., raw material of gunnery — give increased efficiency with present gear?

Answer 2.–My knowledge of the Service system is confined to deductions from the instruments employed. With these I have been acquainted some time, owing to my having been unofficially consulted as to improvements in some of them in 1904, and to having handled them in the *Jupiter* in 1905–6, and in the *Ariadne* recently.

But I have no practical knowledge of their use with guns. Any account I give, therefore, of the Service Fire Control system must be based on *a priori* reasoning.

As I understand it, the procedure that would be relied on to hit an enemy, if war broke out to-day, may be divided into two parts, A and B, viz.:

(A) *The attempt to get the raw material:*
  The enemy would be continuously ranged by rangefinder.
  The corrections for 'error of the day' would be prepared.
  At a suitable distance firing begins at the corrected range.
  The impact of misses in the water are observed and their relation to the enemy's position noted.
  These observations would permit of the range being corrected until

a hit is made, and a *guess* made possible at the enemy's *speed* and *angle of course*.

The estimated speed and course, and speed of own ship, are put on an instrument. A limb of the instrument is kept pointed at the enemy to get his bearing.

(B) *The method of using the raw material:*
When set to *correct speed, course, bearing* of enemy, and own speed, the instrument indicates a rate of change.

The ascertained range and this rate are put on a clock, which now shows an increasing or decreasing range.

A range that this clock will apparently indicate in half a minute or a minute is transmitted to the sight indicators.

This range is put on the sights.

When the sight range and the clock range coincide, the order to fire is given.

I have underlined two data that are *guessed* and one that can only be imperfectly got — the yawing of the ship would make keeping an accurate bearing impossible.

But these three data are the unknown factors in change of range. Therefore part A of the Service Fire Control system is palpably superseded by A.C.

But part B is theoretically sound, in that if all works together as rapidly and as accurately as the circumstances require, and, if the right data are employed, the right range will be on the gun sights at the moment of firing.

As A.C. supplies the correct data to part B, it must add to the efficiency of the Service gear; in other words, the answer to question 2 is 'Yes'.

But this is not to say that the Fire Control system sketched above is the best conceivable. I think a much better one could be substituted for it. But it answers your question to show that part B of the present system does give correct results if it is supplied with correct data; that part A does not include means of getting right data now; that the A.C. system will supply right data; that therefore, within the limits of any other imperfections it may have, part B's efficiency will be increased by the difference between results being always right, because always based on facts, instead of only sometimes right, i.e., when based on correct guesses.

Question 3.–Is such increased efficiency worth [the agreed sum]?

Answer 3.–I do not need to argue that, if there is an increase in efficiency it is worth having. The question is: Is it worth having at [the agreed sum]?

Sea power owes its origin and continued existence to Battle Fleets.

Armoured ships carrying guns that can destroy other armoured ships are the units in Battle Fleets, and exist only to carry guns.

Guns exist only to make hits.

The efficiency, therefore, of navies can be increased by the increased capacity to make hits with battle guns.

Therefore the justification for building £2,000,000 ships like the *Temeraire,* instead of £700,000 ships like the *Renown,* is that £1,300,000 more worth of hits can be got with the one than with the other.

Hits are thus a measure of value.

Hits are valuable in proportion as they are distant – frequent – and formidable.

A gun that always makes two hits under conditions in which another can make only one is worth two such guns.

Hence a ship that can hit twice as fast as another is worth two such ships.

But a ship that can hit further off than another is worth an incalculable number of such ships.

If, then, the A.C. system gets the right data for hits at longer distances than they can now be got, and if it enables the present Fire Control system to get these data to the guns, it must add an efficiency to these ships that they do not now possess. If that efficiency means that more hits are made in a given time at a given range, it increases the value of such ships in proportion. If it enables sustained hitting to be maintained at a range now impossible, it increases the value incalculably.

The cost of the battleships and armoured cruisers, of the first line, built, and building, is over £100,000,000. [The agreed sum is a small fraction] of one per cent of this sum. Therefore, unless the increased efficiency got by the A.C. system is equal to an increase of [so small a fraction] of one per cent in the present hitting capacity, the A.C. system is not worth buying for [the agreed sum].

But if the A.C. system results in hits being got at unknown ranges, with a regularity unattainable by any other way, and therefore increases hitting efficiency incalculably, [the agreed sum] is not an exorbitant price.

Certain other points I will state quite shortly. Will you ask me to explain them if they seem obscure?

In the A.C. system, the rangefinder can be completely armoured, and the communications from the rangefinder to any central station below the armoured deck can be similarly protected.

The operator of the rangefinder, and one other person to observe the first shot (who can also be under armour), are the only members of the necessary *personnel* who need be above the water line.

The only personal skill involved being the skill of the rangefinder operator, all guess work, instinctive judgment, etc., having been eliminated, it must follow that the results shown throughout the Fleet will tend to equalise, instead of showing the astounding variations that the present tests, even with the unreal and unduly favourable conditions of battle practice, exhibit; in other words, it should practically secure mechanical repetition in gunnery results, a thing that cannot happen when gunnery results depend upon personal skill of a very high and rare order, which must be unequally possessed by different people.

Lastly, the number of people employed being very much reduced, so many separate sources of error are eliminated.

One of my reasons for thinking that a better substitute for part B of the present system could be established is that the method I should propose would be automatic.

I must apologise for sending you so long a letter, and shall be glad to hear if there are any points which I have not made sufficiently clear, and on which you would wish me to send you a further statement of my views.

     I am, Sir,
      Your obedient Servant,
        A. H. POLLEN.

Admiral of the Fleet,
SIR ARTHUR K. WILSON, V.C., G.C.B., G.C.V.O.

XI
EXTRACTS FROM A LETTER TO CAPT. REGINALD H. S. BACON, CVO, DSO, ROYAL NAVY. DATED 27 FEBRUARY 1908
(FEBRUARY 1908)

On 25 February 1908, Pollen wrote to Captain Reginald Bacon to inform him that he had had a mechanism designed that would compute the target's course and speed from range and bearing data obtained by observation. Such a device, Pollen proposed, would allow the range and bearing observation instruments to be connected directly to the change of range and bearing machine, which by eliminating the plot would mean that the entire fire control process 'other than the actual range-taking will be purely automatic' and would thus further minimise the danger of human error.[1] Bacon replied to Pollen on 26 February 1908 in a letter in which he questioned the desirability of mechanised fire control in principle.[2] Bacon's letter prompted Pollen to respond the next day with a lengthy letter in which he argued the case for replacing human with machine action wherever it was possible. Portions of this letter were reproduced in a paper entitled 'Extracts from a Letter to Capt. Reginald H. S. Bacon, CVO, DSO, Royal Navy. Dated 27 February 1908', copies of which are known to have been sent to Sir C. Inigo Thomas, the Admiralty Permanent Secretary and a personal friend, and Lord Tweedmouth, the First Lord.[3]

I have to thank you for your letter of yesterday. . . . Don't think me very impertinent if I make the last part of your letter an excuse for putting certain considerations before you with regard to the A.C. system, which I have not yet had an opportunity of doing by word of mouth. I know how busy you are, but I cannot say how much I have regretted that I have never had an opportunity of explaining any of the principles or history or mechanisms of the A.C. system to you.

But you touch the heart of the matter when you say 'how far mechanisms should supersede the human brain involves serious practical considerations.' How serious those considerations are, when you are dealing with fire control, cannot be over-stated. Because it is to make hits that guns exist, and to carry guns that make hits that ships exist, and to work, command, and organise ships to carry guns that make hits that the entire structure of the Navy, with its subsidiary material and the whole of its personnel, from the powder monkey to the Board of Admiralty, has been called into being. . . .

I think I was the first, by a great many years, to point out that fighting meant fighting a moving target, that the Fleet had no provision for engaging a moving target, and that a moving target could only be engaged by a system of surveying or plotting, and I was the first to produce proposed means for such surveying at sea.

Later, I realised that it was no use dealing with the gunnery problem unless it was dealt with in terms of battle; the kind of gunnery wanted was not battle practice gunnery or target practice gunnery, but battle gunnery, i.e., when you are being shot at, and not only shot at, but being hit.

If, then, a system of gunnery had to be evolved, both more complete than the present system (in that it had to deal with a moving target instead of a stationary target) and fit for battle conditions and not only for peace conditions, it became obvious that two things would be required.

First, you would have to provide for getting exact knowledge of the speed and course of the enemy, for this is the clue to all hitting when there is change of range.

Next, you would have to see that this knowledge was continuously supplied under the extremest conditions of stress, because the time when that knowledge is most important is exactly the time when the ship needing it is at the greatest fighting disadvantage.

Now, if I am right in supposing that the gear tried in the *Vengeance* was a plotting gear, there is no question that the Service has accepted my contention of eight years ago, and that so far, whether it is my system that is taken or not, at any rate I have the pleasant satisfaction of knowing that my work has directly or indirectly advanced the art of gunnery.

But no one, I think, has ever disputed that plotting could be done by rangefinder and compass. The only dispute there ever was on that was whether such a method was battle-worthy.

The *Vengeance* gear must employ a great number of people, and must employ them on work where the human brain has to work accurately both for a definite figure result, i.e., an exact figure for range and an exact figure for bearing, and also for an exact time result, that is, one must get synchronism between bearings and ranges; and, lastly, it must work graphically with exactitude – in short, the plotting must be *truly* made from *true* data *synchronously* got.

I do not see, if there are four people to each rangefinder, that you can get the plotting done without employing at least nine persons, viz.:

4	for the rangefinder.[4]
2	for the compass.
2	to receive the ranges and bearings from the rangefinder and the compass.
<u>1</u>	to plot.
9	

so that, looking at the question simply and solely from the necessity of getting exact data properly plotted, your sources of supply are equivalent to at least 7 sources of error for the prime data, with 5 sources for the combination errors. What this works out to as a total source of error I do not know!

Against this the A.C. gear requires only one operator to make the chart. There is therefore only one source of error.

And then we have to look to battle conditions. The first condition is that the tension will be terrific, and therefore the probability of error from each source will be much higher.

The second is that one part at least, i.e., the compass and its operators and communications, will be under fire, and therefore liable to casualties.

Consequently there is first an enhanced probability of error in the results as long as they can be got at all, and, next, a constant liability to all results being stopped by casualties either to men, or instruments, or communications.

Thinking people did not need the *Hero* experiments[5] to teach them that the present fire control system could be knocked out by any chance shell, or fragment of shell.

It was these reflections which made your predecessor disregard the compass-plus-rangefinder system in 1906; and experiment with what I was urging. It was my contention that the operation of getting data for the moving target could never be satisfactory unless it was made both error-proof and, as far as humanly possible, shell-proof. Consequently I had evolved a scheme for making the operation of getting the data automatic from beginning to end, except for the necessary human operation of adjusting the lenses of the rangefinder for the range.

The *Ariadne* experiment was, as you know, from the first limited by set policy to making a chart. It was never intended that the *Ariadne* instruments should be treated as a complete gunnery system, or should be tested for anything but what they were built for, viz., to *get* the data for gunnery automatically (except for the rangefinder) and accurately in all conditions.

I had always wished to carry the thing much further than this. My change of range clock was designed before we left the *Jupiter*, and the sight setting mechanism, to be run straight from the change of range clock, very shortly after.

But there was an obvious difficulty about allowing me to build it for trial with the *Ariadne* instruments.

No change of range clock could be built to deal with a Service gun unless the trajectory of the gun was known to me, as also the formulae for the different corrections which are necessary.

Your predecessor took the line (which undoubtedly was right) that, until the Service had made up its mind about the A.C. gear, I had to be regarded as a person who might at any moment be making his knowledge and experience of gunnery available to the world; consequently no further information could wisely be given to me than was necessary for the purpose of the experiment the Board of Admiralty decided I should make. This was the chief reason why the change of range clock and the sight-setting mechanism were not built in time for the *Ariadne* trials. But this fact makes it a little unreasonable to be 'horrified'[6] because the *Ariadne* instruments were not a complete system of gunnery.

What the *Ariadne* instruments were built to do they did with absolute perfection – that is, we got the data for gunnery on the measured tests well within the limit; and we got them from first to last in all weathers with a regularity and accuracy which is very surprising, considering the small numbers of hours of experience and training which the rangefinder observers had. And we *got* them with the work of *one* operator only.

I have discussed both the change of range clock and the sight-setting mechanism with a good many gunnery officers; and I think that many of them would tell you, if you wished to know, that the probability of these being entirely successful is certainly much higher than the probability that existed of our being able to do what we have actually done with the *Ariadne* instruments. . . .

It is, of course, plain that, if you could have an automatic fire control system, it would present an enormous economy over your present arrangements. . . .

Where the economy would come in, of course, would be in the immense saving of wages. If you take the average pay of a lieutenant to be £200 a year, and the annual pay of a man, with his victuals, and clothing, and so forth, at getting on for half that amount, and assume that at least twenty men and officers per ship are saved – fire control operators and sight-setters – and take the capitalised value of a man earning wages to be about twenty times the amount of the wages, the saving should be £3,000 per year per ship, or capitalised nearly £50,000 per ship, quite apart from what would be saved in battle casualties. Automatic machinery, at almost any cost, is so obviously an economy that the thing need hardly be argued.

But the real argument for automatic machinery is less the saving effected in the displaced labour than the vast advantage of regularity in result; because the elimination of error and the securing of exact repetition is the aim of all ingenuity.

The real issue, then, as to the Admiralty taking or rejecting the A.C. system, turns on confidence and no confidence in machinery. The struggle is as old as the dawn of exact science. Galileo had far fiercer opponents in the astronomers than in the College of Cardinals.[7] Harvey never got anyone over 40 to believe in the circulation of the blood.[8] Henry Bessemer's patents ran out before he could find an ironmaster with intelligence enough to adopt his principles.[9] The inventor of aniline dyes failed to find a capitalist to back him, and so a British invention became the foundation of a vast German industry.[10] Lord Kelvin could not be brought to believe in wireless telegraphy.[11] Telephones were in use for twenty years before the London police heard of them. And the Navy has some odd records of its own.

Still the Navy of to-day is very different from the Navy of 50 years ago. It is far more in touch with scientific thought, far more in touch with civil life. The officers are infinitely more accomplished in mechanical and electrical and scientific knowledge, and in a thousand ways the old barriers between the seaman and the landsman have been broken down.

But, for all this breaking down, the prepossessions, and the accomplishments, and the habits of mind of a naval officer are still totally different from those of the landsman. He is brought up under discipline, and the bulk of his activities are to instil discipline into, and to exercise authority over, others. The major part of his life is still the carrying out of evolutions, and one of the most important of his accomplishments is the power to organise these, so that they shall be carried out with exquisite mechanical precision. It is no wonder, then, that he has an immense preference for doing things by means of the human brain and human skill, drilled into the semblance of a machine, and a correlative dislike for seeing any part of such activity superseded by machinery. If you analyse it out, it is really the same spirit that kept masts and sails in ships for fifty years after steam was invented; it is the spirit that inspired the remark that the screw propeller was an ingenious toy of no practical utility; exactly the same habit of mind that thought the monopoly of the Whitehead not worth buying – at a fraction of what was afterwards paid annually – because a well-handled spar torpedo was 'just as good', and the machinery of the fish torpedo could not be trusted to work.[12] It is a form of conservatism, like many another, based upon confidence in a capacity that is an exclusive accomplishment.

But if for a moment one could get away from the special circumstances, and deal with the thing in the abstract, and propound the question whether exact calculation and exact combinations are preferably done by a body of drilled men, or by an automatic machine, and the questions were put to anyone who has had practical experience in similar problems in other fields, such as manufacture, civil engineering, building, scientific pursuits, etc., there can be no possible hesitation in saying what the answer would be.

I have often been warned against making my instruments too complicated. I am not sure that you did not warn me yourself on this subject. But the answer surely is simple. If there is a complicated operation to be done, it must either be done by a multiplicity of people, or a machine having as many functions as there are functions to be performed.

Now which is the best to have – a lot of complicated people with a lot of complicated heads, or one complicated machine? The errors of a machine can be ascertained by experiment, and they repeat themselves exactly, and therefore can be allowed for in the results. As far, therefore, as instrumental errors are concerned, they can be said not to exist, because they can be ascertained and counteracted. But human errors are in their nature incalculable. . . .

Any number of things which are now done by mechanisms could be done by organised individuals, if the experiment were worth making.

Musicians tell me that an experienced conductor can play a symphony through time after time in exactly the same number of minutes and seconds, so perfectly developed is his sense of time and rhythm, so perfect the drill of his orchestra. Yet no man in his senses, if he wished to measure off an hour, would hire Mr Henry Wood,[13] with the Queen's Hall Orchestra, to do so for him, by beating time and playing the first act of 'Parsifal'.[14] He would buy a 3s. 6d. American watch, which would do the thing perfectly.

Is not the analogy between the automatic fire control system and the Service fire control system exactly similar, with this difference – that on the exact working of the Service fire control system, and on its continuous working, there may at any moment attend issues so vital as to defy expression?

The issue between A.C. and no A.C. is, I think, exactly as you state in your letter, and I cannot help thinking that in a matter of such crucial importance as getting hits in battle, it is quite inevitable that, as in every other sphere of human activity, the future must be with the machine. The history of progress is epitomised in the phrase 'scientific method'. . . .[15]

XII

REFLECTIONS ON AN ERROR OF THE DAY
(SEPTEMBER 1908)

In the first week of March 1908, Pollen persuaded Lord Tweedmouth, the First Lord, that the trials in January had been inadequate and that additional testing was required before a decision could be made with regard to the acceptance or rejection of his instruments. An offer of new trials was made officially on 6 March 1908, but only four days later, on 10 March, the Admiralty Permanent Secretary informed Pollen that no further trials would be conducted, on the pretext that Pollen had refused the offer of 6 March — which was in fact not the case — and that his instruments were therefore rejected for service on the basis of the results of the trials that had already taken place. Pollen's appeals to the Admiralty were to no avail, and in April he was forced to accept a settlement.[1]

The Admiralty's decision to reject the Pollen system was based upon the belief that the fire control problem had been solved more cheaply by the manual virtual-course and time-and-range plotting methods. In late February or early March 1908, Admiral of the Fleet Sir Arthur Wilson had submitted a report to Captain Reginald Bacon, the Director of Naval Ordnance, in which he recommended the adoption of the virtual-course and time-and-range plotting systems that had been tried in competition with the Pollen gear in January and subsequently modified slightly after further testing. While admitting a number of difficulties, Wilson maintained that the Navy's fire control requirements had been essentially satisfied by the complete system of observation, plotting, calculation, data transmission and sight setting that he described in great detail in his report. He observed,

> The methods by which hits can be obtained at long ranges in practice firing are now generally well understood, so that, now this primary difficulty having been more or less solved, greater attention can be directed to the problem of how the matériel and personnel employed to ensure good long-range shooting can be best protected during an action . . .[2]

In March 1908, Bacon had forwarded Wilson's report to the Board of Admiralty along with a supporting memorandum in which he observed,

> The actual obtaining of change of range, which was the sole object of the *Ariadne* and *Vengeance* trials, forms but a small feature of this report, and is a comparatively simple matter, but the interpretation of the results obtained; the estimation of deflection; the calculation of the rate of change, both by the time curve and plotting methods;[3] the application of the ballistic corrections, and the corrections due to wind, course and speed of the enemy, and hence obtaining the accurate gun range, is a piece of splendid work, and cannot fail to mark a very great advance in accurate practical gunnery.[4]

Bacon thus recommended that Wilson's report be printed for restricted circulation, that time-and-range plotting equipment be fitted in all warships that were equipped with range finders 'as quickly as possible', and that the manual virtual-course plotting gear, as well as the time-and-range plotting equipment, be fitted in four battleships and two armoured cruisers of the latest design for further trials.[5] Bacon's recommendations were approved by the Controller, First Sea Lord, and First Lord by the end of March.[6]

Wilson's unfavourable report of 31 January 1908 on Pollen's instruments and later favourable report on his own and Dreyer's system were read by Captain (RMA) Edward Harding shortly after they had been submitted to the Ordnance Department. Harding, whose official connection with fire control matters had been severed by Bacon, but who had understandably remained deeply interested in the fate of the Pollen system, later recalled that Wilson's recommendations were

> so wholly inconsistent with the plain evidence of the results obtained in the test runs that I was wholly unable to understand how anyone, even rudimentarily familiar with the problem, could have come to such a conclusion. It seemed to me it was not a case of the *Ariadne* results being merely better than the *Vengeance* results, for it was not a matter of degree at all, but of kind.[7]

In March or April 1908, Commander Frederic Dreyer presented the case for the manual virtual-course and time-and-range plotting systems in a lecture to the officers in training at the HMS *Excellent*, the Royal Navy's gunnery school at Whale Island near Portsmouth. His remarks were greeted, however, according to one witness, by 'heckling' and queries about the Pollen system,[8] and by May 1908 Dreyer had himself repudiated both the manual virtual-course and time-and-range plotting methods.[9] During the battle practices conducted in the summer and autumn of 1908, in which the formerly stationary targets were replaced by towed targets in accordance with the revised battle practice rules that had been promulgated in November 1907, the time-and-range plotting method that the Admiralty had recommended for general use in the fleet on the strength of the advice of Wilson and Bacon, failed completely.[10]

The failure of the time-and-range plotting method in battle practice prompted a number of naval officers to write to Pollen for an explanation of his methods, which the inventor provided. Pollen then took portions of his replies, combined them after revision to form a single text, and in September 1908 had it printed with the title 'Reflections on an Error of the Day', which also included a mathematical critique of the time-and-range plotting scheme written by Harold Isherwood, Pollen's design engineer (not included in this volume). This paper was widely distributed,[11] and in general appears to have been well received.[12] Dreyer, however, was outraged. On 18 October 1908, he wrote to Captain Constantine Hughes-Onslow, who was then engaged in writing a study of fire control for the Royal Naval War College at Portsmouth, and complained that 'Mr P. has just issued a scurrilous pamphlet . . .' After repudiating the time-and-range plotting method that he had advocated the previous spring, Dreyer then maintained that Pollen had 'stirred up some agitators to believe that his auto-system is the best, but a searching analysis I think reveals the fact that the simple kitchen table methods[13] are better than the complicated machinery game and produce the same results as the latter only does when in adjustment.' 'The simpler we keep our ships,' he asserted, 'the better.'[14]

FOREWORD

These reflections are mostly extracts from, or summaries of, letters addressed to naval correspondents, in reply to innumerable questions as to the Ariadne trials, the possibility of getting A.C. gear, etc. My replies did not lend themselves to print in their original form, and it is feared that the summaries are not too coherent. The responsibility for their being printed must rest with those who have demanded its being done.

I – OF SOME COMPETING PLOTTING SYSTEMS

The appended memorandum, on the attempt to ascertain future ranges, or the rate of change of range, by plotting successive ranges at known time intervals, and entitled "The Relationship of 'Change of Range' to 'Time' ", has been drawn up, at my request, by Mr Isherwood, to enable me to satisfy the demand of several naval correspondents, who have asked my opinion of the value of this method, and for suggestions as to the best way of carrying it out.

It was known to a good many of my friends that we had prepared applications for patents, a long time since, on appliances and devices for trying to get the data for gunnery in this way. We were first put on enquiry into this method by one of Admiral Sir Percy Scott's lectures on gunnery, delivered, as far as I can remember, in 1902, or two years after I had first developed and advocated the necessity of plotting speed and course.[15]

But, after examining this scheme carefully, my colleague and I abandoned its further investigation or pursuit; and did not proceed to apply for patents on our inventions, for the reason that it seemed to us both unscientific and impracticable. In the light of the recent resuscitation of the scheme, the reasons for our giving it up may be of interest, and at any rate will serve as an excuse for our being unable to help those who have so flatteringly asked our assistance. They may be summarised as follows:

(1) This method necessitates the averaging of observations by a curve. A very brief series of experiments convinced us that it was certain that considerable error would be involved in averaging the curve, and a far greater error in trying to continue it. If, instead of attempting to continue the curve, a second plot is started, to get a curve of rate, the errors in the first curve will be repeated in the second to a sufficient extent to vitiate results. Unless the observations are so numerous and so accurate as not to need averaging at all, every curve is the result of a more or less fortunate guess.

(2) Next, it was clear that the observations would have to be continued for a considerable period, certainly for several minutes, before sufficient observations could be got to reduce the error to negligible proportions. The delay in getting the requisite data for gunnery would therefore be greater by this method than by a graphic reproduction of the enemy's course, when the plotting of ranges, bearings, and time can be made synchronous with the target's movements. This delay is accentuated if a second plot is started from the first.

(3) Third, it is extremely objectionable that this method gives no instant information as to the enemy changing either his speed or his course. It has to be remembered that all observations taken in the ordinary manner at sea tend to be exceedingly irregular. If they are plotted in their correct relation to each other by the bearing and time interval being noted, the averaging of them by a straight line is a check on this inaccuracy. But, if a range curve is plotted, a change of course on the part of the enemy changes the factors that govern the curvature; the observations being irregular, the point of departure for the new curvature will not be obvious; and so, after a change of course, several minutes must once more elapse before the proper pattern of the new curve can be detected.

(4) Lastly, we abandoned both this and the method of plotting by range-finder and compass, as also that by range-finder with the bearing taken from the keel line, and corrected by gyroscope, for the reason that all methods for getting at the data for gunnery, involving manual, vocal, aural, and mental operations, seemed to involve an accentuation of what careful investigation into the subject shows to be the root weakness of naval gunnery to-day – viz., that it is dependent upon the co-operation of too large a number of people, and consequently can only give right results if no one makes a mistake, and can never give prompt results, because the succession of operations consumes time, and this, in action, may be decisive of national existence.

'Five minutes,' said Nelson, 'make the difference between victory and defeat,' i.e., between getting the data for hits on the gun and so hitting first, and being hit first.

However bad this system may be, the fact that it has been officially recommended denotes that the old policy of leaving the speed and course of the enemy to be guessed is at last finally abandoned. After labouring the extreme fallacy of trusting to guess work for over eight years, one must admit that this is a step away from the old position, even if one cannot admit that it is a step in the right direction. Note that of the two methods now being tried, official sanction is still given to the one into which guesswork enters most largely. So hard is it to break with tradition altogether!

II – THE REQUIREMENTS OF A PLOTTING SYSTEM

The following is an extract from a letter:

> In working out the A.C. System, the first, or elementary object, was to break down the supposition that change of range could be got in action as it was formerly got in battle practice. It always has seemed fundamental that the distinction between a stationary and a moving target should be recognised. It seems that this could not be hoped for, until the Fleet had not only tried to *range* a moving target, but to *fire* at it as well. I had supposed that *argument* would suffice to show that the things were different. But the Navy, perhaps wisely, prefers to learn by experience.

However this may be, guided by reason alone, it sufficed for me that in action the target would be moving, and, after exhausting other methods, I was driven to the conclusion that the only practicable method of ascertaining the speed and direction of the target's movement – viz., the plotting of its course – must be made the basis of naval gunnery, instead of reliance being placed on guesswork. The second object was to get this plotting done by a system that would stand the stress of battle.

Just as it has always seemed fundamental that battle gunnery meant moving target gunnery, so has it for a long time seemed fundamental that it is only absolute incapacity mentally to visualise what battle means, that can excuse the retention of a gunnery organisation which obviously cannot survive in action conditions.

The respects in which these conditions change the problem may be grouped into four divisions.

First, it will be impossible to choose your weather for fighting – that will be settled by circumstances. It follows then that the fire control system must be one specially thought out to deal with the greatest amount of roll, pitch, and yaw in which guns can conceivably be used. The present rules about weather conditions for Battle Practice have perhaps of necessity introduced a fair weather standard. For example, if there is too much sea on for boats to get to the target, or to lower or hoist in boats and target, the practice is postponed. To-day the standard is still further lowered to permit of the target being towed! What is the consequence? To begin with, there is no experience in fire control in rougher weather than that in which Battle Practice is now carried out. To go on with, men get the impression that shooting is impossible in such circumstances. But if in war we are liable to fight in any conditions, an elementary realisation of what warfitness means would compel preparation for such an emergency. Some officers have this fair-weather standard so ingrained that they cannot realise that rough weather even makes a difference. But with nine degrees of roll, six or eight of yaw, and seven to nine of pitch, there must be a compound angular movement on the range-finder of as much as three degrees per second. With a 9ft. range-finder on the standard mounting in a fighting top, the taking of any reliable readings becomes impossible in these conditions; and yet in the same conditions it is not improbable that many guns with modern mountings could be used. But as it is axiomatic that guns without the right data for range are useless, it follows that the ship could not fight.

Nothing done with the *Ariadne* gear has so impressed the practical men as the fact that with a movement on the ship that made mere standing difficult, quite untrained observers were able to send down ranges and bearings with perfect regularity and accuracy, and so, by one operation, plot the target's course with imperturbable consistency.

It is, I cannot help thinking, much to be regretted that the official tests were limited to fifty minute experiments in a flat calm. Had manual plotting been tested against the automatic in such rough weather as that in which the *Ariadne* gear gave such satisfactory results, the contrast must have been conclusive.

Secondly, there is the great difference in the speed of relative movement that may be expected in battle, and the altered fire control problem that arises from this difference. In the old battle practices the target was stationary: the speed of the firing ships never exceeded 15 knots. At the usual angles of approach or withdrawal, the rate of change seldom exceeded 400 yards per minute. With a towed target it can similarly seldom come up to 600 yards per minute. But with two modern 25-knot ships it may exceed 1,600 yards per minute.

Again, with a stationary target no complication arises by the target's altering course or speed; with a towed target such changes must be both slight and gradual. But in battle conditions such changes may be both so rapid and so great, as to vitiate all forecasts, based on hitherto accurate data, in a few seconds. It follows then that a control system to be battleworthy must be instantaneous in getting results both to enable it to plot the target when both ships are on a steady course, and to detect promptly any change of either course or speed while firing is going on.

And, withal, the results given must be perfectly accurate. Now rapidity, in getting observations plotted, is an important element in securing accuracy. You now know better than I do how fast it is possible to plot manually – I mean how many observations per minute can be recorded correctly. I hear four per minute is considered extraordinarily good. But with the *Ariadne* gear we plotted regularly in unfavourable conditions at from 12 to 15 per minute, and at times did so continuously – i.e., drew the enemy's course as a continuous line. This extreme rapidity is obviously vital if we assume the ships to be moving rapidly: because the side that is able to get sustained hitting first will get so overwhelming an advantage as to make any hitting at all by the other side exceedingly unlikely.

Thirdly, it must be remembered that in action, apart altogether from casualties, which will be dealt with later, there will be an element of tense excitement which must inevitably affect everyone. Is it to be expected, with shells bursting and guns firing, that complicated operations and delicate calculations will be steadily carried on with the imperturbable calm of a peace drill? Why, is it not an explanation frequently given of a bad battle practice that some vital unit in the chain has become excited and muddled his functions? Will not the chances of these blunders be infinitely greater in action? Does it require exceptional power of imagination to realise that the whole operation may be made impossible by one link in the chain snapping, and that the betting that it snaps is 100 to 1? If this reasoning is sound, it follows that any serious preparation for war demands that every weak link should be eliminated. Now the weak link in fire control is the human agent, with his many frailties. Every unnecessary man on whom you rely is more than a possibility – he is almost a guarantee – of disaster; and so I made it my ideal to reduce the human agent to the smallest conceivable minimum; and, by making the mere operation of taking a cut the only human act in making the plot, reduced the personnel, in ascertaining the data for hits, to a single operator.

Finally, there is the most obvious of all the differences that war is going to make – viz., the certainty of being under fire. I put it last because, though the one thing that that happiest of critics, the man in the street, is able to understand, it is really not the most important of the differences. Personally, I would rather go into action with an unarmoured, well-gunned ship, with an unprotected but efficient fire control system, than have everything amply protected, but without guarantee that I could hit my enemy. Offence is so plainly the best defence at sea. But the fact remains that when science has done all that is possible, the element of chance comes in, and if any part of the ship is vital and should have all the protection that brains and reasoning can supply, it is that part of the organisation that changes it from an inert but perhaps interesting product of the shipbuilders' and ordnance manufacturers' arts, into the most formidable engine of destruction the world has ever seen. The fire control of a ship is its eye and its brain: and it is better to keep your head and sight in battle than anything else. Achilles, with only a vulnerable heel, succumbed at last. Had it been his head, could he have been the hero of a hundred victories?

The masts make a brave show: they are a noble target. A shell bursting anywhere may send a fragment that will cut some wires, dislodge some instruments, disable some man on that lofty and completely unarmoured structure. From the first casualty the entire fabric of your fire control is in ruins. A 6in. shell at 12,000 yards is good enough for the purpose. It need not hit the mast at all. It has only to burst anywhere in its neighbourhood. Is it really necessary to have *Hero* experiments to realise so obvious a danger?[16] Is not reliance on so fragile a structure the feet of clay of this idol of naval gunnery we are asked to worship? Even supposing that it has a head of gold in the shape of peace efficiency, I cannot conceive any sane man taking the responsibility for this when the alternative of perfect protection is at his hand.

Because, after all, not the least valuable by-product of reducing the data-getting part of fire control to the work of a single man at a single instrument, is that that individual and instrument and communications can be more efficiently armoured than any other part of the ship or its crew. You would not then have £400,000 worth of armour in a £2,000,000 ship with the thing best worth armouring unprotected; but, on the contrary, the brain and eye more secure than anything else.

The requirements of a plotting system may be summed up in these words. It must be:

> Weather-proof,
> Fool and flurry-proof,
> Shell-proof,
> Instantaneous,
> Accurate.

The *Ariadne* gear was all five.

III – MANUAL METHODS JUDGED BY RESULTS

Hitting Reduced to Chance

The following facts, which can be got by an analysis of the last three years' Gun Layers' Test and Battle Practice returns, seem to confirm, in a rather singular and unexpected manner, the view expressed above – viz., that between the hitting efficiency, where only the skill of the gun layer is involved, and that where the fire control organisation is interposed, the disproportion is intolerably great.

It is evident that the skill of the gun layers, as shown by the order of merit in which the ships are placed after the Gun Layers' Test, is not reflected in the order of merit as shown by the Battle Practice.

During the last three years, the ships at the top of the list in Gun Layers' Test have tended towards a steady loss of place, and the ships at the bottom have tended towards a steady gain. This tendency repeats itself with regularity each year.

If this loss and gain of places is stated graphically by making a vertical column of the ships to represent their order of merit in Gun Layers' Test, and constructing a line at an angle to it, by connecting into a new line mean positions, for loss of place on the right, and for gain of place on the left, on the actual average loss and gain of the three years we shall see that every ship in the list tends to gravitate downwards or upwards to the centre position. A re-arrangement of the vertical line by pure chance would obviously give an absolute tendency to the central position, so that the pure chance line would be diagonal to the old. The diagram on p.192 shows the extent to which the actual re-arrangement in Battle Practice differs from the theoretical pure chance re-arrangement.

It is thought, I know, by some authorities, that the flurry of the gun layers in Battle Practice is the chief cause of upsetting the 'public form' of the Gun Layers' Test. This may possibly explain a ship like the *Prince of Wales* in 1907 losing 53 places, but it cannot possibly explain the *Essex* gaining 64. Are we to believe that good gun layers are flurried into missing and bad gun layers are flurried into hitting? And if gun layers are flurried, will not the innumerable fire controllers be flurried as well?

It would seem to be a more rational explanation to suppose that, between the skill of the gun layer, and his object, the target, there has been interposed an obstacle which tends to reduce all skill to a level and all results to pure chance. A skilful man, to show his skill, requires an accurate tool. The tool in Battle Practice is a correctly set and deflected sight. The best of billiard players is floored if he has to play

> On a cloth untrue
> With a twisted cue,
> And elliptical billiard balls![17]

The lesson, therefore, that I draw from these returns is that latter-day fire control reduces hitting to chance.

Hitting Efficiency Reduced to Almost Nil

In making my estimate of hitting efficiency in Battle Practice, based on the pre-1908 results, it must be borne in mind that the data for setting the sights were known from the start. The target was moored. It had no speed or course. The Dumaresq indicates the rate of change with absolute correctness for the entire time. All that is involved, then, is getting this known rate of change put upon the sights.

'Any fool,' says the cheerful St Barbara, 'having once found the range, can go on hitting a target. Mathematical instruments do that for him, provided he can guess or calculate the speed and course of the enemy, which is a matter of practice only.'[18]

If this lively writer is right, it must be exceedingly gratifying that the results of Battle Practice show that there are no fools of this stamp in His Majesty's Service.

Let us next examine the relative hitting efficiency. If we take the best shots in the Fleet with each gun in Gun Layers' Tests, we find that their hitting efficiency amounts for each nature to the following figures for hits per minute per gun:

> For the 12 in. 2.5 hits per minute per gun
> For the 9.2 5.5 hits per minute per gun.
> For the 7.5 6.3 hits per minute per gun.
> For the 6 in. 11.0 hits per minute per gun.
> For the 4.7 and under 9.0 hits per minute per gun.

Let us call this Standard A.

The average efficiency of the active Fleet in the same test was:

> For the 12 in. .40 hits per minute per gun.
> For the 9.2 2.01 hits per minute per gun.
> For the 7.5 1.58 hits per minute per gun.
> For the 6 in. 3.32 hits per minute per gun.
> For the 4.7 and under 2.38 hits per minute per gun.

Let us call this Standard B.

If we compare Standard B with Standard A, we shall see there is the following leeway to make up to bring the whole Fleet to what, for purposes of discussion, we may call the absolute potential of gunnery efficiency.

> Standard B with the 12 in. gun has 84.0% leeway to make up.
> Standard B with the 9.2 gun has 63.5% leeway to make up.
> Standard B with the 7.5 gun has 75.0% leeway to make up.
> Standard B with the 6 in. gun has 69.8% leeway to make up.
> Standard B with the 4.7 and under 73.5% leeway to make up.

Battle Practice in 1907 consisted in firing at a moored target at an average range of about 6,000 yards. Roughly speaking, a thousand guns took part in this test. I understand that the time allowed was eight minutes per ship, and that both broadsides were fired. In estimating the number of hits per minute per gun, therefore, something less than eight minutes must be taken as the firing time. We will assume that five minutes is a fair estimate. It is difficult to disentangle from the results of the marks published what was the actual number of hits made. The returns of some half-dozen ships appeared in the papers, and, taking their marks as shown in the order of merit, it is possible to arrive at a formula that seems to explain the position of each ship in the list fairly satisfactorily.

Working through the whole list on these principles, it would seem probable that about 1,200 hits were made. If my assumptions are wrong, it will be easy to make the consequential modifications in the result.

If we assume that the target in Gun Layers' Test in 1907 was approximately equal, angularly, to the target used in Battle Practice at the mean range, a comparison can be made between the hitting efficiency of the Fleet in the two tests. The guns employed in Battle Practice were roughly as follows:

      130 12 in. guns
       60 9.2 guns
       50 7.5 guns
      670 6 in. guns
       90 4.7 and under.

Had these guns attained to Standard A, they would have got in one minute's firing the number of hits given in the centre column. Had they attained to Standard B, they would have aggregated the hits in the last column.

12 in.	325 hits	52 hits
9.2	330 hits	120 hits
6.5	315 hits	79 hits
6 in.	7,370 hits	2,234 hits
4.7 and under	810 hits	214 hits
	9,150	2,699

That is to say, had the Fleet shot up to Standard A, its efficiency, measured by hits per minute per gun in Battle Practice, would have been 9.15. Had the shooting equalled Standard B, it would have been 2.699.

If, in point of fact, 1,000 guns in five minutes only made 1,200 hits, its efficiency was actually .24.

Let us call this Standard C.

Standard C compares with Standard A and B, then, as follows:

    C is 2.62% of A, or 97.38% inferior to it.
    C is 8.88% of B, or 91.12% inferior to it.

The leeway to be made up, therefore, is enormous.

Modern fire control, then, fails in securing hitting efficiency by over 90 per cent, because it cannot get the known change of range on to the sights.

It is finally worth noting that, if we omit the top and bottom two ships in each list in 1907, as possibly the recipients of exceptionally good or exceptionally bad luck, we find that, in Gun Layers' Test, the average efficiency of the top five is to the efficiency of the bottom five roughly as 8 is to 1. Adopting the same principle with the Battle Practice order of merit, we find that the efficiency of the top five is to the efficiency of the bottom five as only 4 is to 1. This is exactly what we should expect if we introduced some factor into the test which was in the nature of pure chance, that is, something which prevents skill being reflected in the results.

IV – WHY HAVE MANUAL METHODS FAILED

The question is: What is this factor? What is the quality of Fire Control that exercises so devastating an influence? The factor is undoubtedly the reliance on the co-operation of a multitude of persons who have to make calculations, work dials and indicators, announce, hear and repeat figures, read results on indicators, set them on sights, etc., and do all these things precisely at the right time.

It is clear that the more links that go to make up a chain of communications of this kind, where each link is a human being, the greater the chances of error in the residual result. Each individual man is liable to make a mistake, either in the actual repetition of what he sees or hears or knows, or in misunderstanding the message or task given him. In the excitement of Battle Practice many such mistakes are made. They will be infinitely worse when their results will be most disastrous – *i.e.*, in battle. Every mistake is fatal. Nor is this all. The range is always changing. It is no use having a range on the gun unless it is both rightly calculated and set on the sight synchronously with the period of its correction. Time exactness is therefore quite as important as figure exactness.

The experience of those engaged in the management of a factory, where uniformity and punctuality in getting results is the first criterion of success, will bear me out in saying that it is an industrial axiom to eliminate the unnecessary man, and never to employ labour, whether skilled or unskilled, where an automatic machine can get you the result. The commercial man, of course, has the incentive of economy in pursuing this path; but it would be a prime mistake to assume that the economy he reckons on is limited to, or indeed consists chiefly in, the immediate saving in wages effected in having an operation done automatically. The chief economy lies not in this direct saving of wages in the particular operation, but in securing uniformity in the result and punctuality in getting it. The man of affairs knows from bitter experience that numbers do not further, but frustrate, his objects.*

It was this train of reasoning that drove us, after the *Jupiter* trials, to realise that the data for gunnery could not be got either with the required rapidity, or with certainty of exact repetition, unless all transmission and recording were automatic; and that, given the exact data for gunnery, a

* In the culinary art, the deplorable effect on broth of a superfluity of cooks, is proverbial.

number of hits proportional to the gun layers' skill throughout the Fleet could not be got unless, in their turn, the making of the calculations necessary for the right elevation and deflection, the actual transmission of these calculations, and, lastly, the actual setting of the sights from second to second, were not equally done by automatic processes.

Modern fire control has broken down, therefore, not because it cannot get the data for hits – for this it has never been asked to do until this year – but because it *cannot get the known data to the gun*.

I am aware that there are still officers – and one or two of them with actual gunnery experience – who have not yet been brought to see that so revolutionary a remedy as the adoption of an automatic system is not only desirable but inevitable, and yet they would be greatly shocked at being considered either reactionary or narrow-minded in the matter of progress. I would respectfully point out to them that if the absolute potential of hitting efficiency is 38 times as good as are the results actually got under the nearest approach to action conditions that the Fleet has yet tried, something far more drastic is called for than mere tinkering with the question by improving a little bit of gear here or there; and I would ask them to consider whether the principles, shown by experience in every other field of modern activity to be right, are likely to be wrong when applied to naval gunnery?

The overwhelming success of the *Ariadne* experiment seems to show that the automatic principle, as applied to naval gunnery, raises no difficulties, and therefore is bound to succeed; but this year's Battle Practice, when firing at a moving target is for the first time to be attempted, will no doubt throw further light upon this problem. I confess the only light I expect to see thrown upon it is a confirmation of the conclusions already lamentably obvious.

V – AS FIGHTING EFFICIENCY MAY DECIDE THE NATION'S DESTINY, THE MATTER CALLS FOR DRASTIC ENQUIRY

The matter is certainly one sufficiently grave to call for the united efforts of the best brains and the best wills. Proposals that have any prospect of success cannot, in view of these facts, be dismissed with such threadbare arguments of prejudice as, for instance, that automatic gear is too complicated or too costly. The reason it is complicated is that the operation is so; the test of gear being too complicated is not its complexity, but its being reliable; and, to prove whether gear is reliable or not, trials must be made extending over sufficient time, and over a sufficient number of conditions as to weather, distance, rapidity and variation in change of range, the detection of change of course, etc., to enable the true facts to be known, and prevent condemnation being come to on baseless assumptions.

As to the argument of *expense*, it is to be remembered that the nation pays on an average between 30 and 40 millions a year to maintain a navy, and all it gets in exchange is the fighting efficiency of the Fleet that it subsidises. Can any sane person pretend, when the absolute potential of the hitting power of modern artillery is over 9 hits per minute per gun, and the achieved result is less than 3 per cent of this, that the nation is getting value for its money? And, if it is not getting such value, surely a greater portion of this expenditure should be applied to attaining a better result?

It is perfectly useless multiplying, at fabulous cost, the number of ships and guns unless those guns can make hits in battle. £2,000,000 will give you to-day eight more 12 in. guns in the line of battle. We do not know that these guns would make any hits at all in action. In five minutes' firing, however, with the present standard of hitting efficiency, this will result in 10 hits in Battle Practice. These 10 hits will therefore cost you £2,000,000, or £200,000 each.

Supposing, with an automatic system, you could raise the efficiency of the Fleet from .24 to, say, one hit per minute per gun, that would give you, with the ships you already possess, 760 more hits per minute. No one can suppose that to introduce a completely automatic system would cost £2,000,000, or, if it did, deny that it would be the best investment of such a sum that the nation had ever made. For these hits would cost £1,600 apiece only, instead of £200,000. And a sound system should bring the hitting, not to one per minute, but to over two and a half – and, with more thorough training, nearer still to the absolute potential of Standard A.

Not the least advantage of the automatic system is that it makes all conditions equal, which means that there would be no greater difficulty in hitting a target moving at a virtual speed of 40 knots, and even changing range at this rate, than in hitting a stationary target. If automatic machinery is correctly designed for the work it has to perform, it can be safely trusted to make no blunders.

No ship in the British Fleet to-day could get the rate of change in the conditions just indicated; and if the control party did get the range and the change, it would be powerless to get the known range for a particular moment set upon the gun at that moment.

This is a state of affairs that calls for enquiry.

A.H.P.

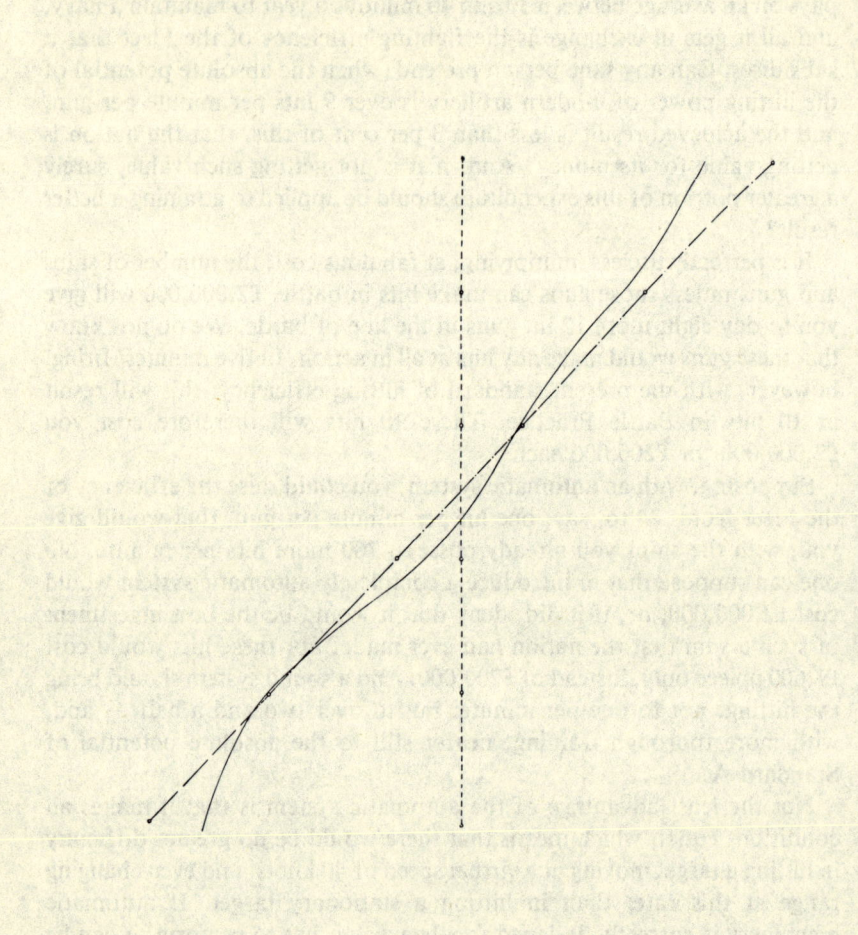

NOTE

In the diagram opposite, the vertical line represents the ships in order of merit after Gun Layers' Test.

The diagonal represents their order when subjected to re-arrangement by pure chance – viz., the first ship tends to lose 50 per cent, the last to gain 50 per cent and the middle ship to keep its place.

The continuous wavy line represents the re-arrangement actually shown to have occurred in the last three years' battle practice. The extent to which it does not coincide with the pure chance line, assuming fire control to be approximately equal in all ships, is a measure of the effect of superior skill.

The extent to which it departs from the vertical is a measure of luck, on the same assumption.

The wavy line has been constructed, by placing a mark to the right of each ship when it lost places in battle practice, and to the left when it gained places. The distance of the mark corresponded with the number of places lost or gained, using the same measure as in constructing the vertical line. The diagrams for each year were superimposed, and the centre of gravity taken of groups. The connecting of these centres gives the line reproduced.

A.H.P.

XIII

NOTES, ETC., ON THE *ARIADNE* TRIALS
(*APRIL 1909*)

At a meeting on 31 March 1908 the full Board of Admiralty had voted to confirm the rejection of the Pollen instruments tried in the *Ariadne* in January, but to compensate the inventor for his losses to date and to award him a fee in recognition of his contribution to the development of fire control theory.[1] Pollen had accepted the Admiralty offer after several weeks of negotiation over the terms, and on 23 April 1908 the Admiralty Permanent Secretary had informed him that he would be paid £11,500 and given back the instruments tried in the *Ariadne*, including the range finder. In return, Pollen had agreed not to divulge the secret of his system to any foreign power for two years, during which time the patents of his instruments would remain in the possession of the British government.[2] Pollen remained convinced, however, that his approach offered the only practicable solution to the fire control problem, and he therefore used the funds and instruments received from the settlement to continue the development of his system.

By the end of 1908 Pollen had founded the Argo Company to back his efforts. Writing as its director on 5 January 1909, he was able to inform the Admiralty that the new firm was prepared to offer for trial a complete fire control system that consisted of an improved gyroscopically stabilised range finder and bearing indicator mounting, an improved automatic plotter, a change of range machine whose computation of the target range and bearing would allow for the continuous variation in the change of range and change of bearing rates, and an automatic system of sight setting.[3] Pollen was unable to gain the support of Admiral Sir John Fisher, the First Sea Lord, as he had in 1906,[4] but he retained the backing of Rear-Admiral Sir John Jellicoe, who had returned to the Admiralty in October 1908 as Third Sea Lord and Controller, which made him the immediate superior of Captain Reginald Bacon, the Director of Naval Ordnance.[5] After an extended official correspondence between the Argo Company and the Admiralty, and a number of private exchanges between Pollen and Bacon, Jellicoe and Reginald McKenna, who had replaced Lord Tweedmouth as First Lord, the Admiralty Permanent Secretary informed the Argo Company on 21 April 1909 that the Navy would purchase a complete set of fire control instruments for trial.[6]

Apparently emboldened by this success, Pollen then drafted a lengthy and bitter attack on the past actions of Bacon and Admiral of the Fleet Sir Arthur Wilson with regard to his fire control system, which he had printed in late April 1909 for the Argo Company with the title 'Notes, Etc., on the *Ariadne* Trials'.[7] In addition to a history of his relations with the Admiralty from 1901 to 1908, Pollen included an appendix in which he provided the reader with excerpts, and in many cases the entire text, of official and unofficial letters to and from the Admiralty. Pollen's intent in putting forward such a work, it would seem, was to bring the issue of past injustices into the open in order to prevent their recurrence in the forthcoming trials of his system. 'What I want to do,' Pollen wrote to Lieutenant George Gipps on 17 May 1909, 'is to precipitate – or rather have Bacon precipitate – a row with me, on which I can produce a précis that I have drawn up of the *Ariadne* fiasco. He must either take this lying down, or face the music of the whole thing.'[8] The correspondence in the Pollen Papers gives no direct evidence that Bacon provided Pollen with the desired pretext, or of the extent to which the paper was circulated, if, indeed, it was circulated at all. It should be noted, however, that the accusatory style of Pollen's squib could not have pleased Jellicoe, whose relationship with Pollen in April or May 1909 did take a sharp turn for the worse.[9]

INTRODUCTORY

(1) A TALE OF CROSS PURPOSES

The singularity of the following facts may be thought a sufficient excuse for recording them. The author's object, however, is neither to preserve certain unhappy incidents from the oblivion to which they would better be consigned, nor to indulge the indignation they most certainly justify – but solely to throw into relief the inevitable consequences of a false position having been created.

(2) A REACTIONARY INSTRUMENT OF REVOLUTIONARY POLICY

The issue is a simple one. The A.C. System was, after patient enquiry, recommended to the Board of Admiralty in 1906, as urgently calling for experiment, on the ground that its success would prove monopoly to be of national importance. When the system was ripe for trial, the new D.N.O. held a diametrically opposite opinion.[10] He was, then, in a false position in having to arrange for the trials and prepare the consequent recommendations. There is apparently a danger of this oversight being repeated. To re-impose these duties on an officer who, however great his ability and distinction, is disqualified both by his professed opinions and his previous actions, from even being impartial, is to prejudice the result. To purchase costly instruments for trial, to secure absolute secrecy, and arrange for obtaining monopoly, can only be consistent with a hope that the experiment will succeed. To entrust the experiment to one who is committed to its failing, is going far towards ensuring failure. The writer believes that success is vital to war-fitness; that if the trials are intelligently, thoroughly, and fairly conducted, this success will be ever clearer than it was in the *Ariadne*; he therefore desires respectfully to suggest the only way such trials can be assured. And with this object, he proceeds to set out, as simply as possible, what appear to be the elements of the question.

(3) THE STRENGTH OF NAVIES IS THEIR STRIKING POWER

He would premise, then, that the strength of the Navy does not chiefly lie in the numbers, size, speed, or disposition of its constituent ships, nor yet in the discipline and organisation of its personnel; but so overwhelmingly in the capacity to make hits in battle, before the enemy can hit back, as to make gunnery efficiency the Alpha and Omega of war fitness. He does not mean that numbers, discipline, organisation, etc., can safely be neglected; but he takes it to be indisputable that if any very marked superiority exists in the use of artillery, a considerable inferiority in other respects may be comparatively innocuous – whereas no perfection in these can compensate for the inability to strike hard when the test of battle comes.

(4) THE BURDEN OF ARMAMENTS

By common consent, the time may not be distant when to maintain a numerical surplus over the combined fleets of foreign rivals must impose a burden the nation will find it difficult to tolerate – it is therefore of the first importance to examine with the most sedulous care any proposal to increase naval power by better methods.[11]

(5) THE DEPLORABLE INEFFICIENCY OF PRESENT-DAY GUNNERY

Now, in the elementary conditions of battle practice, the whole fleet hardly averages 20 per cent in hitting efficiency, and a third of the battleships fail to reach 10 per cent. It must be remembered that these results, humiliating as they are, are only obtainable when every ship is strung to concert pitch by months of patient drill and practice. Heaven only knows what would happen if the fleet were suddenly called upon to produce good gunnery in an emergency that found the required organisation unprepared. And it is a disquieting fact that for a large part of the year, the required organisation does not exist at all. It is impossible then to suppose that the art of gunnery, viewed as a war accomplishment, is not in a rudimentary stage. This being so, is it not clear that it is money far better spent to improve naval gunnery than to multiply inefficient guns?

(6) FIGHTING AND FINANCE: (1) ADDING STRENGTH BY SHIPS COSTS £2,000,000 PER 3 PER CENT

When the ships of this year's programme are completed, there will be 312 12in. and 10in. guns afloat. If the art of gunnery is beyond betterment, this force can only be increased by adding more guns – and as each gun put into the fighting line in a *Dreadnought* costs roughly £200,000, and, as guns can only be added in groups of ten, every 3 per cent increase in power will cost £2,000,000.

(7) FIGHTING AND FINANCE: (2) ADDED STRENGTH BY RIGHT GUNNERY SAVES £6,000,000 PER 300 PER CENT

It can be demonstrated to the satisfaction of any body of experts, that it would not cost a half of £2,000,000 to add, *not 3 per cent, but 300 per cent*, to the striking power of the fleet, by a rational system of gunnery, which would ensure *every* ship being *always* ready to fight, and, while doing this, *save over £200,000 per annum* – an economy that is equivalent to a capital saving of nearly seven million pounds sterling.

(8) GUNNERY TECHNIQUE THE TOUCHSTONE OF ADMINISTRATION

The technique of gunnery is therefore a matter of the supremest national moment, as on its right development the nation's present comfort and future freedom will ultimately rest. No more vital problem can occupy the attention of the distinguished officers and laymen to whom the destinies of the Navy are committed. It is quite obvious that to-day's practices have been worn threadbare in the effort to avoid any drastic reform, and a new departure of some kind is required. The author believes that the ensuing narrative will show that it is a subject that should have been entrusted to the care of an authority, whose qualification was a knowledge of the subject, a capacity for impartial investigation, a lively realisation of the wretched insufficiency of present methods, and an unprejudiced anxiety to find a practical, because scientific, solution of the problem. And it seems clear that this directing authority should not be an individual, but a permanent council of experts.

(9) THE NECESSITY OF RIGHT DOCTRINE

In matters of special knowledge, it is always dangerous to trust to any single individual, nor is the danger sensibly diminished if he is himself an expert. Safety lies in the balanced judgment of many experts – and no advance is possible unless the requirements of the problem are analysed and understood, and invention and solutions looked for in the direction that analysis has shown to be necessary. In other words, the promise of sane progress can only be founded in the establishment of right doctrine, and in resolutely testing alleged novelties by the light of it. Had any doctrine prevailed in 1907–8, it would not have needed a prolonged agony of failure to demonstrate the fallacies on which the time and range method was both based and recommended, for the reason that an elementary analysis, made and proved long before 1908, had already shown its unscientific as well as its impracticable character.[12]

(10) A STABLE POLICY IMPOSSIBLE WITHOUT KNOWLEDGE

Finally, he would ask if it is too late to put the question of automatic fire control beyond the reach of individual caprice and prejudice, into a region of reasoned certainty? Although in one form or another it has been before the Board during the last six years, the importance — or otherwise — of the writer's system has never been thoroughly investigated, or the truth concerning it finally established. Consequently, official decisions have too often been made on the authority of whichever individual is for the moment the more influential, and never on the authority of an acknowledged doctrine. Hence successive acts are the outcome of the conflict of inconsistent views, first one, then another predominating; and cannot therefore be consistent. The worst of it is that, in the absence of thorough knowledge, both of the principles on which the system is based, and of the means by which it is proposed to realise them, neither advocates nor opponents are sure of their ground. For instance, there is not to-day at the Admiralty anyone who has seen any of the gear at work, or is acquainted even with the outward appearance of any part of it. No report by a specialist, either on the gear already tried, or on that of the 'latest approved'* pattern, is available for the guidance of the Board or the instruction of their advisers. The offer of the complete system was made several months ago, and, as yet, not even a sketch has been asked for. For the forthcoming trials to be undertaken on no better information than exists now, is to risk a partial repetition of the *Ariadne* fiasco. This should be avoided at any cost — and there is only one certain way of avoiding it.

(11) THE WAY OF SAFETY

The difficulty in constituting a commission competent to examine the problem, define its requirements, select between competing solutions, and direct future developments, cannot be excessive. The deplorable inefficiency shown in the recent battle practice is surely a sufficient occasion for such an enquiry, and the value of the lead its report would give would be incalculable. Not until this is done will the vast increase in efficiency to be got by rational methods be apparent, nor the great economy of eliminating wholesale the superfluous persons drawing pay and victuals, be acknowledged. Without the definite laying down of doctrine by this

---

* In the Admiralty letter of 21st April the writer was asked to tender for gear of the 'latest approved' pattern. In the circumstances the epithet seems curiously chosen.

or some equivalent machinery, it is hard to see how official action can cease to box the compass of conflicting policies, or how it is to be ensured that any future trials that may be ordered, shall be carried out with impartial intelligence and judicial fairness, and in the sympathetic spirit necessary for scientific investigation.

THE *ARIADNE* FIASCO

I – GUNNERY PROGRESS AND THE BOARD OF ADMIRALTY, 1905–7

(12) THE CHARACTER OF THE INVENTION

The author of the A.C. Battle System pointed out in 1901 (1) that long range firing could only be successfully developed if it was based on knowledge of the target's *future* movements; (2) that this could only be deduced from knowledge of its *past* movements and present position; and (3) that this in turn could only be satisfactorily obtained by the construction of a simulacrum, plan, chart, or plot, of the actual courses, both of the firing ship and target.[13]

(13) THE BIRD'S-EYE VIEW

He believes this was the earliest statement that a bird's-eye picture was an absolute necessity – and the means he suggested, the first feasible method of simultaneously plotting both the path of a moving target and that of the moving platform from which the plot is made. The essential operations – viz., the use together of synchronous ranges and bearings free from yaw error, by plotting them in correct time sequence, on a paper which is moved from the plotting centre in a correct relation to the ship's speed through the water, or, what is the same thing, from a plotting centre that is moved across the paper in a correct relation to the ship's speed – distinguish the writer's inventions from those of the late Col. Watkin, and have all been made the subject matter of various patents granted between 1904 and 1908.[14] It is, of course, immaterial, in considering plotting, as a process necessary to gunnery, whether the ranges are obtained from a single instrument or by a two-observer system, or whether a true plot of both courses is made, or a virtual plot of one only. The A.C. patents permit of either being made at choice.

(14) THE BASIC TRUTHS ACCEPTED

After 1905, the date of the first experiment with the A.C. System in H.M.S. *Jupiter*,* all the practical experts, cognisant of the results obtained, were agreed that the author's contention of 1901 was incontestable; and after the moving target Battle Practice in 1908, the comparative failure of alternative methods of getting the data for forecasting change of range, resulted in a general acquiescence that the 1905 verdict on the theory of the A.C. System was right. As a consequence, there is at last a practical unanimity that naval gunnery must be founded on the three propositions laid down nine years ago, and in one form or another, the British as well as the German Navy, have adopted the general principles of plotting invented and patented by the writer.

(15) HOW TO GET THE PLOT IN ACTION

This narrative is concerned with the disagreement as to how, in action, this bird's-eye view can, most usefully, be obtained. The author, after some months' practical experience with them in H.M.S. *Jupiter*, became convinced that in anything except the easiest conditions, polyanthropic methods – which involve reliance on a multitude of individuals to act together both quickly and correctly in all cases – were doomed to failure and disappointment. It appeared to him that, as in all other fields of human activity, the multiplication of agents involved waste of time and money, and the generation of error, and where the motion of the ship was considerable, or the change of range rapid, the intrinsic difficulties of the required operations would be too great for human agents in any circumstances, and consequently to rely on such agents would be to court disaster.

(16) THE PERIL OF POLY-MANIA

At the time alluded to no moving target gunnery had been attempted – nor was firing practised in the rougher sorts of weather. If, then, the standard of efficiency showed such astonishing variations from ship to ship as the battle practice at a stationary target indicated, what was to be expected when two new unknowns – viz., the speed and angle of course of the target – were added, when the weather had to be taken as it might come, and when all organisation would be at the mercy of the noise, flurry, and casualties of action? It seemed clear that a method that did not emerge

---

* See letter of August 27th, 1906, in Appendix.

with credit from the make-belief pastimes of peace, would certainly be a very feeble stand-by in the tremendous realities of war.

(17) AUTOMATA THE ONLY HOPE

The foregoing reflections could result in only one conclusion. Only a system that reduced the personnel below the drill level, that equalised conditions by the mechanical elimination of difficulties, and that secured the highest conceivable rapidity and accuracy in results by making each operation instantaneous, all of them synchronous, and as many as possible self-acting, could meet the situation. In short, to get the data for war gunnery, an automatic system was a simple necessity.

(18) POLYANTHROPIC ALTERNATIVES

The advantages and disadvantages of the respective methods — the polyanthropic and automatic — were very fully canvassed with the D.N.O.'s Department in the spring and summer of 1906. The two manual methods principally discussed were the use of rangefinder ranges and bearings taken by compass, and the use of a rangefinder mounted with a bearing plate, and correcting the yaw by gyroscope. There was, it is true, a third possible method — namely, the forecasting of change of range by constructing a curve in which the ordinates should be ranges and the abscissae time intervals, and guessing its continuance. He and his associates prepared certain patent applications on improved ways of generating the desired curves — but abandoned them on close examination of the subject, as they found this method held out no promise of useful results. It was not therefore even mentioned as an alternative.*

(19) THEIR USELESSNESS ALWAYS OBVIOUS

Of these methods it may be said that had either a magnetic compass free from irregular deviation, or a continuous gyroscopic yaw indicator, been available, the first two, while not practical, would at any rate have been scientific. The third could in no possible event be either one or the other.

(20) A.C. PRINCIPLES ACCEPTED

In the late summer of 1906, designs for the new A.C. instruments were produced, and, after a perfect galaxy of gunnery talent had been called into consultation, the case for the automatic system was acknowledged by the D.N.O. to be completely established, and after lengthy negotiations the Board of Admiralty finally directed a set of instruments should be built for trial; while the inevitable failure of the polyanthropic method

---

* A full analysis of the time and range system has been printed in Pollen and Isherwood's 'Reflections on an Error of the Day', 1908.

seems to have been so confidently foreseen that even a trial of it with existing Service instruments did not, officially, get beyond the stage of being suggested.[15]

(21) THE INCALCULABLE VALUE OF SUCCESS

It must not be supposed that the partisans of the new gunnery had an easy task in securing official acquiescence in their desire. Not that the vital importance of success was doubted or even disputed; on the contrary, it was candidly admitted by the very highest authority, as appears from the subjoined correspondence, that success would prove the A.C. System to be 'invaluable'. It was the success itself that seemed illusory. Even so late as June, 1907, there was very little hope that the money voted would prove a good investment.

(22) THE ADMIRALTY ENDORSEMENT OF A.C. PRINCIPLES

This endorsement of the A.C. System by the Board of Admiralty, initiated in the spring of 1905, and confirmed in August, 1906, was completed in May, 1907, when the Tactical Machine – the last unit of the complete Battle System – was adopted. This instrument was designed for the twofold object of facilitating the study of the new developments that would be inevitable in naval tactics, so soon as it became certain that a ship's artillery could be used with effect, no matter how rapidly the range might be changing, and as a guide in action when the anticipated gunnery revolution was achieved.[16] The instrument is practically useless for any other purpose. Since the stultification of the old policy by the rejection of the A.C. System, the Tactical Machine has, the author understands, been used as an aid in manual plotting – a very singular instance of the naval genius for adaptation.

II – THE STULTIFICATION OF THE 1905–7 POLICY AND THE VICTORY OF THE POLY-MANIACS

(23) THE VERDICT PRECEDES THE TRIALS

The A.C. instruments were completed in November, 1907, and the experiments directed by the Board were to have taken place in the *Ariadne* in the following January. That the trials took the course they did, was for some time almost inexplicable. It appears, however, from the Navy Appropriation Account, 1909, page 121, that some time before November 26th, 1907, it had already been determined at Whitehall that the Board should somehow be induced completely to reverse its policy of the previous three years, and neither to adopt the automatic A.C. System, nor to prevent Foreign Powers having it; and *this decision was come to* not only before the instruments were tried, but *before they were even inspected or installed*![17]

(24) THE FUNDAMENTAL ERROR

Just as in 1906 the polyanthropic system was dismissed as useless, after the most painstaking and exhaustive analysis, so in 1907 the uselessness of the automatic system was incontinently taken for granted, without the faintest knowledge of what the machines were capable of doing, or slenderest enquiry into their purpose or design. Polyanthropic gear to plot, not the course of the target, but the rate of change, by the time and range method (see pars. 18 and 19), was thereupon installed in H.M.S. *Vengeance*, and tried concurrently with the A.C. gear.

(25) THE BOARD NOT INFORMED

For the fortnight of the preliminary practices, the author, who of course at that time did not know that his fate was already sealed, had the satisfaction of seeing the gear work day in and day out, in the roughest of weather and with unvarying success. He had the high honour of being most cordially congratulated on his triumph by several members of the Board, who were as ignorant as he of the part they were to be dragooned into playing.[18] It was not until the very end of the trials that any inkling of the incredible truth broke upon him.

(26) THE VERDICT BEING SETTLED, EVIDENCE WAS UNNECESSARY

As the evidence had to square with an apparently prejudged verdict, and the tests be completed without disillusioning the distinguished Admiral who had been selected to conduct them, it is not surprising that it was decided that no gunnery or mechanical specialist should be allowed to assist in the *Ariadne* on behalf of the Admiralty, nor to see the gear at work. Nor is it surprising that the trials were limited to 50 minutes in a flat calm – nor that the Admiral did not even carry out such trials as he had laid down as necessary.[19]

(27) THE FINAL STRUGGLE

Early in the month of March, the writer had an interview with the First Lord, and in consequence of certain representations, fully set out in the appended correspondence, Lord Tweedmouth promised to let him have a copy of the Admiral's report, and said he would consider the question of further trials. On March 6th, fresh trials were ordered. Some difficulty however, not difficult to guess, arose. The Lords were, quite untruthfully, informed that the writer had objected to further trials, and they were countermanded; and Lord Tweedmouth was not allowed to keep his promise in the matter of the report.

(28) THE LETTER OF MARCH 10TH, 1908

The adequacy of the polyanthropic time and range method was (as might have been expected, had the November decision then been known) officially accepted; the said decision was somewhat precipitately endorsed by the Board; the author was informed that his system had been superseded by one vastly superior; and a last masterly touch was put to his discomfiture and disappointment, *by his being invited to re-purchase his gear the better to prosecute his alleged desire to sell his inventions to Foreign Powers.*

III – PUBLIC DUTY – OR PRIVATE GAIN?

(29) WHAT TO DO?

As he had been now for some eight years at work on this system, in the belief, amply confirmed by a consensus of expert opinion, that, if it realised its purpose, its war value would be inestimable, and as his one determination throughout had been that, at all costs, his own country should alone benefit by it, so long as it could be kept secret, it was natural that, at the moment of its complete triumph, he should be deeply wounded by the insulting allegation of such unpatriotic motive, and most unwilling to consent to the defeat of the main object that had inspired him to produce it.

(30) NAVAL ADVICE

In this dilemma, it was pointed out, by several officers eminent in the naval service, who were exceedingly indignant over the strange course pursued, that in taking his inventions abroad he would be immune from any imputation of want of public spirit, for the plain and unanswerable reason that he had the permission, nay the encouragement, of the highest authority in so doing. It was known that he had embarrassed himself financially by his long attempt to make his ideas prevail; and it was urged, with equal delicacy and force, that while to be more solicitous for the Navy than the Navy itself, might verge on the ridiculous, by persisting in it, he ran the risk of losing whatever commercial value might attach to his inventions if delay resulted in their secrets being discovered before they were protected abroad.

(31) WHY IT WAS GIVEN

The material advantages in accepting the invitation to go abroad were obvious and attractive. But the author felt that in urging him to do so, his service friends had been chiefly moved by sympathy, and a feeling that they had no right to ask him to make further sacrifices. He did not believe that the case for his system had been fairly put to the Board of Admiralty; and he could not face, with any easy equanimity, the abandonment of what had so long appeared as a plain obligation of citizenship − so he cast about to add up the more favourable elements in the situation.

IV − ELEMENTS OF HOPE

(32) THE PROVED SUCCESS OF AUTOMATISM

First, the complete and indisputable success of so much of the gear as had been tried, had been proved beyond doubt or cavil; every principle essential to the complete Battle System had been tested; the success of the whole, with its corollary of a tactical revolution, was no longer conjectural.

(33) A QUESTION OF AUTHORITY

Next, the authority of those who had pronounced so confidently on the war value of success − and their confidence was unshaken in adversity − seemed immeasurably greater than that of those who had failed to appreciate it when it was achieved. *'Securus judicat orbis terrarum.'*[20] These men had given more months to the study of the principles of the system, than their opponents had devoted minutes to preventing any demonstration of its merits. On one side was every single gunnery specialist of the Fleet who had studied the subject, and every student of the problems of modern war, known to the author. On the other were three − it is true, very distinguished − officers, of whom, however, only one was a gunnery specialist, and he was never in the *Ariadne* at all, and was reputed to be the protagonist of the rival system. The only one who had seen the gear at work admitted he did not know for what purpose it had been ordered; he wished to test the gear, not as an all-weather, quick change of range plotting system − which it was − but as part of the Service fire control system − which it wasn't; and as he was satisfied that the competing method, polyanthropically worked, was 'vastly superior', before he could be informed as to its performances in the only test made at sea, it did not appear to the writer that the conclusions of the Naval Ordnance Department of 1906 were materially weakened by this very eminent officer's adhesion to the other camp.[21]

(34) THE POLY-MANIAC'S FOOLS' PARADISE

Then, too, it was one thing to mesmerise the authorities into the hallucination that a precipitately produced panacea, for which no immediate payment was required, was 'vastly superior' to a laboriously worked out system that must, unfortunately, be purchased — but it would be quite another thing to get decent results with it in the Fleet. It was clear that escape from the chaos that had overwhelmed gunnery affairs, would surely be found when the disillusioning test of practical experience should come. The absolute failure of polyanthropy was inevitable, and in the end the common sense of the Service would both make itself heard and would prevail.

(35) A MORATORIUM ARRANGED

If only then the financial difficulties of the position could be overcome, without the sacrifice of secrecy, the author's duty to lie low and wait for better times seemed to him to be both clear, and even if invested with real risk, not without some compensating advantages. He accordingly addressed a reasoned protest against the character of the so-called trials, and the verdict that had resulted from them, to the Board of Admiralty. He was fortunately successful in inducing a more sympathetic mood, and after some correspondence, a solution at once more agreeable to the writer and more flattering to his work was found. A further period of secrecy was arranged for, the past expenses of the development of the system were liquidated, and the author's services to naval gunnery were both acknowledged and remunerated.

(36) V – THE MORAL

At the risk of repeating a thrice-told tale, the writer will venture to point the moral of these curious transactions.

(37) SLAP-DASH INFALLIBILITY

It seems quite absurd to maintain that the story told above shows that during the last eighteen months, gunnery technique has been recognised to need the best attention of the best experts. The writer does not doubt for an instant that those responsible for smothering the truth about the

A.C. gear have acted quite conscientiously throughout. They started with the assumption that the system was nonsense – and they never studied a single one of its principles or details long enough to undeceive themselves. They were quite sincerely convinced that there was no system worth the name to test, that automata were intrinsically vicious, that naval gunnery had far better pursue its polyanthropic path, and not get mixed up with the disturbing influences of modern thought and methods. And these opinions they quite frankly hold to-day.

(38) DID INFALLIBILITY FEAR THE TRUTH?

To encounter such opinions is the common fate of all who, in any walk of life, busy themselves with trying to alter the existing state of things. What has amazed the writer most in the present case, is that the officials concerned were not sufficiently confident to have the courage of their convictions. Had their prejudices amounted to certitude, they surely could not have adopted a course so little consonant with the chivalry traditional in their Service. Generosity is not stretched to the breaking point by giving a fair trial to an opponent you are bound to beat. If they were *positive* they were right, they could have afforded to *prove* the author wrong. Even to *seem* to tamper with the decencies of justice, was almost to admit that they feared the truth.

(39) A GUELPH IN JUDGMENT ON A GHIBELLINE[22]

The author would, then, first call attention to the unpleasant fact that the decision was not merely left to those who were ignorant of the case for automata; it was committed to the unqualified mercies of its avowed opponents. The fate of the policy of having taxi-motors on the streets would hardly be doubtful if its destiny were in the hands of the hansom-cab men. The sailing masters of the three-deckers could never have been impartial judges of the value of steam power – nor stage coach drivers of the benefits of railroads. But to leave the investigation of automatic gunnery to those obsessed by a passion for polyanthropy, was not one whit more consistent with common sense, than these supposititious cases. The obscurantism of vested interests is just as potent where personal predelictions are affected, as when financial disadvantages are concerned. Indeed, it is hardly a hyperbole to say that an impartial verdict in a matter of this kind cannot be looked for, unless those who judge ardently desire success. For it is clear that only those conscious of the inadequacy of existing methods, and therefore looking for better, will go to the expense of experimenting with new things. It is the *wish for success* that alone explains there being any trials at all to judge.

(40) THE PROCESS OF SECURING A VERDICT

Had it been recognised that gunnery was the be-all and end-all of naval war, it could not have been permitted that the expert knowledge of the Service should have been excluded in deciding on so momentous a matter. Had general staff principles – even without their organisation – been acknowledged, it must have seemed preposterous that a single admiral, not a gunnery technician, and admittedly uninstructed as to the history or purpose of the gear, should have been entrusted with choosing between the productions of a layman whose objects were unknown to him, and the inventions of a naval officer, whom he had every reason to trust blindly in gunnery questions. This officer had, by a unique genius for polyanthropic organisation, kept the Admiral's flagship at the head of the Fleet for two years, in the old-fashioned exercises which depended entirely on a form of skill that automata would make unnecessary. He had been promoted Commander at an unusually early age, owing to the Admiral's favour, thus very properly earned. There could never have been any doubt as to the side towards which his superior and patron must lean.[23]

(41) A DRUMHEAD COURT MARTIAL

Can it be wondered at that the supposition arose that the Admiral in question was selected, less for his gunnery knowledge than for his dominating personality? That it was known he would view the competing system with a friendly eye, and that his unparalleled ascendancy made him the one man who could overawe the Admiralty into stultifying a whole catena of connected acts of policy, and registering as their own a decision *made for them five months before the question came before the Board?* The supposition may be baseless – but it is difficult otherwise to account either, for the trials, that should in the ordinary course have been, part, a scientific enquiry, and part, a judicial proceeding, assuming, without disguise, the character of a drumhead court martial, or for the further trials being countermanded and the First Lord being unable to keep his promise in the matter of the report.

(42) THE MISUSE OF A GREAT NAME

While the author regrets that the most eminent sea commander of our time should have been betrayed into casting the halo of his great name over a system so inept, as to be doomed to failure from its inception, he would add that it was always obvious that the Admiral never considered himself entrusted with the duties either of an investigator or of a judge.

His clear understanding of his instructions was plainly shown both in his words and his acts. He knew the young commander's invention must be superior;[24] the author's was, therefore, not worth trying. His duty was to deal summarily with a grotesque imposition; and he discharged it without any thought of disguising either what his opinions were, or that they were preconceived.

(43) THE NEMESIS OF POLYANTHROPY

It must not be forgotten that the ostensible ground for jettisoning the A.C. System, was that the time and range method was 'vastly superior'. Can it be supposed that if the gunnery experience and talent of the fleet had been reflected in official action, that the collapse of the substitute would not instantly have been followed by retrial? Instead, the failure of the 'superior' system was lightly passed over, as if the reputation of those who had procured the condemnation of the automatic system, were in no way tied up with its success; and they forthwith proceeded, in a final effort to justify the proceedings of 1907–8, to adopt the hand-worked copy of the author's system, rejected as impracticable three years before.[25]

(44) FIAT JUSTITIA

To the writer, this evading of the issue hardly squares with the accepted canons in parallel cases. If the only witness in a successful prosecution is subsequently proved to be an impostor, the sense of justice of all mankind demands the instant quashing of the conviction. It is, at the least, unusual for those who have pledged themselves to the 'vastly superior' character of the detected sham, to set about looking for a new witness, whose testimony will justify the maintenance of the condemnation.

(45) INJUSTICE THE OFFSPRING OF IGNORANCE

Of the two obvious aspects of the proceedings on which the author is so bold as to comment, the trifling with the war efficiency of the fleet is so much the graver, that he apologises for introducing a consideration that may be thought merely personal to himself. And yet a private injustice, if it is the result of any defect in public organisation, is a legitimate matter for common concern. It cannot be doubted that the author was led to suppose in 1905, 1906, and 1907, that if he would forego the protection of his inventions by patents in foreign countries, he would be assured of a fair trial and equitable treatment at home. To deny him this was to be untrue to a plain obligation, and to do him a poignant wrong: and while the writer is glad to acknowledge that this wrong was to some extent remedied at a later date, he makes bold to say, that had the matter of the A.C. System been throughout handled in the light of expert knowledge of gunnery, neither the grave risk of the system going abroad could have been incurred, nor the injustice to himself have been committed.

A. H. POLLEN.

APPENDIX

NOTES ON THE ORIGIN OF THE A.C. SYSTEM

'My Lords, ... have no intention of questioning your claim to be regarded as the sole inventor and originator of the system for obtaining the data for gunnery by the use of ranges and bearings taken simultaneously and used in conjunction with a chart gyroscopically corrected.' – Admiralty Letter, September 21st, 1906.

1901 – First Principles

The writer first saw practice with ships' guns at Malta, in March 1900. The range was 1,400 yards. He was given to understand that the difficulty in ascertaining the range prevented practice at longer distances. As in South Africa the same guns had been used with effect at 10,000 yards, the problem of approximating sea and land conditions seemed worth investigating. The preliminary analysis of naval gunnery requirements, out of which the A.C. System grew, were commenced in the spring and summer of 1900. The results were indicated in a communication to the Admiralty dated January 26th, 1901.

The following extracts show how the A.C. system was originally brought to official notice.

> Let us suppose the Captain of a cruiser to find the mast of an enemy's ship in sight on the horizon, by means of this device he would, within a few minutes, know her exact distance from him, and, inferentially, her exact place upon the chart. On the observation being repeated every few minutes, he would soon be in possession of the knowledge of her exact speed and direction.

Later in the same letter occurs this passage:

> We claim that were this machine correctly used it would be possible for any ship provided with it to ascertain the exact distance, course, and speed of any vessel in sight within a few minutes of seeing either a mast by day or a light by night; and to open upon her with a more effective fire at 10,000 yards than is now possible at 5,000 yards, and that without the faintest possibility of her being able to reply effectively.

The increased efficiency foretold would be due to the future position of the target being known.

Incidentally, it may also be worth observing that it was contended in the same letter that the movements of the ship would have to be counteracted by the use of gyroscopes:

> Further than this, however, important inventions have been made, on which we have recently effected considerable improvements, for combining the telescope for observation with an arrangement of one or more gyrostats, by means of which it is considered that very much more accurate observations could be taken on board ship than have been customarily effected. These devices we shall be happy to submit and consider as part of the inventions now submitted to the Admiralty.

The accuracy alluded to was, of course, accuracy in observed bearings.

Official Discouragement and Service Encouragement

This letter was sent to the Admiralty in January, 1901, and various interviews and certain correspondence took place, with the result that the writer soon realised that it was perfectly hopeless attempting to go further into the matter, as it seemed to be clear, from the nature of the communications made to him, that the whole of his representations and schemes were treated as if he was suffering from some harmless form of insanity.

The suggestion that the patents should be kept secret and the matter investigated only elicited the reply that the system did not seem to be of value to His Majesty's Service, and that secrecy need not be observed.

For the following two years, therefore, he made no further representations to the Admiralty in the matter. But in April and May, 1903 – a year before the *Venerable* experiments[26] – when he was once more in Malta, he had the opportunity of seeing a gun layers' test carried out in a battleship. A great deal had happened in the years that had intervened between his first contact with naval gunnery, and his second; and he found, amongst gunnery officers, so much interest in his general ideas that, on his return to England, in the month of June, he put in hand the manufacture of the calculating machine (which was an integral part of the original scheme) and set to work on the general development of the other mechanisms that would be necessary.

The Second Attempt

In the summer of 1904 he brought the matter, for a second time, before the Admiralty; and after receiving first a formal refusal to examine or experiment with the System, succeeded in getting a suggestion that the matter should be reopened when the designs were complete.

This summer he wrote a memorandum which was headed 'Memorandum on a proposed system for finding ranges at sea and *ascertaining the speed and course of any vessel in sight*;[27] in which the following passages occur:

> The system consists in the application to a moving base or a moving object of the ordinary principles of surveying, where practically all the operations, except reading of the angles and compass, the charting, and deducing mean range, are automatic and instantaneous.

In the same pamphlet he describes the operation of making a chart, showing how observations will be recorded in series so that the median line will show the exact course and speed of the enemy, and consequently those in control of the charting will be in a position to forecast what the range will be some seconds ahead with extraordinary closeness.

Again later on he says:

> An accurate knowledge of range at any one moment is no help to forecasting an accurate range at a future moment, unless the rate of change is known. To ascertain this rate of change, it is necessary to know the course steered and the speed of both ships. This can be obtained by the proposed system, and apparently in no other way with certainty. Assuming that the actual course and speed of an enemy is recorded in the graphic and easily understood form of a chart, a rapid rate of change instead of a slow rate would introduce no new difficulty in forecasting what the future range will be.

This memorandum was elaborated in a rather fuller pamphlet entitled 'Fire Control and Long-Range Firing';[28] and after the third Admiralty rejection of the scheme in January, 1905, this was issued to a limited number of sympathisers in the Service. In both the memorandum and the pamphlet the vital importance of *ascertaining the speed and course* was emphasised above every other possible consideration.

It will be noted that all the essential objectives of the matured A.C. Plotting System are set out in this memorandum and the letter of January, 1901. The only material advance was not in the analysis of the desiderata for gunnery, but in the mechanical processes for obtaining them with the required ease, speed and certainty.

The Fatal Omission at the First Trial

In the discussions the writer had with various gunnery officers between August, 1904, and the actual commencement, in May, 1905, of the manufacture of the instruments, afterwards tried in the *Jupiter*, he was led to suppose that the movements of ships at sea, under conditions which would be called gunnery conditions, were so considerably less than he had supposed, that it would not be necessary either (*a*) to facilitate the taking of observations by gyroscopic elimination of the movements of the ship, or (*b*) to correct the bearings by that or similar means, in order to make the chart accurate; and it was in consequence of the omission of these important safeguards that the instruments tried in the *Jupiter* failed.

This gyroscopic control was an integral part of the scheme submitted in 1901. It was abandoned in 1904–5, or, rather, postponed, partly because, as stated, he was advised that it was not indispensable, partly because it was obvious that the amounts, if any, that the Admiralty would be willing to spend on the trials would not allow of the prolonged and therefore costly experiments that the development of such a control would require.

It is quite clear now that, if any advantage could be got by ascertaining ranges at distances far beyond the capacity of a one observer instrument, a two observer system could be built under gyroscopic control, which would give practically perfect accuracy to the limit of sight.

The results of the *Jupiter* trials were inconclusive. After the experiments were carried on with the help of gyroscopes both the feasibility and the value of plotting – whether true or virtual, as circumstances made the one or the other preferable – were, the writer thought, conclusively proved – and, what was more important, the experience gained indicated the future line of advance.

The vital lessons of the *Jupiter* experiments were, therefore, the elimination of yaw, the making of all necessary human operations easy, and the widest possible adoption of the automatic principle.

Dates of the Fundamental Principles

The following are the dates at which the main features of the Aim Corrector system were decided upon as necessary to make it a war gear.

(1) That knowledge of speed and course was the only conceivable foundation for naval gunnery.

> The summer of the year 1900, and brought to the notice of Admiralty in January, 1901. Universally accepted by the Service specialists, 1905. Adopted as the gunnery method of the whole Navy (except the Channel Fleet), 1908.

(2) The necessity for facilitating observations and the automatic elimination of yaw by gyroscopic control.

> In December, 1900, and brought to the notice of the Admiralty in January, 1901. Experimentally tested in *Jupiter*, 1906. Successfully shown in *Ariadne* gear, 1907.

(3) The necessity for armoured protection for the entire system.

> In October, 1904, and brought to the notice of the Admiralty in February, 1905.

(4) The necessity for automatic operations to exclude both personal errors, co-ordination errors, and time lag.

>In January, 1906, and brought to the notice of the Admiralty in March, 1906. Incorporated in *Ariadne* gear, 1907.

(5) The necessity for converting the one-observer rangefinder from a difficult instrument to use to an instrument lending itself to continuous results in all weathers.

>In January, 1906, and brought to the notice of the Admiralty in the summer of 1906. Shown to Admiralty representatives June, 1907. Completed November, 1907.

(6) The desirability of making the rest of the fire control processes automatic, 1906–7–8–9; submitted to Admiralty in same years.

That these main points were original to the writer, in that he did not borrow them from anybody else, will probably be conceded. It is, however, obviously possible, that many, or, at any rate, some of them, may have occurred to others concurrently with, or antecedently to him, or may have been common knowledge.

Should the question of the originality of the leading ideas be debated, however, it may be remembered that until 1908, the official method of ascertaining the speed and course of the enemy was to guess it.

A curious confirmation of this is to be found in the contribution of a lively writer in *Blackwood* of February, 1908. The writer is evidently a naval officer.

>Having once found the range, any fool can go on hitting a target. Mathematical instruments do that for him, providing he can *guess* or calculate the speed and course of the enemy...[29]

THE WATKINS SYSTEM

The writer has been told that his idea of a surveying basis for naval gunnery is a plagiarism of the Watkins gear. His ignorance may be inexcusable, but, as a matter of fact, he had never heard of the Watkins' inventions – they are still officially secret – until 1906. He does not now understand how they work, or their exact purpose. That plagiarism has been suggested indicates that plotting is one of the objects aimed at – it is possible two observers are employed. The writer's own account of the genesis of his ideas, quoted from the 1904 Memorandum – viz., that they are an adaptation of the principles of surveying to a moving base and moving object, should be a sufficient answer. He has never, and does not now, attach great credit to those who have good ideas. To have good ideas is no more a virtue than to quote Scripture. The difficulty is to live up to them. If credit is to be claimed, it must be won by perseverance, willingness to make sacrifices, and hard work in making those ideas effective, both by the production of suitable means, and in breaking down the opposition of those who, either from natural conservatism, from want of opportunity or ability to understand, or, as seems sometimes to happen, from sheer perversity, make progress so painful, so dilatory, and so difficult.

The writer would have seen no reason to be less proud of his work, such as it is, if he had borrowed his ideas from Col. Watkins: and had any of the innumerable officers, who must for years have been familiar both with the Watkins' invention and the difficulties of naval gunnery, solved the latter by an adaptation of the former, he would not think the fact of such adaptation would detract from the credit due to a successful solution.

As a matter of simple fact, he had never heard of Colonel Watkins in 1900 – and could not now give an intelligent account of his system, as he does not know more about it than that it must resemble the A.C. System in some particular; but as he is ignorant of the point of resemblance, he cannot throw further light on the matter.

NOTE TO LETTERS OF AUGUST AND SEPTEMBER, 1906

The following two letters explain themselves. It may be noted that, while the effect of the expert advice given to the Board is reflected in Lord Tweedmouth's reply – there had yet been no *independent and authoritative* investigation into the claims of the A.C. System sufficient to be an abiding guide to official policy. The recommendations that inspired Lord Tweedmouth's recognition of the *invaluable* nature of success, were *departmental* – not *general staff* recommendations. The consequence was that, with the retirement of Admiral Jellicoe, a year later, and the advent of a new departmental head, who neither knew, nor wished to know, anything at all about it, the entire case for the system was forgotten.

It may be interesting to observe that in 1906, there was no disagreement between the writer and the Admiralty experts. The disagreement was in reality between the experts and the sceptical members on the Board. It was the chief of these who was the first to congratulate the writer, when the December experiments in the *Ariadne* showed that his promises of what the A.C. gear would do, were more than realised.

The letters show that the writer, quite mistakenly, supposed that a demonstration of the success of the instruments would have set all doubts at rest. In view of the absolute unanimity of the specialists, no sane person could have anticipated the contingency that the gear could succeed and not be understood. Consequently it was not provided in the agreement that the trials should be either fair or thorough. It is clear, therefore, that the investigation suggested by the writer is even more needed to-day than then.

COPY

188, Fleet Street,
London, E.C.,
August 27th, 1906.

Dear LORD TWEEDMOUTH,

I hope the importance of the matter may excuse the unusual step I am taking, in sending you a long letter about the negotiations now pending. My reason for writing is that it seems highly desirable that a statement of the facts should be before you.

May I preface my remarks with the assurance that, while I should very highly value the honour the Admiralty would do me should they associate themselves with my work, I perfectly realise that the matters now in dispute are essentially business matters, and must be decided on principles and considerations appropriate to such affairs. I have the less hesitation in premising this from the fact that no one representing the Board has so far suggested that any other motives ought to weigh. It follows, then, from this that the discussion is purely impersonal, and, whatever the upshot, can give rise to no feeling of grievance or annoyance to one side or the other. The decision in the matter rests with the Admiralty, and if I write fully it is because I am anxious the decision should not be made in ignorance of any essential facts.

The present situation arises out of the following circumstances.

As stated in the official letter of the 21st, I am the inventor of a new system of gunnery, consisting, briefly of semi-automatic means for reducing the difficulties in making hits on a fast moving target at long range from a ship steaming at high speed to the simple proportions of hitting a stationary target with a stationary gun.

I first brought this system to the notice of the Board of the Admiralty in the year 1900, and, during subsequent years, on various occasions, urged various considerations to show that it was worth examination and trial. For some time, however, the importance of hitting at long range at all was not easy to establish, and I was altogether unable to arouse sufficient interest in the subject to make an 'a priori' case in its favour. The only response I received consequently was a repetition of the statement that my system would be of no value to the Service, and that I was at liberty to patent it abroad and develop it in any way that I saw fit.* I held on, however, till 1904, when the great developments in the art of gunnery

---

* The writer has, in all, received seven official intimations that his gunnery inventions were of no value, and need not be kept secret.

within the Service had given rise to new systems of fire control, and made the employment of artillery at long range seem, not only possible, but necessary. An enquiry was made into my system, and finally, in May, 1905, an agreement was entered into by the Admiralty to pay me a sum of £4,500 to supply instruments for a trial, and hold back the taking out of foreign patents until its conclusion. This sum of £4,500 was supposed to consist of £2,000 towards refunding part of what I had spent on developing the system in the previous five years, £1,500 the estimated prime cost of the instruments, and £1,000 for my services in supervising manufacture and assisting at the trials.

The trial took place; but was a partial failure, for the reason that I had designed my instruments for a slower and a more limited yaw than as a fact was found to exist. On the other hand, the trials gave ample evidence – and the Committee's Report, I think, admits this – that, had the problem of yaw been eliminated, the system could have been worked successfully.

I have now invented a method of eliminating yaw; and greatly improved all the weak places in the system, and greatly simplified it by substituting one-observer for two-observer range finding. It is not surprising, therefore, that those familiar with the *Jupiter* experiments are now confident that, if the yaw correcting mechanism works, the new instruments must succeed, and Professor Boys, a very high mechanical authority, is equally confident that the yaw correcting mechanism will work.*[30]

The *Jupiter* trials brought me into close contact with a great many officers of the Fleet – men of the highest eminence as experts in their various branches whom the Admiralty had selected for many of the most important administrative and technical posts. It can readily be believed, therefore, that it was a very great relief to me to find that the confidence I had always reposed in the soundness of my system was amply confirmed by the practically unanimous endorsement of so large and important a body of expert opinion.

If, therefore, now I put forward a statement of the reasons why Aim Correction is of epoch-making importance, I have the satisfaction of knowing that their cogency does not rest upon my authority, but upon that of a set of men best capable of judging in such a matter, of any persons in the world.

If I may paraphrase and connect together the various statements made to me verbally and in writing, I would say that, if it succeeds completely, it is unique in the history of the evolution of armaments because it promises an immediate extension of the effective fire zone, without the substitution of anything in the way of new artillery; that it makes so fundamental a change in the technique of gunnery as to make probable an instant and

---

* The Professor's judgment was proved to be right.

abrupt revolution in men's ideas of the desiderata for ships and the best methods of fighting them; that it is consequently of paramount, and indeed vital, necessity to the Navy, not only because its influence might be decisive in any war in the near future, but because the development it will make inevitable in material and methods must be so radical that it will be of prime financial importance that their control shall remain with this country, and the ruinous expenditure avoided that must follow from the competitive pressure the possession of the system by rival powers would entail.

That all this follows logically from the system succeeding is self-evident; and it is, therefore, not unreasonable, when I am given to understand that the Admiralty wish to treat for the exclusive use of it, that I should assume that the basis for negotiation will be a recognition of its potential value.

It would no doubt be far more satisfactory to me to enter upon this negotiation for a monopoly after the instruments have been proved to work successfully; for that matter it would probably be more satisfactory to the Admiralty to postpone negotiations until after all uncertainty has been removed, for the reason that it would be easier to obtain sanction to comparatively high terms for a system whose value was not in doubt than low terms when success is only hypothetical.

Unfortunately, however, the fact that I have come to the end of the resources that I can devote to this work, and must arrange to get the further capital required elsewhere, makes this postponement very difficult. Not that there is the slightest difficulty in getting the necessary funds, but because the admission of partners would rob me of control.

As the whole *raison d'être* of my negotiations with the Admiralty was that this control should not pass from me except into the Admiralty's hands, I put forward terms which, on the assumption that the system was of overwhelming importance, was common ground, would, I hoped, be recognised as exceedingly moderate.

My terms were £8,000 for bare out-of-pockets and the cost of the trial instruments; and, for the monopoly—

(1) A royalty of £1,000 per ship.
(2) An annual royalty of £200 per ship.
(3) A commission on the cost of the instruments.
(4) A guarantee of twelve ships per annum for five years.

These monopoly terms were proposed as the utmost limit I could get. But I suggested that they should be subject to the revision of independent arbitrators should experience show that this limit was not justified.

An examination of these terms – on the basis of the instruments costing £5,000 per ship, shows that, for the sixty ships guaranteed, a total of £255,000 was payable in the course of some sixteen years – an amount equal, I am told, to a present worth of rather less than £170,000.

Remembering that, unless the system succeeds, nothing at all is paid, the question is: Does the additional efficiency to the offensive gained by the system justify this amount?

The offensive – i.e., the hits a ship can make with its big guns, is the primary reason for building her. If one of the new class of battleships cannot hit with her broadside of eight 12in. guns at a range when the circumstances make hitting important, the two millions each approximately cost is so much money thrown away. Hits then are a measure of value. If the Aim Correcting System should double her hitting efficiency, the value got for the two milllion pounds would be doubled too. If it enables her to hit at ranges and under conditions when getting hits is inconceivable by any other method, then the increase of value got cannot be assessed at all.

This, of course, is an elementary and self-evident truth, and I now proceed to apply it to the matter in hand. Suppose sixty ships to cost on an average £1,250,000 apiece at least, and £75,000 a year to keep up. These are figures well within the mark. The guaranteed ships, therefore, over sixteen years represent a minimum outlay of

£75,000,000 first cost.
£45,000,000 upkeep.

£120,000,000 total.

Assume the system to give a ten per cent increase of efficiency, and there will be £12,000,000 better value. Obviously, if thoroughly successful, there should be a betterment, not of ten per cent, or a hundred per cent, but one that cannot be measured in percentages at all.

But supposing only twelve million pounds better value. Is £255,000 an extravagant amount to pay for this? It is only two per cent of the increased value, leaving aside altogether what it may be worth to possess a military advantage over a possible enemy.

If my reasoning is correct, it is difficult to understand why these terms are considered so preposterous as not to be worth discussion.

At the conference on Thursday, the 9th inst., it was said that the Admiralty had to be guided in this matter by what had been paid in similar circumstances. For instance, I was told that the late Mr. Whitehead had been paid £125,000. But, unless I am greatly mistaken, he had in addition a very profitable manufacturing contract which has subsisted for the last thirty years. The fact, moreover, that the then Board of Admiralty declined to purchase the monopoly of his invention, and so enabled him to make many millions by dealing with all the other nations, seems to rob this instance of any particular bearing on my case in this connection at least, however instructive it may be from another point of view.

Again, it is currently said that a five per cent royalty is paid for the use of the Krupp patents on cemented steel armour. If this is so, something like £750,000 must have been paid by the Royal Navy directly or indirectly to Herr Krupp and his executors in respect of the ships built in the last ten years or now building. Here, again, there is not only no monopoly, but Krupp steel can have no admittedly supreme superiority, because in the same period a great many ships have been built that are protected by Harveyized or nickel steel.[31] Other instances could easily be multiplied, such as the Marconi wireless telegraph system, turbine engines, etc. I allude to them only to show that neither the principle of royalties nor the payment of royalties of a very large amount are without precedent; and, in all the quoted cases, there has been not only no monopoly to the Navy; but, except in the case of the Whitehead torpedo, the thing that has been so liberally paid for is by no means the sole and only way of getting the advantage sought.

Perhaps the fact that there is no parallel to the present negotiations may explain some of the difficulties that have arisen in dealing with it. But, at first sight, the fact that an absolute and exclusive monopoly is being sought would seem to justify a predisposition to offer unprecedently high terms.

The actual offer made to me by the Admiralty is £6,500 for the trial instruments, and £90,000 – to be spread over a series of years for all my rights for all time in the Aim Correcting system.

If we suppose the period fifteen years – the life of the patents – this offer would have an approximate present worth of £63,000. Supposing it to include a repayment to me of what I have tied up in the business, say £5,000, and say a £1,000 (per annum) in respect of my seven years' work, the nett profit to me on the transaction would be £50,000 present value.

What I fail to grasp is any logical connection between the view that I had supposed common ground that my system might prove to be of paramount national importance, and therefore of a potential value hardly to be assessed or calculated and the amount of the offer made. £50,000 is, it seems to me, neither commensurate with the possible military value of the system, nor with its commercial value if independently developed.

Suppose the question of monopoly had never been raised, or the monopoly is not worth acquiring, and it succeeds completely, and I developed the business on normal industrial lines.

[The writer then discussed its commercial possibilities.]

This being the situation, no thoughtful person can for one minute suppose that the offer which the Admiralty has made to me is one that I shall be at all likely to accept. It is quite true that my perhaps exalted view of what my system will achieve makes me exceedingly loth to take any step that would make it impossible for the British Navy to enjoy a monopoly of it. But, when all is said, the Admiralty have to-day just as good reasons for being sure of their ground as I have; and, if it is their deliberate opinion that the offer they have made me is the limit of what the system may be worth, it would be the merest quixotism for me to affect a greater solicitude for the country's interests than their Lordships. The possession of a secret or monopoly worth only £90,000 spread over fifteen years, a matter of some £6,000 a year, cannot be of very vital moment.

It has occurred to me that a possible explanation of the apparent incompatability between the thing offered for sale and the price suggested, may lie in some fundamental misunderstanding. Possibly it may be thought that it would never be necessary to purchase from me whole equipments for fire control, but only certain essential features, such as the yaw correcting mechanism and the charting table. These, it may be supposed, should not cost more than £700 or £800 per ship. Again, it may be thought that the system would only be applicable to very few ships, and, in this way, £6,000 a year might appear a liberal royalty on a small number of instruments of moderate cost.

But, of course, this train of reasoning is based upon a complete ignoring of the pivotal fact of the situation, viz.: − that what I have to sell is not instruments, but a system, the embodiment of certain laws of gunnery which I have been the first to codify. My instruments as at present designed will no doubt soon be superseded by others that embody the essential features of my system more completely. The monopoly of instruments is only incidental. It is the knowledge of the system which they make workable, and the exclusive knowledge of it, to which high, possibly supreme, value attaches.

I have put in two alternative offers in reply to the Admiralty offer. The first, which I intend to apply to all my inventions, is to leave future terms, if necessary, to arbitration, my offer of August 9th being the limit that can be awarded.

The second is embodied in my letter of August 24th.

May I, without impertinence, make a final observation? To secure an option on the monopoly, the Admiralty will have to incur the extremely small liability of £6,500. Not another farthing need be paid until the Admiralty is quite convinced that it will get good value for its money. The alternative is to let the thing go: to risk losing the chance of monopoly once and for all. A secret that is told cannot be recalled.

If it succeeds, and there is ample material for forming a judgment on the probability of this, certain results are very plain. First, the navies that adopt it step at once into a position of eminence in gunnery more markedly superior to ours, than ours is superior to theirs to-day. There will at once be an imperative and instant necessity for the British Navy to have it, too. The Admiralty must either deal with me for the instruments that practical experience has taught me to produce, and buy them at a price that commercial experience has shown to me to be reasonable, or use my system without agreement with me and without the benefit of the improvements which are certain to be made.

I am well aware that the law gives the Crown the power of so dealing with a recalcitrant and unreasonable patentee.[32] But, while this power exists, and has existed for many generations, it is notorious that it has never been exercised. The circumstances connected with my developing this invention and bringing it to the Admiralty's notice, would, I cannot help thinking, give the attempt to use this latent right so arbitrary an aspect as to make it inconceivable that any government would employ it against me.

On the other hand, it is equally inconceivable that a fair market price for my inventions, after they have proved successful and been sold largely abroad, would not imply an ultimate expenditure by the Admiralty very much greater than any acceptance of my terms would involve.

It may be of interest for me to add that since I completed the designs for the system tried in the *Jupiter* till to-day, I have in no way increased or modified my terms, although of course the probability of success is immeasurably higher to-day than it was two years ago.

In conclusion, may I say that I shall be very pleased, if any further explanation from me is wanted, to attend at any time and any place.

With many apologies for trespassing so long upon your attention,

I am,
Your Lordship's obedient Servant,
(Signed) A. H. POLLEN.

TOMICH HOUSE,
BEAULY,
September 3rd, 1906.

Dear MR. POLLEN,
Thanks for yours. I am most interested in your range and course finding experiments, thoroughly appreciate how invaluable a successful result from them would prove, most anxious they should succeed, and that our Navy should reap the full advantage; and you have all my good wishes and sympathy.

But potential values are hard things to assess, and so I and my advisers have to be cautious, though it does seem to me our offer was a liberal one.

Sincerely yours,
TWEEDMOUTH.

NAVY APPROPRIATION ACCOUNT, 1907–8, p. 121

An Admiralty letter of 26th November sets out (par. 4) the short provision, amongst other matters, that falls on Sect. II. of Vote 8, amounting to £290,000, and continues:

> From which an abatement can be made of £50,000 in respect of payments to Mr. Pollen, in connection with his aim-correcting apparatus, which, though provided for, cannot be made this year.

The trials were closed in January – the decision was made in March. Unless therefore it was already settled that the system would not be purchased, there was no ground for saying that the payment *could* not be made within the financial year. It would be interesting to know if in the Estimates for 1908–9, drawn up *before* the Board's decision of March 10th, it was assumed that the verdict passed in November would be confirmed.

NOTE TO THE 1908 CORRESPONDENCE

The correspondence that passed between March 2nd and March 25th, 1908, calls for a word of explanation. As appears from the Admiralty letter of March 10th, the Lords were bound under the agreement to decide by March 14th. On March 1st, no decision having been communicated, the writer called on Lord Tweedmouth, put certain considerations before him, and received the promise of being shown the substance of Admiral Wilson's report. Lord Tweedmouth also undertook to consider the question of further trials. This interview and promise the writer confirmed by the letter of March 2nd.

On March 9th he received the letter dated the 6th, in which the decision to carry out further trials was conveyed. His reply of this date does not, it will be noted, justify the statement that he refused these trials. He saw Lord Tweedmouth for a few minutes on the evening of the 9th, and was invited to call again the next day with any suggestions he wished to make with regard to the trials. But on the 10th Lord Tweedmouth was not able to see him. He accordingly wrote thanking him for his good offices, reminding him of his promises and enclosing the memorandum of the proposed trials he had been asked to bring.

This letter crossed the Admiralty letter of the same date, announcing that the fresh trials would not be held, and saying, first, that the new trials had been ordered out of 'consideration' for the writer, which was nonsense, and that he did not wish for them, which, as is clear from the correspondence, was untrue. The letter further contained the following words in par (4):

> As my Lords ... from your letter to Sir A. Wilson of 24th January understand you *wish* to enter into negotiation with certain Foreign Powers, etc.

The passage alluded to was as follows:

> Certainly, after coming successfully through the Torbay tests for exactness in results, seeing the A.C. gear at work in the roughest weather, and ranging so many moving ships up to 10,000, 11,000, and 12,000 yards, with uniform ease and constant precision, I find it difficult to suppose – until any competing system is tested against mine to the point where one or the other breaks down – that it is valueless to the British Fleet, or negligible in the hands of any other navy.

The fairness of the interpretation is on a par with the statement that the writer's letter of March 9th showed that he did not wish for further trials.

The writer's entire conduct for nine years showed that he had laboured incessantly, and at great loss and risk, to compel the British Navy to preserve monopoly and secrecy. If he had *wished* to negotiate abroad, he might have availed himself of many opportunities. It is significant of his sincerity that the only press allusion to his work that has ever appeared with his name connected with it, was apparently inspired by an anxiety to prove that the *Vengeance* system was superior!*

The writer left England for some days, and on his return saw Lord Tweedmouth again, and wrote the two letters of March 25th, one addressed to the Secretary, and the other addressed individually to each Lord. He did this as it had been intimated to him by Lord Tweedmouth that his references to Admiral Wilson had given great offence. The writer wished to make it clear that no offence was intended, but that as Admiral Wilson had been completely deceived, he felt bound to inform the Board of what was a demonstrable truth.

In the fourth paragraph of the letter of March 10th to Lord Tweedmouth, the writer meant that had the trials been thorough and *proved* that the gear was either useless or unnecessary to the Service, he would not have minded its being rejected. He could then have developed the thing commercially or not, as he thought fit.

---

* NOTE. – It has been objected that the writer behaved with both want of patriotism and impropriety, in suggesting, in a recent communication to the Board, that it might be dangerous to allow the A.C. System to go abroad without further trial. He fails to see how, in discussing a course in which he had the offer (for a consideration) of official assistance, his patriotism is in any way compromised; or, that in doing so, he can be held to be employing a threat. If there had been no other way of securing monopoly to the British Service, he holds that almost any course would have been justified that in fact would prevent what, in his optimism, he would consider a national disaster. Fortunately, quite different considerations have produced the desired result, while the letter quoted absolves the writer of any imputation of unworthy conduct.

COPY

188, Fleet Street,
London, E.C.,
March 2nd, 1908.

My Dear LORD TWEEDMOUTH,
I beg to thank you for your prompt accession to my request that so much of Admiral Wilson's report as deals with matters of fact in connection with the A.C. gear should be communicated to me, so that any explanations I may be able to offer should be considered by the Board before a decision is come to.

In brief, my reasons for asking for this are as follows:

(i) From a public point of view, it seems desirable that the Board should not take the risk of deciding, in so grave a matter, on grounds that examination may show to be mistaken.

It is to be remembered that Admiral Wilson had no mechanical or gunnery assessor in the *Ariadne*, or was instructed before the trials, or presumably assisted after them by anyone familiar either with the history of the invention or the working of the gear. On certain mechanical questions he seemed not to have given himself time to understand either the principle or actual working of the instruments.

It is fully to be expected, therefore, that, in giving reasons for his adverse verdict on the system, he will have put forward condemnatory statements – both about the actual machinery and the theory of its employment. These statements should certainly be examined by someone closely acquainted by experience with the subject. Mr Isherwood and I are, in the nature of the case, the only persons who have had this experience.

(ii) My second reason is that, on January 4th, I received a letter from the Admiralty, which said (as a reason for refusing to make me an advance, secured on my patents, to enable me to finance my losses on the experiments) that:

The Admiralty may decide not to adopt the system, in which case . . . the value of your patents . . . might be seriously affected. . . .

If this opinion is well founded, it is clear that an adverse decision, if come to on insufficient grounds, would inflict a very grave injustice on me.

I submit that, in view of the fact that I have for some years, at great personal loss to myself, risked the possibility of protecting my inventions abroad, in order to enhance the probability of the British Navy enjoying a monopoly, and have throughout disclosed all my inventions in perfect good faith, that for the Board to condemn my system unheard, would be without precedent.

I am strengthened in this view by the straightforward and honourable way in which their Lordships treated me on the occasion of the *Jupiter* trials. In that case I was in communication with Admiral Parr and the Committee throughout. I attended the meeting where the criticisms – which were to be the foundation of the report – were tabled and discussed. I was put in possession of the case against me, and asked to reply to the objections made, both verbally and in writing. Finally, the material portions of the report, when made, were sent to me by the Admiralty, in case, on reconsideration, I wished further to criticise the verdict, before it was acted on by the Board.

I am sure, therefore, that my request has only to be understood for it to be considered reasonable.

May I take this opportunity of drawing your attention to two enclosures?

(i) A memorandum on the origin of the A.C. system – a short epitome of the history of my connection with the gunnery problem, your own, and your colleagues, perusal of which would be most gratifying to me, in view of Admiral Wilson having informed me that I was mistaken in supposing myself to be the pioneer in the advocacy of plotting to get gunnery data; and

(ii) Extracts from a letter to Capt. Bacon, D.S.O., R.N., being an argument in favour of exact instruments being preferable to mental, vocal, and manual operations for the making of intricate calculations, the only value of which is their accuracy in the most trying circumstances.[33]

A. H. POLLEN.

COPY
C.P.G. 1699/7578.
CONFIDENTIAL

ADMIRALTY,
6th March, 1908.

SIR,
With reference to Admiralty letter of 14th February last, C.P. 7519, I am commanded by my Lords Commissioners of the Admiralty to inform you that they now propose to carry out further trials with your Aim Correction Apparatus, in accordance with the provisions of the agreement dated 18th February, 1908, before arriving at a decision as to its adoption or non-adoption in the Service.

2. My Lords presume that you will wish to be present at the further trials, in accordance with the last paragraph of clause 2 of the agreement, and they will arrange that you shall be informed at the earliest possible date of the time and place fixed for the trials.

(Signed) C.I. THOMAS.

A.H. POLLEN, ESQ.,
    188, Fleet Street, E.C.

COPY
CONFIDENTIAL

188, FLEET STREET,
LONDON, E.C.,
March 9th, 1908.

SECRETARY,
  THE ADMIRALTY,
    WHITEHALL, S.W.
    C.P.G. 1699/7578.

SIR,
Your letter of the 6th March as above duly received on Saturday morning.

(1) I note their Lordships now propose to carry out further trials of my apparatus. As the apparatus has been left untended in places, in the

Ariadne, of considerable exposure for some seven weeks, it would, I suggest, be desirable that someone acquainted with the machinery should proceed forthwith to get it into proper working order, clean off rust, etc. If their Lordships desire that I should undertake this work, my partner, Mr Harold Isherwood, will be prepared, at a moment's notice, to take my engineers to the ship and proceed at once with the work. Perhaps you will instruct me on this point as soon as their Lordships' pleasure is known.

(2) I shall of course be much interested in any trials that their Lordships may carry out, but I regret that my business engagements put it out of the question my being able to attend continuously. No doubt, before the question of my attending actually arises, I shall be informed more fully what trials are proposed, and how it is suggested they should be conducted.

(3) Will you be good enough to make clear to me your reference in par. 1 of your letter 'in accordance with the provisions of the agreement dated 18th February, 1908'? At the present moment it is not clear to me what you mean by it.

(Signed) A. H. POLLEN.

COPY

188, FLEET STREET,
LONDON, E.C.,
March, 10th, 1908.

Dear LORD TWEEDMOUTH,

I am sorry you were not able to see me to-day. I shall not be in town to-morrow, and therefore I take the liberty of sending to you the memorandum with regard to the forthcoming trials, which I was going to submit to you at the interview.

I have no doubt that you will not have overlooked your promise to have the material portions of Admiral Wilson's report sent to me, and that I shall have an opportunity of seeing them before the form of the new trials is decided upon.

I hope you will not mind my saying how gratified I am at your having grasped the situation so resolutely. I think that, without such a masterful cutting of the Gordian knot, a position ultimately exceedingly painful to everybody must inevitably have developed.

I should not in the least degree mind the Admiralty rejecting my system, but I can hardly be blamed if I resented it being rejected on the absolute parody of a trial which has taken place, and in face of what I know to be the unanimous opinion of every practical Service man who has heard an account of what we have done.

Apart from everything else of course, the rejection would be a particularly cruel libel on my invention; and, as I have worked at a loss for many years over this thing, and risked the taking out of patents, to help the Service, this would have been a miserable sort of return. I am so glad that your strong hand has saved us all from so disagreeable a situation.

(Signed) A. H. POLLEN.

MEMORANDUM

(1) The trials of the A.C. system should have been in the nature of a judicial proceeding, or a scientific investigation. This implies:

   (1) An impartial head.
   (2) Competent expert technical, tactical, and mechanical assessors.
   (3) Thorough comprehension of the intention and object of the gear, and the policy of the Board of Admiralty.
   (4) Exhaustive experiments to vertify or disprove the claims made for the system.
   (5) A thorough consideration of the tactical and other developments that must follow upon success.
   (6) An equally thorough consideration of such improvements as obviously would improve the results or facilitate their use.

(2) The person whose gear was on trial should have been heard with regard to each experiment, and any experiments he asked for should have been made.

(3) The report of the Commission trying his system should have been submitted to the Admiralty, together with the inventor's comments on all finding of and inferences from matters of fact.

The above was the procedure followed in the *Jupiter* case.

COPY
CONFIDENTIAL
C.P. 8383/8026

ADMIRALTY,
10th March, 1908.

SIR,
With reference to your letter of 9th March, I am commanded by my Lords Commissioners of the Admiralty to acquaint you that the further trials proposed in Admiralty letter of the 6th March, C.P.G. 1699/7578, were suggested out of consideration for yourself. By your letter of the 9th March, you do not appear to wish these further trials to take place. In the absence of such further trials, the Agreement between yourself and the Admiralty will expire on the 14th instant.

(2) My Lords have accordingly determined to decide the question on the basis of the trials that have already taken place.

(3) After a full and careful consideration, my Lords have decided not to exercise the option, open to them under the conditions of the agreement dated 18th February, 1908, of acquiriing the sole rights of the 'Aim Corrector System.' The said agreement will therefore become inoperative, and my Lords will take steps to re-assign the secret patents to you at such a date as you may wish (except any which, under the provisions of the agreement relating to Tactical Machines entered into in 1907, are to remain assigned to the Crown and sealed as secret for 3 years if so desired by the Admiralty) so that you will be at liberty to make any use of them that you may desire.

(4) As my Lords do not intend to make further use of the instruments now installed on board H.M.S. *Ariadne*, and as from your letter to Sir A. Wilson of the 24th January they understand that you wish to enter into negotiations with certain Foreign Powers, they would be glad to know if you are desirous of re-purchasing the instruments for use in connection with those negotiations.

(5) In conclusion, I am commanded to convey to you an expression of their Lordships' thanks for having in the first instance offered the exclusive rights of these patents to H.M. Government.

C. I. THOMAS.

A. H. POLLEN, ESQ.

COPY OF LETTER SENT TO INDIVIDUAL MEMBERS OF
BOARD OF ADMIRALTY, 25th MARCH, 1908

SIR,

I enclose you, for your personal perusal, a copy of a letter addressed to-day to the Secretary.

I am writing this, after consultation with many of my Service friends, because it seems certain that the Board have been led to their recent decision by a complete, though undoubtedly sincere, misrepresentation of the details and purpose of my system; and that this misrepresentation itself has been due to the underlying principle of my gear neither having been understood, nor consequently tested.

I share with my naval friends the conviction that it is only necessary for members of the Board to know the facts set out in the letter for the final rejection of my system on demonstrably mistaken grounds to be made impossible.

I go further, and say for myself that as far as terms are concerned, I should prefer to leave myself in the Board's hands – without reservation, unless it were for reference to an impartial tribunal – so long as I am sure that they genuinely know what I had done, and the completeness with which I had succeeded.

The *Ariadne* instruments, I may say, were built to solve a portion of the entire problem only, viz., the making of a plot automatically in any sea or action conditions. The short experience we have had has shown that great and far-reaching improvements are possible.

Further, the issue under the expired agreement was whether you should buy the system outright for £100,000 or let it go.

It seems, then, that if I am willing to have the question of terms impartially decided, and able to submit material improvements, it is possible for the Board to re-investigate the question, without necessarily animadverting on the character or adequacy of the recent trials, or even reconsidering their decision, which was limited to not purchasing for £100,000.

I am most anxious that no member of the Board should think that I wish in any way to reflect on Sir Arthur Wilson. My close connection with the Service for the last eight years, has given me an intimate knowledge of the respect and veneration in which that most distinguished officer is held throughout the Navy – and these feelings my short relations with him have made me share to the full. That he was not informed of the principles I was working on, and consequently mistook the purpose of my invention, is undoubtedly to be regretted – but to point this out does not involve any arrogation of the right to criticise him, in a matter in which he could have had no motive but to act for the best interests of the Service, of which he is perhaps the greatest living ornament.

My eight years work on this problem has not only brought me neither profit nor recognition, but has landed me in heavy loss; but I am more anxious to-day that the Navy should benefit than that I should be rewarded; and I am confident that if the Board see their way to appoint an impartial Committee to examine into the question and report to them, any action this report leads them to adopt, will be both the right one for the public service and just to me.

(Signed)   A. H. POLLEN.

COPY

188, FLEET STREET,
LONDON, E.C.,
March 25th, 1908.

SIR,
I beg to acknowledge yours C.P. 8383/8026, and note that their Lordships decided on Tuesday, March 10th, not to hold the trials ordered on the previous Friday, nor to exercise their option under the agreement with me.

Yours of March 6th, saying the Lords had ordered further trials, did not inform me that they were ordered out of consideration for me; nor did mine of March 9th say, nor can be supposed to mean, that I did not desire them.

My whole contention, since protesting in January, against the character of the Torbay tests, has been that there have so far been no trials at all. That I have refused further trials, is utterly unfounded.

* I understand their Lordships have decided — as a substitute for the guess-work that until now has been the Service method of getting the enemy's speed and course — to equip the British fleet with a plotting gear to carry into effect the A.C. theory of gunnery, and that this gear is virtually the hand-worked imitation of my system, thoroughly discussed, and discarded as useless, by the Naval Ordnance Department in 1906; thus throwing open to the world the armoured, automatic, all-weather system, so largely developed and perfected by public money, and the opportunities successive Boards of Admiralty have afforded me.

In these most unexpected circumstances, I deem it my duty to recall

---

* The writer did not learn till months afterwards that it was the time and range method that had been resuscitated as a substitute for his system. He would probably have expressed himself more strongly had he known at the time of writing. It was only after the complete failure of this system, that the course commented on was adopted. *It is to-day that the British fleet is being equipped with the gear discarded in 1906.*[34]

certain salient facts to their Lordships' attention. I can produce proof of anything herein that may be disputed.

In the year 1900 I discovered the truth that naval gunnery should be founded on plotting the course of the enemy, and in 1901 and 1906 I invented two schemes for carrying this theory into effect. The first was invented when there was no rangefinder accurate at long distances, and therefore included a rangefinding system. The second, when the accuracy of the 9ft. Barr and Stroud was well established, was limited to machinery that converted this instrument into an automatic, all-weather, armoured plotting device.

The truth of my fundamental theory was acknowledged in 1905, and is now so thoroughly established that many do not know that I discovered it.

The war requirements of the system were gradually evolved in the years 1904, 1905, and 1906.

These requirements necessitate that the plotting system shall –

(a) Be workable with accuracy and rapidity in all weather conditions – i.e., the ship's movements must be neutralised, however great or rapid.
(b) Give perfect results when ranges change with the rapidity that may be expected with modern fast ships (that is, at rates greatly exceeding 1,000 yards a minute) – i.e., must be instantaneous in its action.
(c) Reduce the personnel to a minimum, so as to lessen the chances of error in the strain and confusion of action – i.e., must be automatic.
(d) Be so designed that the whole system can be completely protected – i.e., must be armoured.

In 1905, 1906, and 1907, successive Boards of Admiralty, acting on the advice of the Directors, Captains, Staffs, etc., of the following departments and establishments, viz.:

(i) Department of Naval Ordnance,
    D.N.O. and Staff.
(ii) H.M.S. *Excellent*:
    Captain,
    Experimental Staff,
    Senior Staff.
(iii) The War Course College, Portsmouth,
    Captain and Assistants.
(iv) Department of Naval Intelligence,
    D.N.I. and Staff.
(v) The Navigation School.
(vi) The Inspector of Target Practice.

adopted and consistently pursued the unprecedented policy of assisting me to develop this system, on the ground that its successful production would revolutionise naval war, and must be monopolised in the national interest.

This system has now been successfully produced, and fulfils the four conditions set out in paragraph 5.

I am informed that the reason why the Board has now rejected it, is that they are persuaded:

(a) That a plotting system has been produced superior to mine; and
(b) That grave mechanical defects exist in my system.

*I understand that the plotting system in question consists in taking ranges by a Barr and Stroud, with an improved mounting worked by four persons: taking bearings by the standard compass, and transmitting both ranges and bearings vocally to a charting centre where they are plotted. I suppose the system to employ 9 people to make the chart only.

This system, except for the improved mounting of the Barr and Stroud, was considered by the Naval Ordnance Department in 1906, and rejected as not complying with the war conditions set out in paragraph 5.

As the Torbay trials were not carried out to ascertain if either system complied with these conditions, it seems clear that those responsible did not know that it was with the sole object of complying with them that the A.C. instruments were built.

Apparently it was thought I had contended, and the body of specialists referred to had actually believed, that plotting could, in any circumstances, only be done with my instruments; and, consequently, instead of testing my system for its professed purposes, they devoted their energies to saving the purchase price, by producing a rival system that could make an acceptable plot in a flat calm, when there was a trifling change of range or none at all.

The trials were stopped the moment this was proved; but, as this was always self-evident, no proof was needed. Everyone knew in 1906 that in smooth weather and small change of range, there were several ways of plotting with ordinary Service instruments.

**The system superior to mine, therefore, is the self-evident method, officially rejected as unseaworthy and unbattleworthy in 1906.

This summary abandonment, of all the objects that were the sole justification of the Board's policy in the last three years, can possibly be explained by the mechanical objections being thought serious and insuperable.

*See note to par. (2).
**It was even worse!

I understand that among the defects alleged are that:
(*a*) A change of course throws the instruments out of gear. An operator, it seems to be supposed, has to stop the gyroscopes when this takes place, and release them again when the ship is on her new course.
(*b*) The gyroscopes creep, and so must cause an error in the gunnery data given.
(*c*) The system requires a large personnel to work it – 4 or 5 at least.
(*d*) It covers too small an arc – 180 degrees only.
(*e*) The instruments were shown to be unreliable.

That these criticisms are due to a fundamental misapprehension, both of the functions and working of the machinery, can fortunately be demonstrated without any further trials at all. I am prepared to prove to the satisfaction of any consulting expert their Lordships may appoint – with very little more evidence than is available at the Admiralty today –

(*i*) That the gyroscopes do not need the constant attention of an operator.
(*ii*) That the instruments are not thrown out of gear by a change of course.
(*iii*) That the gyroscopes, after such a change of course, automatically select and keep the new course.
(*iv*) That in this operation no error in the gunnery data can possibly arise.
(*v*) That the gyroscopes do not and cannot creep so as to cause any error in the gunnery data.
(*vi*) That the gear, from the beginning till the end of both series of trials, never once broke down, or failed in a single particular.

The instruments in the *Ariadne* were designed for making the plot only; they actually made this as rapidly as observations were made. We left the designing of rapid means of reading it until the trial showed us the best way. This way is now clear. I will therefore further demonstrate:

(*vii*) That a simple modification will make the reading of the results instantaneous, so that, in any change of speed or course by the enemy, the rate and direction of his movement will be reported within a minute, and checked from minute to minute.
(*viii*) That the system does not need more than one person to make the chart.
(*ix*) That, with the aforesaid modification, it can never need more than one other person to translate the chart into gunnery data, and to

give this result verbally or otherwise, to whatever organisation or machinery is employed to transmit it to the guns.

(x) That it is just as easy to make the charting table to work over 270 or 360 degrees, as over 180 degrees.

(xi) That my system is more rapid in getting results, more reliable and accurate in all conditions, and employs fewer people than any man-worked system; that it can be completely protected by armour at small expense; and is exactly what it was represented it would be in 1906.

The above are immediately demonstrable matters of fact, and I submit that if their Lordships have been influenced in their decision by such mechanical objections, they can satisfy themselves, beyond the possibility of mistake, that these objections simply do not exist. They will then be in a position to decide whether or not official trials are desirable, to see whether my gear has or has not fulfilled all the objects of the policy they followed from 1905 to 1907, on the unanimous recommendation of their expert advisers.

(Signed) A. H. POLLEN.

To the Secretary,
The Admiralty.

EXTRACT FROM A LETTER FROM THE ADMIRALTY TO
MR. A. H. POLLEN, DATED 23rd APRIL, 1908

'I am commanded by my Lords Commissioners of the Admiralty to inform you that they have given full consideration to your letter of 25th March last, and subsequently correspondence relative to your Aim Corrector System, and that, while not desiring to make further trial or use of your system in H.M. Navy, they recognise that your prolonged labours in the elaboration of your invention have been of material value to H.M. Service.'

The arrangement finally concluded was that:

(i) £11,500 was paid as a full and final settlement.
(ii) The patents were to be kept secret for 18 months.
(iii) Mr. Pollen was bound to secrecy for the same period.
(iv) The Admiralty undertook not to work, without agreement, any of the secret patents.

XIV

MEMORANDA AND INSTRUCTIONS INTRODUCTORY TO THE USE OF POLLEN'S TACTICAL INSTRUMENT
(*MAY 1909*)

In 1907 or 1908, Pollen had delivered twelve of his tactical machines to the Royal Naval War College at Portsmouth, where they were intended for instructional use.[1] The teaching staff complained, however, of 'mechanical defects, and inaccuracies in the machines'. The accepted tactical doctrine of the day, moreover, called for courses that were parallel and speeds that were equal to that of the enemy, which meant that little change in the relative position of the two fleets was expected.[2] The War College instructors thus could see no point in using a machine that was designed to study the ways in which the relative positions of two engaged fleets might change over time. Upon hearing of these difficulties, Pollen wrote to Rear-Admiral Lewis Bayly, the President of the Royal Naval War College, on 20 January 1909 to explain the purpose of his tactical machine. Pollen then revised and expanded his letter and had the text printed in May 1909 with the title 'Memoranda and Instructions introductory to the use of Pollen's Tactical Instrument'.[3] (The 'Instructions' have been omitted from this volume.) This paper was probably circulated at the Royal Naval War College and among a few of Pollen's correspondents.[4]

It appears that there has been considerable difficulty in realising exactly what the utility of the Tactical Instrument may be. The general line of criticism appears to be that the machine is of no very great practical value, for the reason that all it can do can be done, and with sufficient rapidity, with pencils, suitable rulers, and compasses; and that in one set of circumstances – viz., where there is either no, or a very small, relative movement of the opposing fleets, the machine, being unable to portray it, is altogether useless.

To explain the intention of the machine is the best answer to these criticisms; and it is first necessary to state that it is part of a system of instruments, designed to carry into effect a certain theory of fighting for ships of war. The complete system has been known as 'The Pollen A.C. Battle System'.

To make the Tactical Instrument the first subject for discussion is, in point of fact, to discuss the means of dealing with the finished product of the system — a difficult task, if its nature and object are not first understood.

The Tactical Instrument might be rhetorically described as a map and guide to a new country; the A.C. System, as means for getting into that country. If there is no means for getting into the new country, the maps of it are not of more than academic interest, and certainly cannot be useful elsewhere.

Carrying on this metaphor, the new country, proposed to be made accessible by the battle system, is that in which sustained hitting can be ensured, in any conditions of change of range that can be set up by the fastest fighting ships.

The A.C. Battle System has grown out of an analysis of the processes required for establishing and continuously maintaining the right vertical and horizontal angular relationship between the line of sight and the axis of the bore, so that correct aiming and gunlaying shall always result in a hit as far as the accuracy of the gun permits. It consists in a series of mechanical contrivances designed to synthesize these processes in all conditions of weather, and is based on means for getting exact knowledge of the position, speed, and course of the enemy, irrespective of his speed.

Theoretically, therefore, it neutralises the difficulties arising from *movement* altogether, and so should tend to reduce the problem of securing hits, to the proportions of hitting a stationary target with a stationary gun, no matter what the speed or course of that target or of the ship carrying the gun may be. . . .

The history of tactics on land and sea is a history of weapons, the means of using them, and the methods of disposing and moving the forces that carry them — *e.g.*, the missile tactics of the bow differ from the shock tactics of the lance, and the tactics of the gun differ from the tactics of the ram. Similarly, the tactics of mounted infantry are distinct from the tactics of pure infantry, although they carry identical arms; and the tactics of the universally dirigible, constant speed steamer differ from those of the sailing ship, with its limited choice of courses, and speed dependent on the strength and direction of the wind.

It seems to be an interesting fact, and illustrative of the argument, that almost the first tactical development resulting from the adaptation of steam to ships, was to regard the new command of mobility as constituting the ship itself a weapon of precision. The cult of the ram died hard, and only as the dirigibility of the gun came gradually to be realised; and it is curious, nevertheless, that an eminent naval protagonist of modern gunnery should, as late as 1903, have held that ramming was still one of the contingencies of battle. This shows that any belief at all in the effectiveness of gun-fire, even at the short range that makes ramming an unthinkable absurdity as a tactical resource, is of very recent growth; and this want of faith can in turn be traced to the consciousness that naval gunnery, at that date, was not based upon any analysis that made the promise of future progress at all assured. For that matter, although the analysis is now more generally understood, its corollaries are not yet accepted.

For the tactical ideas of those who command fleets and ships at the present day, general reference must be made to conversations with such Admirals and Captains as have discussed this matter with the writer, and to such publications in English, German, and French as have printed papers on naval tactics since the very recent development of long range firing. This extension of range has not affected tactics, except to increase the probable initial range. It has introduced, that is, no new principle.

The broad conclusion from these sources of information is that the tactics of the day are the tactics of identical direction and parallel courses.

Given two fleets approaching each other with the intention of fighting, each would try to get into a position approximately parallel to and in the same direction as his enemy; and the possession of superior speed would only be used to draw ahead of the enemy's line, with a view to securing a larger number of guns bearing at a given range, and forcing the head of the line round or cutting across it, and thus throwing him in confusion. It has been well argued that this and the 'choice of range' – the only pretended tactical advantages of speed – cannot be effective against a vigilant opponent. Whatever the merits of this controversy, however, it may be admitted that the present object of ideal tactical handling is to procure and maintain an almost constant range.

It was round this conception of sea fighting that the whole controversy on the tactical value of speed centered when the Admiralty policy, with regard to the *Indomitable* class and the *Dreadnought*, was under discussion a year or two ago; and this formation has been taken as typical, and constant range as the ideal, because undoubtedly the present technique of artillery does not permit of guns being used, with effect, in conditions where ascertaining change of range, and getting it effectively on to the sights, are more exacting than they were, say, in the recent battle practice.

During the past year, the first experiment has been made in the British Navy in engaging a moving target; and it is understood that the greatest change of range set up in any one minute in this practice has been about 175 yards.

It is said that the number of hits does not exceed, if it equals, that made with the moored target, when the change of range, though easier to ascertain, was in reality greater. If this is so, it would appear that an average of not more than 17 or 18 per cent of hits is got even in conditions that are so little removed from those set out above. Those, therefore, who have accepted parallel course and identical direction as the inevitable formation for effective fighting to-day, have apparently excellent reasons for so doing.

But with guns carried at a speed of 25 knots (or 833 yards per minute), and with a target moving at the same speed, a change of range of 1,666 yards per minute might be set up. The extreme conditions now experimented with, and the extreme possible conditions, are therefore roughly as 1 to 10. The difficulties in obtaining hits to-day, which consist largely in the fact that finding the enemy's speed and course, and converting that knowledge into correctly-set sights, has to be done by polyanthropic methods, have been held to vary as the square of the velocity of the change of range. It is not necessary to accept this formula, however, to realise that the present system cannot afford a final solution of the problem of control.

If the exigencies of gunnery technique have limited the tactics of fighting to these dispositions, any extension of that technique must similarly lead to an extension of choice in the selection both of formation and of course in any given circumstances; and the logical outcome of this principle is that if the new technique extends hitting efficiency to the limits of mobility – another way of saying that if a gunnery system is produced which will ascertain the change of range in any circumstances up to, say, 1,666 yards per minute, and keep that change of range correctly on the sights, then the tactical developments that must follow upon this will be at least as revolutionary as that gunnery system itself.

In these circumstances, the ideal conditions for the tactical employment of artillery at sea, will not be the maintenance of a nearly constant range, but the setting up of the most rapidly varying range that other considerations may make permissible, in the hope that, as the efficiency of the tactician's own fire will not be thereby diminished, the enemy may be surprised and embarrassed by a manœuvre for which he is unprepared.

And as the command of the highest conceivable speed must contribute to making it easier to increase the rate of change, it does not need further demonstration to show that the possession of such speed will be a tactical advantage of the highest order, although possibly for this reason only.

Just as change of range is the crux of naval gunnery, so is a right estimate and forecast of the enemy's formation and manœuvres the crux of tactics. But this is only to state the same thing twice over, once in the parlance of the artillerist, and again in the language of the tactician. The essential problem is knowledge of the enemy's movements, correctly obtained and stated or used in a form effective for each of the two purposes to the serving of which it is necessary.

The A.C. System and the Tactical Instrument aim at solving the movement difficulty by eliminating it altogether – *i.e.*, in the gunnery system by reducing all problems, as far as human agents are concerned, to stationary proportions, and in the tactical instruments by exhibiting all situations as geometric figures.

The Tactical Instrument is, therefore, a means for showing, with the very great rapidity necessary if it is to be of any use, the virtual future course of an enemy when the factors that decide that course are known, and to afford to those who, in the expected conditions, will be engaged in the tactical handling of fleets, an ever-ready forecast of the position relationship which is being set up by the courses and speeds that are partly under their control.

With only 175 yards change of range per minute as the maximum to be expected; the alteration of relative position that can be set up in eight minutes cannot be very great, nor, for tactical purposes, very difficult to estimate, and therefore a means of forecasting it will not be very important.

On the other hand, if there is to be a change of range that may run to well over 1,000 yards a minute, the change in relative position in a very short time will be enormous; and means for knowing it beforehand, and of knowing instantly and exactly how it can be modified by each change of own course, will be a vital necessity.

In these circumstances, if reference may be made to the criticisms quoted, no work with ruler and compass would be of the faintest use; just as, with a constant range, no guide to future position could be of the slightest value.

If, when the new artillery technique is an achieved fact, this instrument is to be a necessary adjunct in the management of fleets in action, it will also afford, perhaps, the best means of anticipating, working out, and studying these conditions, as a preparation for war.

It would appear as if there were two noticeable possibilities which may make exhaustive investigations into the potentialities of the new tactical order of particular importance and interest.

The first is the inevitable development of attack from two quarters – *i.e.*, either the case of two or more ships engaging a single vessel, or the use in conjunction, with the single object of destroying the enemy, of two squadrons, not necessarily working together in one line – say, a fast squadron of armoured cruisers and a slow squadron of battleships – such as has already been discussed by certain foreign writers.

The suggestions for this attack have not, so far, been based upon the supposition that any extreme change of range will be set up; but rather with the idea of the fast squadron taking up successive positions well ahead of the enemy's line, with a view to raking him with an end-on fire.

A gunnery technique that permits of an effective attack, by a high speed squadron going across the line of the enemy's course, whether by a traverse of the van, or, as would seem preferable, of the rear, must, however, greatly multiply the possibilities of such combinations; and consequently make the rehearsel and preparation of such manoeuvres of imperative urgency.

The second is the natural development of the torpedo. It is stated that the latest torpedo can already operate at a range of 3,000 or 4,000 yards further than the present type, and at a much higher speed. Whatever the effective limits of this new weapon may be – 6,500, 7,000, or 7,500 yards – it would appear as if this distance would always represent the minimum range to be maintained in action by any considerable squadron in line ahead, in conditions when the enemy can forecast their approximate positions at the end of the time of travel of the torpedo. Clearly some guide that would give ample warning, several minutes ahead, that ships are approaching this minimum, would be desirable for this reason only.

But, further, this increase in range, speed, and power of the torpedo may, not improbably, make the employment of large squadrons in a continuous line exceedingly disadvantageous, for the reason that such a formation would present too favourable a mark in any circumstances.

Once more, then, the development of weapons will have modified the art of tactics; and should the divided attack already alluded to be made imperative for this reason, and thus the further subdivision of naval forces become the rule, the necessity for a guide to forecast the future position-relations of these separate squadrons, both with regard to each other and to the enemy, will be greatly emphasised.

XV
[Pollen] TO REAR-ADMIRAL THE HON. STANLEY C. J. COLVILLE, CVO, CB
(*JULY 1910*)

Trials of Pollen's gyroscopically stabilised range finder and bearing indicator mounting and automatic true-course plotter, which had been installed in the armoured cruiser *Natal*, began in October 1909 under the supervision of Captain Frederick Ogilvy, a close associate of Vice-Admiral Sir Percy Scott and one of the most respected gunnery experts in the Royal Navy. Ogilvy was greatly impressed with the performance of both instruments. He was convinced, however, that in battle it would be necessary to be able to plot the position of the target while one's own ship was under helm – that is, while one's own ship was turning. Pollen's plotter patent of 1904 had contained a description of a means of providing such a capability,[1] but he had subsequently been discouraged from developing the idea by the service insistence upon economy and mechanical simplicity. Ogilvy's insistence upon the importance of 'helm-free' performance, however, induced Pollen to have his engineers design modifications to the existing plotting instrument that would allow plotting to be carried out during the turning movement of one's own ship.[2]

While the design work was in progress, trials of the instruments in the *Natal* continued. On 10 November 1909, however, Pollen and his engineers were ordered off the *Natal* at short notice in accordance with orders issued from the Admiralty. The Admiralty justified this action on the grounds that the presence of non-naval personnel aboard the *Natal* during the trials that were planned for that day would jeopardise service gunnery secrets, but as a result of the absence of Pollen's engineers the equipment malfunctioned in such a manner as to be rendered inoperative for several days.[3] In response to this incident, Pollen obtained an interview with Rear-Admiral Reginald Bacon on 13 November, at which time he discussed 'the possibility of putting our relations with the Admiralty on a basis at once less likely to be humiliating to ourselves and better calculated to favour the interests of His Majesty's Service.'[4] Bacon, surprisingly in light of his past actions, promised greater co-operation, agreed to the necessity of adopting Pollen's system, and stated that the only question remaining was the matter of financial terms.[5] At a meeting of 10 December between Pollen and representatives of the Admiralty, tentative agreements with respect to co-operation, the purchase of the gyroscopically stabilised range finder and bearing indicator mounting, and the trials and possible purchase of the modified plotting instrument were reached.[6]

Pollen's improved relations with the Admiralty were, however, to be short-lived. By a cruel twist of fate, a gift of oysters that he had presented to the officers of the *Natal* to celebrate the success of his instruments in the trials proved to be contaminated with typhoid.[7] Ogilvy was taken seriously ill and died on 18 December 1909, a loss that deprived Pollen of his strongest and most influential advocate.[8] Worse was soon to follow. On 25 January 1910, Admiral of the Fleet Sir John Fisher was replaced as First Sea Lord by Admiral of the Fleet Sir Arthur Wilson, who remained an inveterate opponent of Pollen and his system. The adverse effect of the Wilson succession was further compounded by the fact that the just-converted Bacon had retired from the service in December 1909 and was replaced as Director of Naval Ordnance by Captain Archibald Moore, who proved himself to be as obstructive to Pollen as his predecessor had been.

On 18 January 1910, Pollen was informed that no order for the gyroscopically stabilised range finder and bearing indicator mounting would be forthcoming, as he had been led to expect at the meeting of 10 December, until after further trials.[9] Although these trials, held in March 1910 in the *Natal*, were successful, disagreement over terms resulted in an Admiralty refusal on 11 April 1910 to place an order.[10] After further negotiations, the Admiralty on 29 April offered to purchase forty-five mountings from the Argo Company at a higher price than it was willing to pay previously, but Pollen in return was asked to void the contract that had been signed in February 1908.[11] The Admiralty order was smaller than Pollen had hoped, and the voiding of the 1908 contract meant that the Admiralty was no longer committed to pay Pollen either an award for having conceived of his system or royalties for its use in the event that his system was adopted. He nevertheless accepted the Admiralty offer in the belief, undoubtedly, that the purchase of a portion of his system would lead to profitable orders for the remaining instruments after their successful trial.[12]

By June 1910 the modified plotting instrument, the change of range machine – which was called Argo Clock Mark I – and range and bearing data receivers for the gunners, substituted for Pollen's proposed automatic sight-setting system at the request of the Admiralty, were ready for testing in the *Natal*. Preliminary testing revealed serious defects in the 'helm-free' arrangements of the modified plotting instrument, and in the design of the range and bearing receivers, both of which required the constant attention of Pollen's engineers to be kept operable.[13] The understanding on co-operation that Pollen believed had been established in his discussion with Bacon the previous November, moreover, proved to be non-existent, and on 15 June Pollen's engineers were once again ordered off the *Natal*. In the subsequent official trials, which were evaluated by a committee headed by Rear-Admiral Stanley Colville, the commander of the First Cruiser Squadron,[14] the absence of Pollen's engineers exacerbated the problems that were caused by both the mechanical imperfections of the trials instruments and the inexperience of the naval personnel assigned to operate them.[15]

On 17 June 1910 Pollen wrote to the Admiralty to protest against the removal of his engineers two days before. The Admiralty Permanent Secretary informed Pollen on 20 June, however, that the orders barring the Argo Company engineers from the *Natal* would stand.[16] On 1 July, by which time the official trials had been concluded, Pollen wrote to Rear-Admiral Colville to present his view of the results and included a copy of his letter to the Admiralty of 17 June. Pollen had the letters printed, and they may have been submitted to Colville in this form in the manner of Pollen's report to Admiral of the Fleet Sir Arthur Wilson of 24 January 1908 (included in this volume).[17]

There is no record of Colville's reaction to Pollen's letters, but his committee's report of 1 July 1910 was highly favourable to the Pollen system in spite of its vicissitudes in the trials. While the committee rejected the range and bearing receivers as unsuitable for service use, they concluded that the Pollen automatic true-course plotting gear was unequivocably superior to the manual true-course and virtual-course plotting methods that had been tried at the same time. They therefore recommended that additional trials be carried out on the automatic true-course plotter and the clock, after both had been modified in light of the experience gained in the recent trials.[18]

188 FLEET STREET,
LONDON, E.C.,
July 1st, 1910.

Dear ADMIRAL COLVILLE,

As appears from the annexed copy of a letter to the Secretary of the Admiralty, I did not hear, until after the trials of our gear in the *Natal* had actually begun, that this Company's representatives were to be excluded from the ship.

It has been common ground since November last that our gear is not yet ship or sea worthy, and I suggested – as you will see by my letter – that our engineers should be in attendance, as much to prevent as to remedy the minor breakdowns that were obviously to be expected. I feared, and my fears seem to have been realised, that some runs would certainly be abortive if this suggestion were not acted on. I was counting, further, on our engineers being able to explain personally to you and your Committee certain points that need elucidation. As we have been disappointed in this, I beg leave to submit these explanations in writing.

(2) The gyro-controlled rangefinder mounting having been already ordered by the Admiralty, it is not necessary for me to explain to you the points on which it is seriously defective, nor to specify the improvements and alterations which we propose to make. The trials now going on are, I understand, particularly directed towards testing the plotting table, the change of range clock, and the indicators. It is therefore to these only that I shall address myself.

(3) May I begin by reminding you that, owing very largely to our own mistakes, we never got the entire gear into running order before the trials began?

Next, no ammunition was allowed for preliminary firing under the conditions in which I understand your trials will be made.

These two facts put the gear at a double disadvantage. In the first place it has to be tried without our supervision, when it has never been run under our supervision; secondly, it is being tested for firing without any preliminary practice on the part of the officers who have to carry out that firing.

(4) The plotting table was originally designed so that the paper on which the chart was to be made should be pinned upon the plotting table, and the plotting pen traversed over it. This form of plotting obviously limited the representation of our course to periods when its being treated as rectilinear, introduced no material error into the results.

To plot a turn, or to plot during a turn, was consequently impossible. We had long been of opinion that this was a grave omission, but the strong objections so constantly made to our gear that it was already too complicated had deterred us from attempting to rectify it; although our first patent, taken out in 1904, contained a claim for means of doing this. The means we proposed at that time would certainly not have given us the result we wanted, but the fact that we did claim these means is evidence that eight years ago we considered plotting during a turn as one of the prime necessities of the situation.

(5) The late Captain Ogilvy, almost as soon as the gear was installed in October last, pointed out to us that the military importance of plotting during a turn seemed to him to be overwhelmingly crucial. The saving of time was the essence of all military operations, it was vital therefore to be ready to fire at the earliest possible moment in all contingencies. A ship under helm, ought never, if it could be avoided, to be out of action in the matter of getting her guns rightly sighted.

Further, and on this he was most emphatic, he pointed out to us that, when manœuvring, no ship under the necessity of keeping station could be on a steady course for any useful length of time. To limit plotting, therefore, to when she was on a steady course, was to limit the occasions for getting the guns ready to exceedingly brief periods, and those far apart. He urged us therefore to ignore all the complication objections, and adapt the table forthwith, and as best we might, to the means we had some time since devised for making plotting independent of helm.

(6) These alterations we have carried out, but they have been carried out without previous experiment, and only with such knowledge of ship's conditions as we had acquired in the brief period between the installation of the table and our first expulsion from the *Natal* in November last. It is not surprising, therefore, that several material points escaped our observations. To explain these it is necessary for me to go into some little detail.

(7) I have told you that the table was originally made to have the paper on which the chart was to be made pinned on to it. The table top was therefore constructed of thick soft wood, so that inserting drawing pins should be easy. It was not until the table had been subjected to the high temperatures in the lower conning tower for some months that we discovered that it warped so considerably as to prevent the regular and consistent marking of the pens while the instrument was plotting. Unfortunately by this time we had already adopted the method you have seen in the *Natal* of dealing with the turn. You will remember that we have now got the plotting head fixed,

and propel the board carrying the paper under this by means of two spiked wheels, which by differential gear are made to reproduce at the point where the pen marking our course is situated, the angle of course of our ship at the speed the indicator shows her to be going.

(8) The misfortune of the situation is that we had not, when this design was decided on, a sufficient knowledge of the conditions in the lower conning tower to anticipate that the wood would warp to the extent that it does, and we have now found that not only does the soft wood of the original plotting table warp so as to introduce serious friction at some points more than at others, and so set up varying resistances to the board that is being propelled, but that these plotting boards themselves have also warped, with the result that the prickers fail at times to get a proper hold. When this occurs the ship's turn is of course not reproduced, as the board does not turn through the full angle. If the clock were set to data so defective, the range could not be changed correctly.

(9) And in this connection a further point may be mentioned. Limited, as we were, to driving the board that carried the paper from below (it being impossible, with the present design, to have put pressure rollers to hold it down from above), we thought it necessary to make the boards of sufficient weight to set up an adequate friction with the pricker wheels, the better to ensure that every movement of the wheels was communicated to the board. What we did not realise, however, and here again our inexperience of the sea was the cause of our undoing, was that, if there was any motion on the ship, a heavy board must have a tendency 'to take charge', and swing from side to side as the ship rolled or pitched.

(10) A second source of frequent disappointment during the preliminary trials has been our choice of the marking members of the apparatus, viz.:— pens made on the stylographic principle. These pens have often given trouble, and, although constantly running satisfactorily when they were under the daily care of those who made them, have not always proved satisfactory when this fostering attention has been withdrawn, and on several occasions, by persistently refusing to mark, have been the cause of the failure of the plotting instrument altogether.

(11) Again, in their inoperative position, the pens are too close to the paper, and at least on two occasions the shifting of the plotting board has led to their being broken. This last disadvantage is one of the incidents of the present table being a compromise between two contradictory ideas, and results from our having tried to adapt an instrument for rectilinear plotting only, to helm plotting.

Several suggestions have been brought forward for replacing these pens by a special form of pencil, by a rubber-faced striker with inked pad, or by a metal striker marking through a typewriter ribbon. Anyone of these would give a good result, although the last is probably the best.

(12) A further defect in the table, owing to its being a compromise, is undoubtedly that we have to use a human operator to follow the movements of the gyroscope below by a follow-the-pointer mechanism. In point of results, the gravest disadvantage of this arrangement is that no operator can follow the movements of the gyroscope with exact synchronism, so that there is a failure both in the rapidity and accuracy with which results can be got. A less disadvantage, but still a real one, is that there are several objections to the addition of an operator.

In a new table this necessary part of the functions would be carried out by an automatic relay of a similar character to that employed in the range-finder mounting which has already been approved by the Admiralty. There should also be a relay enabling the speed indicator to control the table drive.

(13) The means provided for reading the course and speed of the enemy, although distinctly better than any that have been used before, are not as good as they might be for similar reasons to those given above for other defects, viz.:—that the device under trial is an attempt to impose new functions upon a table originally designed for a different purpose. At present a portion of the protractor graduations is covered by the bearing arm. It is obvious, therefore, that in a new design the protractor circle should be above the arm, where all portions of it could always be legible.

(14) Experience, too, has shown that the range-finder operator should be warned some few seconds before the elapse of each minute, so that he should be more particularly careful to be on the target for range and bearing when the minute marks are made.

(15) The more serious of the above defects follow from a single cause, viz.:—that the driving power of the pricker wheels, and what I may call their frictional attachment to the board that they control, are utterly inadequate, in view of the weight of the material that they have to drive and the uneven surface over which they have to drive it. Were we designing a new table to deal with the turn we should propel not a board, but a single sheet of stiff paper, or light cardboard, and it would be held quite securely between upper and lower rollers. In this way there could be no slip at the point of propulsion, and with even less power in the propelling gear, nothing similar to what printers call 'registration' error would arise. In this event of course the upper surface of the table, over which the paper would be propelled, would probably be of polished metal. There would be no frictional resistance, and the tendency to take charge when the ship rolled would be negligible.

(16) I think you will find that practically all the troubles likely to result in the plotting table, other than electrical troubles, which I will deal with later, have been due to the features mentioned above. For the sake of clearness I will summarise them:

(1) The warping of the table and boards has given rise to undue and inconstant friction, which prevents the pricker wheels engaging uniformly with the board they drive, with the consequence that the board does not always, as it should, reproduce the course of the ship. But our experiments have shown that when the pricker wheels do engage properly, the course of the ship is reproduced perfectly.
(2) Troubles with the pens owing (*a*) to the pens being of unsuitable type, and (*b*) their too close proximity to the plotting table in their inoperative position, leading to breakages, etc.
(3) The weight of the boards caused them 'to take charge' when there is much motion in the ship, thus throwing out the plot of both ship and target.

(17) I now pass on to the method of getting the data from the table to the clock. The necessities of the case in the *Natal* have led us to put the clock in a separate compartment. It should obviously be alongside of it, and at present it has been found impossible to perfect the drill necessary for getting data from one to the other with sufficient rapidity.

(18) The clock itself has only once, I understand, given any trouble, when the adjusting nut on the ball bearing of the variable speed disc became loose. This had never happened before, and I should fully expect an error in the rate of change to have arisen from it. The pedestal of the clock needs re-designing to make its working parts more accessible; and the clock itself will be greatly simplified by the elimination of the features that make a running calculation of the change of range and bearing during time of flight. This running gear is an unnecessary complication: if these changes are calculated on a triangulating linkage from time to time, it should be sufficient. The new design will also permit of corrections of range and bearing without stoppage.

(19) As to the indicators, it is only fair to us to say that in the first place we have never considered this type of indicator, nor indeed indicators at all, as the proper method of communicating the results got from the clock to the guns. It has always seemed to us that either these results should be communicated by automatic sight-setting – mechanically a perfectly simple device – or by indicators of the follow-the-pointer type,[19] which should be combined with the sight, so that hand setting could be resorted to if for any reason the central control broke down.

(20) Next, it was at Admiralty request that the range and deflection was combined in one indicator instead of being given in separate indicators, so that one could be placed on either side of the gun.

(21) Should, however, indicators of the type that we have installed be desired, there are certain faults which we have found in them which we should remedy. In the first place, they should be larger and their working parts made more readily accessible. By making the frames so that the numeral drums could be removed without dismantling the whole instrument, all parts could be readily reached.

(22) I now come to the electrical troubles that we have experienced. I am informed that you have not upon your staff, for the purposes of these trials, any representative of the torpedo branch; and I therefore make bold to ask that judgment may be suspended, in the event of any electrical defects showing themselves, until the cause of such breakdowns has been ascertained. I say this because, since we have been in the *Naatal*, we have, in addition to the invaluable suggestions of Captain Ogilvy, had the benefit of the criticism of two torpedo officers of quite exceptional attainments,[20] and thus for the first time have had the advantage of co-operating with members of this important and highly expert branch of His Majesty's Service. It has been abundantly brought home to us that, while the general character of our electrical devices seems to be well calculated to bring about the results we have had in view, our actual practice is very far from coming up to what is necessary in a ship of war. If I may give an example, the abutting contacts in our junction boxes have not proved satisfactory, and should be replaced by plug and socket contacts. We have had many electrical failures in the short history of the installation, but the bulk of these can be traced to these junction box contacts.

It is unnecessary, perhaps, for me to go into further details of the electrical failures. But I think an electrician would assure you with regard to such that each single failure can be explained by faults for which the remedy is obvious to him, although such failures might possibly be considered incidental to the scheme by those who were not similarly experienced in electrical work.[21]

(23) There is a French saying that the self-excuser is the self-accuser. I hope you will not consider that it applies in the present case. You can readily understand that for landsmen to tackle successfully the production of gear of a similar complication to ours, and to get all its details to stand the rough and tumble of ship's life, must be a task of no ordinary difficulty. At the time when the designs for the *Natal* gear were made, our sea experience had been of the very shortest, our ignorance of Service practice was therefore considerable. In looking at the results – especially when the

instruments have been working without that supervision by the designers (which in the preliminary trials to some extent concealed or neutralised the defects in practice and design) — it would seem to be essential that the cause of each breakdown should be noted, so as to ascertain if it is inherent in the scheme itself (that is intrinsically a part of it, so that the system could not work if it were altered), or is merely due to the design being defective, because it is extemporised, to inexperience of sea conditions, or to ignorance of the electrical practice which in all other naval gear has been found to be necessary.

(24) If instead of only looking at the failures we look at the successes, may I be permitted to suggest a consideration that seems to arise from what I have seen the gear do? The clock seems to speak for itself, but I make bold to say that the principle of automatic plotting, now that it can be made effective *in all conditions in which the rangefinder can give useful results*, irrespective of the plotting ship being on a steady course or under helm, has shown itself to be realisable without difficulty in practice. It follows from this that knowledge of the speed and course of the enemy has become attainable with a speed and accuracy unthinkable by any other means. To suppose that the most gifted and experienced observer can ever — let alone always — guess these data by watching from aloft, with any similar accuracy or within an equally short time, or detect a change in the enemy's speed or course with a like promptitude, is surely an untenable proposition.

(25) There is a possible war advantage in the self-plotting device that may not be apparent when engaging a target not liable to obliteration by its own smoke. If we suppose our ship to make a turn, and the enemy to be obscured during the turn, so that no observations could be taken, the self-plotting would afford exact data for resetting the clock — and thus the correct range and change of range could be kept on the sights if the enemy did not change course or speed in the interval — and the assumption that he is firing presupposes that he cannot change either materially.

(26) That 'plotting' has now got a bad name is undeniable; but surely the reason is that no one has hitherto seen the thing done in the only way in which those who invented plotting intended it should be carried out. Between the plotting which is possible by the *Natal* gear, and the plotting which the Service adopted in 1908, there is a wide difference. The first is a synchronous graphic record of what the rangefinder is doing, the value and excellence of which must increase with any improvement in that instrument. The plotting table is simply a rangefinder indicator: just as the devices already ordered in place of it, are rangefinder indicators, only the first is graphic

and the second numerical. Each is really not a separate unit, but the logical complement of the range and bearing finding unit. Manual plotting, though aiming at the same result, attempts to achieve it by an organisation more complicated than any machinery, and has been shown, after very painstaking experiments on a colossal scale, to be impracticable when its object was most necessary, and to give only very meagre results when conditions were easy. Nor does it appear that any improvement, either in the rangefinder, its mounting, or in methods of transmitting its data, can greatly simplify its over-manned organisation.

(27) That the Service would prefer guesswork to manual plotting has always appeared inevitable. Manual plotting was sufficiently tried out in the *Jupiter* in 1905–6 for it to be clear that the loss of time and the many sources of error inherent in such an operation marked it from the first as one that no improvement in organisation could ever make of any war value. But the necessity to know the enemy's speed and course remains paramount. This unsuccess of the manually made plot should not blind us to the military value of the plot itself – indeed, the fact that the experiment has been so assiduously persisted in, is the strongest confirmation that this military value is both obvious and incalculable. If we assume that in war minutes are of vital moment, it will not be safe to prefer a slower and uncertain way of getting results if a quicker and surer one is to hand. Are we to go back deliberately to guesswork, or are the difficulties – none of them considerable – in producing means for getting an automatic record of the rangefinder's work to be resolutely faced? I cannot doubt that with the experience we already have, and the knowledge that is available in the Service, there can be no question as to ultimate and speedy success should co-operation to ensure it be recommended.

A. H. POLLEN.

CONFIDENTIAL

188, FLEET STREET,
LONDON, E.C.,
June 17th, 1910.

SECRETARY,
  THE ADMIRALTY.
SIR,
I am informed that on June 15th at 7 p.m. a signal was received by H.M.S. *Natal* saying that the Admiralty did not approve of the Argo Company's representatives remaining on board during the trials, and that accordingly

Mr Isherwood, Mr Landstad, and the two mechanics were transhipped to H.M.S. *Invincible* and brought into Weymouth at 4.30 yesterday afternoon.

It may be remembered that in November last the Company's representatives were similarly sent out of the ship, without previous warning, and, on that occasion, against the expressed wish of the Commanding Officer. Their dismissal was immediately followed by a breakdown, which put the gear out of action for several days and necessitated extensive repairs. This accident would not have occurred had the Company's Engineers been present.

As the present decision may have been come to without either full acquaintance with the local conditions, or this previous experience being in recollection, I beg very respectfully to submit the following considerations:

The present trials may be directed towards either of the two following objects:

(a) To ascertain the military purposes of the A.C. system., *i.e.*, to demonstrate the increased gunnery efficiency to be obtained by its use, or
(b) To ascertain if the devices themselves can be trusted to run without skilled attention, are in all respects suited to ship and sea conditions, and are thus fit for installation in their present form throughout the Fleet.

In the light of the practice exercises recently made, the second object may be regarded as definitely achieved already, in that our experience makes it perfectly clear that, in their present form, the instruments are neither reliable without more skilful attention than they can get in a newly commissioned ship nor in many of their unessential details suited to ship or sea conditions; and consequently are not in the shape in which any extensive installation of them would be desirable.

That many elements in our present designs need radical alteration has already been recognised by their Lordships in the case of the rangefinder mounting, without the necessity of drastic alterations being considered an objection to a considerable order being given. The case for radical alteration is more obvious and stronger in the case of the other instruments.

It appears to me that to remove the Argo Company's engineers at the present moment is certainly to risk, if not to ensure, such a repetition of the November breakdown as may prevent any demonstration of the military purpose for which the instruments have been built, as the time available for the trials is exceedingly short owing to the forthcoming manoeuvres.

It will be clear therefore that when the time and money already devoted to these experiments is considered, to delay still further the demonstration of the purpose of the gear would be very much to be regretted.

I therefore respectfully suggest that the trials should be limited to demonstrating the military purpose of the system, and that, until the moment of this being successfully accomplished, Mr. Isherwood and Mr. Landstad should be allowed to remain on board for the sole purpose of attending to the actual running of the instruments.

No doubt, when the trials are over and the ship goes into manoeuvres, the instruments will be put to severe and regular employment, when the particular defects which render them liable to breakdown will become even more obvious than they are at present. Reliability in machinery is a plant of relatively slow growth, and experience is necessary for the evolution. The knowledge already obtained in the *Natal* should go far towards ensuring future gear being free from the objections on this score of the present, but whether future experiments are worth while or not must depend upon it being made obvious that the military purposes are worth achieving.

(Signed)   A. H. POLLEN.

XVI

THE QUEST OF A RATE FINDER
(*NOVEMBER 1910*)

On 19 August 1910 the Admiralty Permanent Secretary informed the Argo Company that the Admiralty had rejected both the automatic true-course plotting instrument and the clock, and required that improvements be made in the range and bearing receivers that were already scheduled for delivery, as arranged in April of that year,[1] as part of the order for the gyroscopically stabilised range finder and bearing indicator mountings. The Admiralty expressed an interest, however, in evaluating a clock that had been modified to take account of shortcomings revealed in the *Natal* trials.[2] On 25 August Pollen, in reply to the Admiralty letter, insisted that some form of plotting was essential for effective fire control, and stated, therefore, that he intended to carry on the development of both the plotter and the clock.[3] In an interview with Captain Archibald Moore, the Director of Naval Ordnance, on 26 August, Pollen was told that the Admiralty had decided to reject plotting in all forms.[4] Pollen nevertheless continued redesign work on the automatic true-course plotter as well as on the clock. In November 1910 Pollen restated the case for his system and recounted the history of the trials in a paper 'The Quest of a Rate Finder', which was circulated together with a second paper, 'Of War and the Rate of Change' (included in this volume), in early 1911.[5]

NOTA BENE

In the following pages the word 'rate finder' is used to express apparatus for ascertaining the range, bearing, course, and speed of the target ship. A time and range curve would give the present rate: but not the factors for forecasting the range continuously. Means of making this forecast from these materials, and of correcting it for change in time of flight, and of keeping it, when made, on the sights, are − in the opinion of the writer − a need, at once primary and fundamental, of naval war.

FOREWORD

A NOTE ON THE SEARCH FOR A RATE FINDER

No one can be quite insensible to the gentle flattery of consultation. A letter dated from a gunnery school and addressed to me as 'Professor' is, and will long remain, one of my most treasured possessions. Nor am I at all indifferent to friendly enquiries as to the fortunes of my system. But, with the best will in the world to reciprocate such pleasant attentions, it is impossible to reply to every correspondent with equal, nor to any, with adequate fulness.

I have, therefore, selected from recent communications three typical passages, and appended to these an omnibus reply, made up and amplified from the original answers to them, in the hope that the substance of most questions likely to be put to me will be found to have been dealt with.

This reply is practically limited to giving an account of the gear we had in the *Natal*; but my correspondents go much further than this. One of them, for instance, alluding to the unsatisfactory character of the present position, says it is a case of 'as you were 1907'. He might have gone further back, for it was in 1901 that the first representation was made to the Admiralty that by plotting successive positions of the enemy his speed and course could be ascertained. It was in 1905 that the first plotting experiment was authorised, and when, in the following year, the *Jupiter* trials resulted in failure, the problem of how to provide for moving-target control came up once more before the D.N.O.'s department.

The new 9ft. Barr and Stroud seemed so good that the range finding side of the *Jupiter* system was not worth developing. But the plotting or rate finding side remained. That plotting was theoretically the only way of getting a true rate on which to open fire, had been admitted since 1904. Enough had been done in the *Jupiter* to show that, with right methods, serviceable plotting might be feasible. Its actual failure in that ship was known to be partly due to paucity of ranges, *i.e.*, because the range-finding was inadequate; but from the first we all saw that this was neither the only, nor the main difficulty. Its real vice was that it was a manual process: and one experience of polyanthropy was enough. While it was clear that plotting would be impossible unless the ranges were numerous and fairly accurate, and the bearings as numerous and perfectly accurate, it was even more clear that, however good and numerous the observations, they could not be useful unless they were first synchronously taken, and next recorded simultaneously with their being made. The essence of the problem, since 1906, has always seemed to lie in the single word 'Speed'.

It remained to invent methods of satisfying these conditions, and to incorporate them in what would appear to be a working system of gear. Our efforts resulted in the gyro controlled range finder, the system of mechanical reading and transmission, and the automatic table to make an instantaneous graph of what the range finder operator was doing. These combined constituted a rate finder that, except for the actual making of the cut, was entirely automatic.

Before undertaking the experiment, I had to ask for a considerable sum to cover the larger part of the expenses. But so improbable did success appear, owing to the novelty of the design, that there was great and serious opposition to risking a penny piece on it.

It was in July, 1906, that it was first proposed to make a further trial of manual plotting, this time with Service gear, partly so that some results should be available to strengthen the case for proceeding with the more scientific, though more costly, method; partly that a system should be developed for temporary use while the automatic, should it succeed, was being manufactured.

But a scrupulous and thorough review of the lessons of the *Jupiter* trials resulted in the gunnery authorities – after many specialists from outside Whitehall had been called in – being convinced that there was nothing to be expected of manual plotting. It would be equally useless as a temporary rate finder, or as a mere demonstration of the principles of rate finding. I had the honour of submitting a memorandum, which more or less summed up the arguments that had done duty during many informal deliberations over this proposal, the conclusions of which were accepted. It ended with these words:

> The real point to bear in mind is that no charting can be reliable, or give results of the faintest value unless
>
> (a) Range taking, bearing taking, and yaw correction are attained first by one operation; next
> (b) Are obtainable at any moment; and
> (c) Are transmitted synchronously and instantaneously to the chart.
>
> For these reasons, the writer can see no possible good object being served by any experimental trial of Aim Correction by means of extemporised instruments, for the reason that *no test from which the synchronous getting of the data and their instantaneous transmission are absent, could throw light on a system which depends entirely upon them.*[6]

It is worth remembering, then, that the rejection of manual plotting in 1906, was not made on *a priori* grounds. It arose, naturally and inevitably, from the examination of known facts. The automatic method was adopted, not because it would be a better and quicker way of finding the rate, but because there was no other way of finding the rate at all. It was the fact of its being automatic that alone made success possible.

Hence the issue put to the Board in August, 1906, was that the choice lay between automatic plotting and guesswork. It is exactly the choice we have to-day. Guessing, in 1906, seemed, even to the most sceptical, an unthinkable alternative, if there was any possibility that the automatic would succeed; and it was less with the expectation of success than the feeling that no stone must be left unturned to escape so forlorn an alternative, that the grant for the A.C. trial was authorised.

We see thus that since 1904 the ultimate need for a rate finder had been undisputed. How is it, one may ask, that we are still without what, for six years, everyone has hitherto recognised to be the first necessity of gunnery?

The answer is, I think, simple. I have shown that manual methods were not rejected in 1906 on *a priori* reasoning, but on strict proof of their inadequacy. But the automatic method *was* rejected on *a priori* reasoning, before any proof of its failure was possible. And the prejudice of a previous conviction clings to it still. It was not condemned in March, 1908, because it failed when tested, while the manual method succeeded. It was condemned in the autumn of 1907, because the manual method was assumed to be sure of success. The official and public rejection that took place in March, 1908, when the Board of Admiralty for the first time considered the results of the trial in the *Ariadne*, only confirmed the earlier unofficial and private decision made by the Ordnance Department, in November, 1907.

A letter to the Treasury explaining that this change of policy had been the reason for spending, on some different object, the amount already provided for the purchase of the A.C. system, is published in the Navy Appropriation Accounts of 1909, and is dated the 25th November, 1907.[7] But I have good reason for thinking that the decision was come to by the Naval Ordnance Department at least six, if not ten weeks, earlier still.[8] It must, in any case, have been a considerable time before the letter was written, certainly long before the trials were arranged. And the letter is dated three weeks before the *Ariadne* gear was inspected after its first erection, and seven weeks before the trials of the competing systems began. The conclusion is irresistible that the A.C. system was abandoned in November because those who rejected it were absolutely confident that no trials were necessary to show that manual plotting would solve the problem. The decision was made – and unquestionably in all good faith – because the verdict of the trials was a foregone conclusion.[9]

And this is why the tone in which the manual method was commended was one of almost amused surprise that such heavy weather had been made of so straightforward and elementary an affair. In those who thought that manual plotting was an original invention of their own, this attitude of mind was natural. I confess that, before the *Jupiter* experiment, it never occurred to me that our plotting being manual would be fatal to its utility. I doubt if anyone, unless taught by actual sight of the failure of manual processes, could have anticipated that the problems of plotting in exacting conditions must be insoluble by polyanthrophy. Certainly naval officers, whose daily task it is to solve apparently far more complicated problems by extemporised organisation and gear, would be the last to admit that defeat was certain. Even with the 1908 Battle Practice to warn them, many still thought manual plotting to be feasible. Experience has now convinced everyone that they were wrong, just as experience convinced us that we were wrong five years before. Unfortunately, with the change in the direction of the N.O. Department,[10] the sponge of oblivion was so completely passed over the slate of experience that the evidence known to be quite overwhelming in 1906, was not known even to exist in 1907.

But note that the decision of November, 1907, did not in any way imply that rate finding was unnecessary. On the contrary, it confirmed its being absolutely necessary. It was, negatively, a rejection of automatic rate finding, but, affirmatively, it ordained the adoption of manual rate finding. One may think the decision unfortunate, without seeing in it any acknowledgement that anything so unwarlike as guesswork was inevitable. Indeed, if the facts of the 1905 experiment had been known, and hence manual success had been suspected to be even uncertain, can it be doubted that the *Ariadne* trials would have been extended to rough weather, a high rate of change, sudden and large changes of course, and other tasks more formidable than those that confronted us at Torbay? The best proof that it was thought that there was nothing to prove, is afforded by the fact that these trials were limited to tests of 50 minutes duration in a flat calm.

The fact, of course, is that no one connected with the 1908 experiment would have acquiesced in the suggestion that to seek a rate finder was either a superfluous or a hopeless quest. They acted as they did because, not knowing the proofs to the contrary, they thought that a rate finder was both easy to find, and that they had, in fact, found it. Hence to accept guesswork as inevitable now is not the logical outcome of the fire control policy that prevailed either in 1905, 1906, or 1907–8. The logical outcome would be to continue the constructive policy of 1908 – and that policy was based on the known truth that means for rate finding were required, and must be supplied.

The disfavour that plotting is in through the failure of the Fleet to do it manually has given the word a kind of sinister meaning, and I find that many people have really forgotten its fundamental purpose. They recognise it can only be done by rather elaborate machinery, and, when so done, they suppose it is in the nature of luxury, and not a necessity of the case. I sometimes wish that we could drop the word 'plotting' altogether. After all, to get any sort of corrected rate of change, or a range for opening fire, you certainly have to use a range finder, and, as a fact, range-finder mountings have been ordered to transmit, not only ranges, but bearings synchronously with the range. These must be transmitted to indicators. To use them, they must be transcribed, and probably in a graphic form. Now, looked at simply, an automatic plotting table is only a range-finder indicator. You can indicate by figures on a numerical indicator, or you can indicate by a line on a graphic indicator. Mechanically, there is little to choose between them. We are all familiar with the advantages of the barometer that is made to record its risings and fallings automatically. A continuous and graphic record of what the range finder is doing is, in the same way, preferable to figures, which vanish the moment a new figure is required. It saves many necessary and difficult operations, and these consume all too precious time, occupy members of a not too numerous personnel, and breed all too disastrous errors.

Moreover, to indicate by figures that have to give way to others is to limit the speed of observation; partly, because no indicator can work beyond a certain speed; much more because, before this limit of speed is reached, the power of the human eye to read these rapidly-changing figures has gone. Whereas, a graph traced on paper, in precisely its right relative position to a similar graph of our own course through the water, can be made at any speed – and often continuously – for the cut, with practice, can be kept on at the right rate of change. Thus it is both an exact index to the factors that make up the change of range at the moment, and will afford means, and apparently the only means, of averaging out the inaccuracies of the telemetric product of the instrument. I suppose it is too late to get the phrase 'graphic indicator' accepted instead of 'plotting table', but I feel somehow that, if it could be accepted, a great deal of prejudice would be wiped out with the alteration of the name.

A.H.P.

November, 1910.

LETTERS ON THE RECENT TRIALS IN H.M.S. *NATAL*

I

October, 1910.

Dear MR. POLLEN,

I have been wishing for some months to congratulate you on your order and on the preservation of secrecy. I wish the order had been for the system complete. I am truly glad you have not been driven abroad. That would have been a disaster. How are things going with your system? Is it still officially alive? They are not going any too well with us.

Plotting, so long the centre of our hopes, has collapsed. For three years we have been doing our best with it. We have tried it with your gear,* with Admiralty gear, and with no gear at all. We have tried virtual plotting and rate curve making. But all have been equally disappointing, and for the moment we are in a state of chaos.

Disgust with existing arrangements is widespread, and its extent would

---

* This refers to our manual table. [Editor's Note: See Preface, note 8]

surprise you. As far as gear or system is concerned, it is universally felt that we are not ready for war, if war means long range. We have concentrated on this for years, and have not found the right way to do it. Have you ever given much thought to short range shooting?* I don't mean with machine guns, but with the main armament. It may be, and probably will be, forced on us, and at the moment we are certainly not as prepared as we might be. Our only reliance is on the skill and experience of our gunnery staffs. I do not know much of what foreigners are doing, but I doubt if any Navy has so much actual experience as we have. Much of this experience is no doubt wasted because the system is bad. Still I cannot help hoping superiority in practical knowledge should make us give a good account of ourselves in action. But we ought not to be contented with this, if real progress is possible in method.

It is only with better method that we can raise the average – or get an average that can be relied on. To-day we simply do not know that any particular ship would hit at all.

II

August, 1910.
Dear POLLEN,
I have heard very little about the *Natal* trials, beyond the report that none of your gear has been adopted. This surprised me, because I supposed it admitted that as soon as the failure of the manual was acknowledged, we had the automatic method to fall back on. Was not this the whole object of the Admiralty settlement with you in 1908? It is very disheartening to be reduced to guesswork now, when three years ago automatic control was dangled before our eyes.

Is it a fact that the results in the *Natal* were bad? Why? when they were so good in the *Ariadne*? Did the gear break down altogether? If it did, are the faults irremediable? And finally, have you given up automatic plotting as hopeless? . . .

I am certain that if your people and we made a determined onslaught on the problem, and your inventions were developed in the light of Service experience, we should get a method of gunnery much beyond what we have to-day. Did you not tell me, a year ago, that you hoped to agree a basis for such co-operation? The resulting system might not be exactly your method. But then I have never thought you could succeed unhelped by us.

* The requirements of short range and long range gunnery are, in essentials, similar. At short ranges the increase in the danger space, and the decrease in telemetric errors, make absolute accuracy less important, and so obviate the necessity of spotting. On the other hand, the speed of the different operations is more important than at the ranges at which perhaps an enemy could not fire at all.

III

October, 1910.

Dear POLLEN,

... In last year's Battle Practices, the plotted course of the 6 to 8 knot target was something like 14 deg. out on the average. The target's speed was so low that it is difficult to say what the average speed error would be in action; judging from P.Z. and manoeuvre results,[11] the errors would be considerable, and if the speed were higher we might be less likely to get as near as 14 deg. to the course.

Naturally, the common verdict is that plotting is dead. We are therefore in the unsatisfactory position of being 'as we were' in 1907, *i.e.*, thrown back on guessing how to set the Dumaresq. A series of experiments in estimating speeds, etc., were made recently in the Home Fleet. The speed was generally from 15 per cent to 30 per cent wrong, and the course error was said to average about 10 deg. This has been trumpeted around as a victory over plotting, as if to be only 10 deg. out was something like a triumph!

If we cannot get the enemy's course nearer than 10 deg., it means that we cannot begin with the rate right, and therefore cannot bracket for gun range only. To correct rate by plotting, always a very ticklish job, we must spot an immense number of trial shots, and must assume that all through the correcting period, the conditions will be favourable for seeing the fall of the shot. But can we spot at all against a fleet firing?

Anyway, to spot for speed and course as well as for gun range means throwing away tons of costly ammunition, and the loss of what may be much more costly time. Until our guns can be made to shoot more accurately, the gun must, ultimately, be its own range finder at long range; but surely we ought to get a better, quicker, and cheaper way of finding the rate than bumming off round after round from 12-inch guns!

Our battleship types, our system of gun training, and our entire conception of battle tactics are all based on fighting at long range; to sit down under a supposition that we shall always be on an average 10 deg. out knocks the bottom out of this theory. Our present case is not a very logical corollary to the *Dreadnought* policy.

Would it not be better to chuck long-range fighting altogether, like the Germans?[12] We must either do that, or find the rate before the first shot is fired, and keep it afterwards.

REPLY

14, Buckingham Street,
Strand, W.C.
2nd November, 1910.

My Dear ——,

Many thanks for your congratulations: the order for the range finder mountings was an immense relief. It would have been too bad if, after ten years' work, so largely at the public cost, with the full proof in our hands that the greatest of all control problems were solved, and with secrecy successfully maintained, we had been forced to sell our system to foreigners to save ourselves from bankruptcy. All these dangers are now, thank heaven, safely behind us; and as far as future progress is concerned, we get the added boon that we are no longer dependent on official subsidies or approval, before we can put new experiments in hand. It is true that we have no further orders as the result of the trials. But we have been asked to submit an improved clock. *Quoad* plotting, we are neither alive, nor under any formal sentence of death. Let us say we are, officially, in a state of suspended animation, and, unofficially, looking to a new trial with animated suspense.

(2) All plotting, I gather from the letters I receive, is for the present to be ruled out of Admiralty plans; the manual, polyanthropic method, because, I suppose, it has as a fact failed; the automatic, because it has not, officially, succeeded. There is no need to enquire as to the 'why' of manual plotting's decease; and, after so many years of costly experiment, there is, perhaps, no patience to enquire into the 'wherefore' of our failure. I do not fear refusal to try us again, if we come with reasonable guarantees that the gear not only solves the problem, but is of the stuff that ship's officers and men can work and look after, without immoderate demands on special and unusual skill.

(3) But, officially, the period of talk, promise, experiment, and subsidies is apparently over. Far be it from me to question the propriety of this attitude, however much I may, for various reasons, regret it. I have been treated for years with unparalleled generosity, and I can well imagine a Government Department wearying of being dragged for ever in the wake of a light-hearted optimist. It is, however, not difficult to show that the painter has been cut within hail of port. I suspect there is still only a partial comprehension of the exact objects we are driving at; which may, indeed, be more my fault than anyone's. However, so far as the future is concerned, the point is immaterial. We could not make much quicker progress even if our

(4) rate finding system was once more basking in the sunshine of Admiralty favour.

(4) Now to the details of your enquiries. It is rather difficult to answer. I have not seen the *Natal* report myself; really know very little of what actually happened; and can only explain such meagre facts as have been communicated to me by conjecture; but certainly, the plotting done was not satisfactory. It was unfortunate that the gear, being experimental, could not be run without our help, and that we were not on board. I should say that the reason why we alone, who knew the gear thoroughly, could run it, was twofold. It is partly to be explained by the unsuitable construction of the table; partly by the defective state of the electrical communications. Briefly, the origin and details of these defects are as follows:

(5) The table was originally built on the same principle as that used in the *Jupiter* and the *Ariadne*. This principle may be called 'straight line' plotting, in that the course of the plotting ship is assumed to be, and rendered as, a straight line. The chief disadvantage with a table so constructed is that it is impossible to plot during a turn.

(6) So soon as the late Captain Ogilvy – who was so high an authority on fire control, largely because his gunnery ideas were inflexibly tested by the conditions of action – had learnt the history of our somewhat kaleidoscopic adventures, and mastered the military purposes that had guided us, he begged us to free the system of this limitation.

(7) Ogilvy's view of his duty to the Service was not limited to standing mutely by while the wrong thing was being done, and then freeing himself of further responsibility by gleefully recording the particulars in which we failed. In the *Revenge* he had seen more manual plotting than anyone afloat, and his scepticism as to its utility was inveterate.[13] That we should revert to guessing the rate was incredible. It was therefore automatic gear or nothing. He took it for granted this must be the Admiralty view also, for unless the Board had ardently hoped for success, the large sums spent on it would never have been so invested, and the experiment he was entrusted with, never have been ordered. To realise the necessity for success, to hope for it, to spend £30,000 and more to get it, and then to nullify the whole proceeding by denying such little co-operation as was necessary to make it certain, seemed to him to be far too irrational a proceeding to be the policy of Whitehall. He had no instructions to treat us as suspicious characters, and supposed himself selected to co-operate by offering every suggestion that would accelerate the production of what, if successful, must add so greatly to the fighting power of the nation in war.

(8) He had followed with keen interest the tale of the gradual evolution of what I called the war features of the system. How in 1901 I had put

gyroscopic control in the forefront, and been induced to drop it out of the *Jupiter* system, with fatal results to the success of that experiment; how the *Jupiter* experience had shown, once and for all, the overwhelming importance of *rapidity in getting results,* and the hopelessness of expecting rapidity with polyanthropic methods: how we had learned the lesson that a fair weather gear would be useless in war: how, finally, we had in 1904 projected making plotting independent of helm by a very simple, but alas! quite ineffective, device, and had, of late, in sheer terror of opposition, left this addition to the future when the cry, that the gear was too complicated already, would no longer stand in the way of its logical development. And he at once fastened on the importance of the one thing we had left undone.

(9) His view was, that unless the Admiral's tactical initiative was to be crippled, every ship leading a line into action might, at any moment, be under helm; and that any ship having to keep station, whether in action or not, would almost always be under helm. It would be hard enough to keep a steady course when actually firing – these periods of steadiness must not be wasted in getting the rate. To limit rate finding to straight-line plotting was to make it of only occasional utility to fleet ships, but if it could always be used, whether under helm or not, its fighting value should be past reckoning, for the excellent reason that nothing else could conceivably take its place.

(10) His encouragement to go forward with the needed conversion was, of course, a great spur to us; and when, a year ago, negotiations were begun with the Admiralty for finding a *modus vivendi* by which we could co-operate with the Service in the final development of our system, the sympathetic reception with which our proposal to embark on a new experiment was welcomed at Whitehall, removed our last misgiving in putting in hand what seemed the best compromise we could make in the circumstances.

(11) With the gear as it was, the only practical way of reproducing a true continuous bird's-eye view of both courses involved propelling a large board, on which the two courses were to be constructed, over the existing table. This board was to be driven by wheels mounted below the surface of the table, with spikes just showing above the surface so as to engage the board. These wheels, by a differential arrangement, controlled by gyroscope, caused a point on the board, midway between the wheels, to reproduce our movement through the water. The lower plane of the plot board and the upper plane of the table were, of course, assumed to be parallel.

(12) But by February, when the change was to be made, the table, the top of which was of thick soft wood, had warped in the most

extraordinary manner, owing to the heat of the lower conning tower, and, by the time the official trials came on, the boards had warped almost as badly as the table, with the lamentable result that the spikes did not always engage. To make the gear work at all accurately, therefore, involved rather a pretty problem in watchfulness. It was impossible to remedy this warping by planing, on account of the clearance, between the fixed and moving portions of the table, being settled by the protusion of the pricker wheels.

(13) It is quite true that in the preliminary runs, which extended over about ten days, we did some extraordinary good plotting under all degrees of helm, but we only did it by having the engineers and mechanics, who had constructed the gear, nursing it with continuous assiduity. Now as to the communications.

(14) In the *Ariadne* a special house was built to protect the range finder with its mechanical and electrical attachments. The existence of this house rendered it unnecessary to make the transmitting gear at the range finder absolutely watertight. The protection so given explains our freedom from electrical troubles in that ship. But it was not until the last moment that we discovered that no such provision existed, nor could be made, in the *Natal*. We were consequently unprepared in this particular, and, going to sea as we did before the instruments were half erected, they were almost continuously under water for some days, and electrical troubles set in owing to the soaking of the transmitters, cables, deck connections, etc., etc., that recurred again and again during the whole time that the instruments were at work. These troubles can be certainly avoided, by employing the Service methods of running fire control cables, which have stood the test of continuous work in turrets.

(15) You will see from the above that whereas we had originally put our gear into the *Natal*, in 1909, with what seemed well-founded confidence that it would be workable as a ship's fitting, the installation had in fact, by 1910, become purely experimental, owing to the alterations made and the defective wiring. It could only show its professed purpose while in the hands of those who had long been familiar with every detail of its construction; for, to make it work, it was as important to anticipate, as to remedy, all the little breakdowns that were likely to arise, each trifling perhaps in itself, but fatal to the result.

(16) Long before this the establishment of a co-operative basis for future work had broken down, and when the official tests came on, our people were, without notice, ordered to leave. Even had we expected so severe a measure, there would have been no time to train the officers of a newly commissioned ship to do what we had done; much less to teach them what we knew about the gear. Hence any

official demonstration of the success of the new rate finding system became impossible from that moment.

(17) I did indeed try to represent this state of things at Whitehall. My point was simple. The trials could either be run to show the fighting objects of the system, or to prove the actual reliability of the individual instruments, communications, etc. But, as it could not be disputed that the instruments were altogether unreliable, and as no sane person would suggest supplying them in their present form for general use, there was nothing to prove on the second head; while, if our people were excluded, it would be impossible to give that demonstration of fighting purposes, the making of which seemed to be the sole object of the experiment. However, the arrangements already made for the exclusion of our engineers were not altered.

(18) I expect, as I have already said, impatience with an intolerably drawn out experiment explains the course taken. Results, not excuses, were wanted, and Service, not landsman's, results. My assumption that the *pourparlers* of the previous winter implied a willingness to treat the gear as only intended to demonstrate the principle – the details of future construction to be settled in the light of the experience so obtained – was evidently premature. There had been many changes of personnel since the thing had begun, there was nothing in writing to alter the previous basis, and I doubt if the circumstances were presented by me to the Board with convincing lucidity. I doubt too, and for the same reasons, if the novelty of what the gear was designed to do was very clearly apprehended. The official attitude was, I take it, 'if the system is still experimental, we don't want it.'

(19) After all the work, the money, and, what is of infinitely greater importance, the time that has been lavished on the system, it is of course a great disappointment to me that we have not had an opportunity of exhibiting its military purposes. But the facts being what they are, it does not surprise me that the actual tests were unsuccessful and inconclusive, nor that the Admiralty has not asked us to build an improved table. Before doing that, the verdict of the trials must have been in our favour – and how could it be, when the production of any evidence of the purpose of the system was made impossible?

(20) However, it is of course quite vital, if true bird's-eye plotting is ever to be serviceable, to get over the mechanical defects that made the *Natal* table unworkable in ordinary ship conditions, and with this object I have had a model table built which is already running. The methods of surmounting the obstacles to success have shown themselves to be quite simple, and the solution we suggested before the trials began seems to be perfectly satisfactory.

(21) Bearing in mind that we never had any electrical troubles in the *Ariadne*, and deducing from this that the electrical system is right when it is watertight, it seems to me to be a mere question of getting our practice to square with the experience of the Service in other fields, to make certain that a table built on the principle Captain Ogilvy urged on us as imperative will work with continuous regularity. We are therefore going over the drawings carefully with a view to building a new table which will, we hope, incorporate the lessons of our hard and disappointing experience, and be, what does not now exist, a rate finder that can be used in all weather and under full helm. And, whatever it costs, it will certainly be cheaper than rate finding with a brace of 13.5's![14]

(22) The change of range clock has been sufficiently well thought of for us to be asked to get rid of certain working defects, and to submit a new edition for trial. I forget if you ever saw the clock in the *Natal*? If not, I may explain that it is one on which are set the following data:

    (a) The enemy's range.
    (b) His speed.
    (c) The angle of his course, relative to ours.
    (d) His bearing.
    (e) Our speed.

(23) When these data are set, the clock continuously generates the true range, and the change of range in time of flight, and it is this corrected range that is transmitted to the guns. The *Natal* clock further calculated the change of bearing in time of flight (or the theoretical deflection), and was built to transmit this, with the range, to the gun. But the clock worked none the better for these additional functions.[15]

(24) The principal defect in the working theory of the clock is that it runs, like the original plotting table, on the 'straight line' principle, hence the automatic generation of bearing must be stopped while it is being re-set manually as we turn. Another, not less serious fault, is that it is some little time after it is started before it fairly gets to work; *i.e.*, the backlash is excessive. The range curve given by the clock, therefore, tends to lag behind the true range curve. Both difficulties can seemingly be got over.

(25) The official results of the clock runs have now all been plotted out; and, allowance being made for this backlash, its accuracy is truly astonishing. I have very little doubt that, when we have somewhat simplified it, a new fitting of very great value will be available for the Service. It ought to be as good a *rate keeper* as the new table will be a *rate finder*.

(26) If we assume this clock to be adopted, the question arises, how are we to get the first four of the above data for setting it? I am far from quarrelling with your statement that hand plotting is dead: indeed I never quite understood the very sanguine hopes for its success that were entertained three years ago.

(27) But I find that in many quarters the failure of manual plotting is somehow or other supposed to compromise the utility of any kind of plotting, however done, whether manual or automatic, so that the desperate means you allude to − viz., relying solely on guesswork for the speed and course of the enemy, is now thought by many to be the only conceivable way of arriving at these highly necessary data. There is no necessity to criticise the Admiralty, if this is the official view − for official views are the basis of action, and I don't see what practical action the Admiralty could take to remedy present deficiencies, however much it might wish for the success of automatic plotting, until a working system is produced.

(28) But when faith in the ultimate inevitability of guesswork is inculcated, it seems to me that, in the face of what has been accomplished in the *Natal* and the *Ariadne*, this Christian resignation of spirit is, happily, quite superfluous. In both ships when the atmospheric and light conditions were favourable, we have got almost the exact speed and course of the enemy within two minutes of the first observation. It has been done with fair regularity in the roughest of weather and under full helm. With the improved gear now building, and a better range finder, it should, I think, be got quicker at longer distances, and in a wider range of indifferent light. But, at any rate, two minutes, or even three, is a good deal better than guesswork, when it is admitted that the average guess is likely to be 10 degrees out.

(29) There is a good alternative means of getting and keeping rate − alternative, that is, to straight line plotting. When our range finder mountings are supplied, it should be an easy matter to make, semi-automatically, a time and range curve that would give an accurate opening rate. With a time and bearing curve similarly and simultaneously made, the two rates could be resolved into speed and course of the enemy, by a simple linkage we are now constructing, and the clock set from the result, and thus the rate, once found, would be kept right[16]. It is a roundabout way of doing what you want; and will take more time than true plotting − a serious disadvantage. But its worst fault is that, as in straight line plotting, you must be on a steady course while the observations are being made, which brings it within the strictures Ogilvy passed on the straight line table, and these seem to me to be unanswerable.

Hence one is forced to the conclusion that the only 'rate finder' that can possibly fulfil the wants of battle is a table on the Ogilvy principle.

A. H. POLLEN.

XVII

OF WAR AND THE RATE OF CHANGE
(DECEMBER 1910 – JANUARY 1911)

By 1 December 1910 Pollen had completed an essay on the relationship of the development of fire control to that of naval tactics, which was entitled 'Of War and the Rate of Change'.[1] Proofs of this paper were ready by the end of December, and on 1 January 1911 Pollen added a 'Postscriptum' after reading the texts of two lectures that had been given at the War Course College at Portsmouth by Vice-Admiral Sir Reginald Custance.[2] 'Of War and the Rate of Change', together with 'The Quest of a Rate Finder' (included in this volume), were apparently circulated widely.[3]

Pollen was aware that his earlier prints had caused offence, and in January 1911 he sought the advice of Vice-Admiral Sir Reginald Custance on the matter.[4] 'As far as I can see,' Custance replied on 9 January 1911, 'there is no reason why you should not circulate your pamphlets. The arguments for and against your so doing remain unchanged. You are pushing theories which can be advanced only by that means, and you have faith in their truth.'[5]

NOTA BENE

The writer of the following pages was, he believes, the first to point out that successful firing beyond point blank ranges could only be founded on knowledge of the enemy's range, bearing, course, and speed. He has been concerned in developing gear to ascertain these with the requisite rapidity – and speaks here and elsewhere of such gear as a 'rate finder'. A time and range curve would give the present rate: but not the factors for forecasting the range continuously. Means of making this forecast from these materials, of correcting it for change in time of flight, and of keeping it, when made, on the sights, are – in the opinion of the writer – a need, at once primary and fundamental, of naval war.

APOLOGY

At the suggestion of certain correspondents, I have taken the liberty of printing some observations of a general character, on what appears to an outsider to be the inevitable trend of tactical thought. Most readers will probably feel that my views are unduly coloured by optimism as to the future of the A.C. system; and it may well be that this optimism is my only claim to originality. Frankly, I find it impossible to get away from the mental obsession, that if difficulties of ship propulsion and handling were the crux of war in the sailing era, the difficulties that beset the use of modern guns are the crux of naval war to-day.

My own efforts to get over these difficulties have been directed towards making the processes of finding and keeping the rate, and getting the sights set synchronously with the change of range, at once more accurate, more seaworthy, and, above all, rapid to the point of practical instantaneity. In these efforts I have, from time to time, been unexpectedly successful in obtaining official support – though somewhat curiously unsuccessful in retaining it.

It has happened more than once, that experiments ordered when one set of officers were advising the Board, came up for trial when different counsels prevailed. At one time the task imposed on the gear for trial was so easy that the chief reason for ordering the trial seemed to have been ignored. At another, the conditions – by our being excluded from the ship – were made so adverse, that the gear could not exhibit the military purpose for which it was designed, although it was recognised, when the experiment was authorised, that the gear was experimental.

My own experience has, then, made me an admirer of continuity of policy and aim. But, had I not been so lucky in getting the support I have, I might possibly have been found lamenting, and with greater eloquence, a conservatism nothing could shift from the groove it travelled in! So I probably illustrate the old saying that orthodoxy is 'my doxy', and heterodoxy, 'the other fellow's'. Readers must, therefore, discount my partisanship – which, though uncompromising, is, I hope, never truculent, disrespectful, or embittered by disappointment.

It is, of course, quite impossible for a person in my position to be an impartial onlooker in a situation of this kind. If, then, I venture to offer my views, it is, first, with the diffidence that becomes one who knows that he is, beyond conversion, prejudiced in the matter of thinking positive knowledge a surer basis for gunnery than guesswork; and, next, with the still greater diffidence that becomes one whose practical experience is almost non-existent. It is, indeed, a simple historical fact that I have given longer consideration to the problems involved than any other individual; and it has been my privilege, in the course of nearly eleven years' work, to make many warm and intimate friendships in the Service, with some gain, I hope, to my comprehension of the question. But I am fully aware that what a landsman has to say on a sea problem is generally vitiated by much that is unreal, because the innate difficulties of all sea work must be unrealised by him.

My last refuge might have been the proverbial advantage of the looker-on, who is said to see most of the game, but, alas! it is my misfortune never to have been allowed to be the spectator at the firing of a single shot under (what are called!) action conditions.

A.H.P.

1st December, 1910.

1

I find it hard to believe that it is only ten years since my saying that, with proper methods, naval gunnery should be more effective at 10,000 than it actually was at 5,000 yards, caused genuine concern for one's mental condition amongst friends. But at that time hardly a shot had been fired, at sea, beyond a mile and a half; to talk of five miles, or even two and a half, might well have sounded like the vapourings of a visionary.

2

And that such a statement is so little startling to-day is undoubtedly due to the achievements of the naval service in the intervening period. The ground that was first broken by the evolution of the scientific training of gun layers, has since been amazingly tilled by those who have spread over the vineyard in the development of control. Much has been done by individuals — more perhaps by the efforts, first, of the Fleet Committees on Fire Control, that in the Mediterranean and the Atlantic did so much in the critical years 1904 and 1905 to lay the basis of the present organisation,[6] and, later, of the gunnery staffs at Whitehall, Whale Island, and Victoria Street.[7] As one of my correspondents very properly observes, the resultant practical knowledge collectively possessed by the Fleet to-day is immense. If we take only Battle Practice results, the contrast between what we do to-day, and what we did seven or eight years ago, is truly astonishing; and, if we go further into detail and compare year with year, there is also marked progress to be noted.

3

This progress, however, appears to an onlooker rather as an all-round growth in individual skill, availing itself keenly of the fruits of sporadic ingenuity, than as a progress in method, such as could only be the outcome of concerted and organised mental effort. Wherever one touches a centre of gunnery activity, one is more apt to find some detail of gear, or organisation, the subject of care, than the discovery of the basis on which the whole structure should be raised. Hence, as a general rule, the particular performances of their own ship, in the heavy gun tests, interest the officers concerned far more than the general principles involved. This is not unnatural, because a task is annually set by the prescribed course of Battle Practice, and the duty of each ship is to perform it as best it may. But if one compares the nature of the task that is set, and the method adopted for carrying it out, one finds it hard to believe that those who ordained the end can possibly have prescribed the means.

4

Very long range has, until this year, been made the sole rule in Battle Practice; and, as a certain number of ships have always done extraordinarily well, certain quite unproved things have been taken for granted. The first is that what one ship can do, all *ought* to do, and thus hitting at long range has been made the chief object of gunnery training. And, secondly, because a certain method has in fact solved the problem for *certain* ships in *certain* conditions, it has been supposed that it is necessarily good enough for *all* ships in *all* conditions. Again, it is notorious that to hit at long range involves difficulties that do not exist at short range. Hence it is further supposed that short-range gunnery must be easy, and will present no hard problems to those who are proficient at the other. And so neither the wisdom of the long-range ideal, the adequacy of the method of making it effective, nor the problems of short-range shooting, have, as a matter of fact, come in for organised, systematic, and continuous investigation.

5

And this brings me to the curious fact that, in all the advance that has been made, there has never been, so far as I am aware, any attempt thoroughly and independently to enquire into the basis on which our gunnery system rests, unless the informal, though thorough, enquiries made by the D.N.O. in 1906 are an exception. A perfect store of priceless experience is available; but the pity is that the bulk of it has been acquired without examination into fundamentals, or is the possession of those who cannot ensure its teachings being reflected in Service policy. There is no machinery by which it can be turned into a common channel, made a depository of all the known facts, and thus the true foundations of the art be independently and authoritatively analysed. Only the creation of some such machinery can, one would think, bring into being an acknowledged doctrine of action gunnery; and I think it is hard to resist the conclusion that it is the absence of a permanent organisation, formed with this object, that explains a certain lack of continuity of aim that has distinguished our fire control policy in recent years.

6

There should be no occasion for surprise in recognising that these statements are true. Shooting from beyond point blank ranges only dates back some seven or eight years, less than the period of specialist service of a gunnery lieutenant.

The use of guns in battle is still, therefore, in its tenderest youth. The fact seems to be that we have drifted into the practice of an exceedingly complex and difficult proceeding, and done it so quickly that we have failed to recognise that it is an art just as distinct from other naval arts, as the accomplishments of an hydrographer are distinct from those of an historian.

7

Long-range gunnery was not grafted bodily and suddenly on the Service, as were steam power and fish torpedoes. With the introduction of steam, engineers, expert in its use, had to be imported ready made, with the result that their craft has always been a thing apart. With the introduction of torpedoes, selected officers had to be given special training; and hence the torpedo branch is already an old-established service within the Service, with traditions, assured aims, and a policy of its own, which no non-torpedo man would dream of criticising. In both fields *expertise* – that is, a knowledge recognised as special – is also recognised as an indispensable qualification for taking a hand in directing them. Naval officers, who are neither steam experts nor torpedo specialists, whatever their other accomplishments may be, are regarded by those of the craft as laymen. It is quite inconceivable that any Nestor of the Navigation School could be put in command of the *Vernon*, or a Solomon of the Submarines asked to decide on a new type of steam turbine.[8]

8

But gunnery is still supposed to be one of that multitude of accomplishments which every seaman can master as he masters all the rest. But surely the principle of Jack-of-all-trades can be pushed too far. I dare say it does not very much matter if one of the first violins in an orchestra is sent to play amongst the second violins; but there would be some loss of harmony if the leading piccolo were set to blow the trumpet, even if it were his own.

9

Certainly in gunnery administration we have not yet advanced very far in the differentiation of functions. That there is a distinction between the duty of providing guns and the duty of elucidating the principles of using them, cannot be ignored as negligible. There is real danger if the quasi infallibility that always attaches to every Captain of a ship in the affairs of his ship, is always to attach to every head of the gunnery department in all the affairs of that department. An officer may be a heaven-sent administrator, yet not necessarily the best judge of the niceties of an art in which he may be entirely inexperienced. When all is said and done, there is not a greater gulf between the way a ship is fought through a modern fire-control organisation and the way she would have been fought ten years ago, than there is between wireless telegraphy and the methods of signalling of 100 years ago. We should not expect a Trafalgar Signal Yeoman to make much of a hand at running a Marconi installation. How can we expect final wisdom from those whose other duties may have prevented their mastering the art for which they are chosen to legislate?

10

I am far from saying that only the best exponent of an art can safely act as a director of it. There is, on the contrary, a very grave danger, when the exceptional man lays down the law, because he may forget that he is exceptional. He is apt to think others equal to him. Imagine how disastrous it would have been if the Ithaca Territorials had been armed by Ulysses with weapons the counterpart of his own. It was doubtless the best bow ever seen; but useless to anyone but himself, as the candidates for the favours of his putative widow found to their confusion.[9] We run into similar risks when we take the performance of persons of outstanding merit as the standard we are to expect from ordinary men. The plots manually made in the *Vengeance* were, I was told, in every respect as good as, if not better than, ours made in the *Ariadne*. The inability of average ships to equal these performances is an instance in point. You do not need the greatest exponent of gunnery to preside over gunnery policy. But you cannot expect wise direction unless the existence of gunnery technique as an art apart is recognised. And this implies that decisions affecting it will be guided not by slap-dash generalisations, but by the considered opinions of those who possess both the most experience and the best judgment.

11

It looks, therefore, as if the shifting moods of fashion might well be steadied by some organisation, of which the individuals composing it changed slowly, so that it could keep a certain permanence of corporate being without losing touch with the vivifying ozone of the sea. It should be the function of this organisation to embody the experience and co-ordinate the suggestions of the Fleet, the schools, and the various staffs concerned. Infallibility is not within the reach of mortals. But an organisation, small enough to be active and large enough to be impersonal, changing enough to be always drawing in the latest knowledge, and enduring enough to represent the traditions of past experience, should go some way towards making escape from the more obvious forms of error certain. The judgment of contemporaries to some extent anticipates the verdict of posterity. You must make that judgment come from a wide enough circle to be sure that it is representative of the best.

12

I have heard it said that the danger of incorporating specialists in this way is twofold. First, responsibility would be diffused, enterprise checked, and new departures delayed; secondly, the authority of the executive would be weakened. But these are surely chimerical perils. The organisation suggested would have no executive duties – and consequently no executive responsibility. Its functions would be limited to giving expert judgment on specific problems: it would be a body less advisory than informative. It is inconceivable, if it were properly constituted, that it could lack the courage of its convictions – and when its convictions were expressed, its verdict would be a protection to those who acted on it. And, for that matter, Admiralty authority is not enfeebled by being kept in unison with the best Service opinion. If enfeebled at all, it can only be so when in conflict with that opinion. You cannot box the compass of mutually destructive policies, and keep the prestige of having been right each time. Opposing policies may both be wrong. They *cannot* both be right. A business man is not less enterprising because he protects himself with legal or other expert advice. I yield to none in my admiration of the system that makes Whitehall absolute. But the more absolute the rule, the greater its need of authoritative information as to the requirements of purely technical problems. And so long as the information it gets is not authoritative, there is the risk that questions will be treated as open, or not urgent, that in reality are both settled and crucial.

13

Did such an organisation exist, I feel confident that two of the first questions it would consider would be, first, whether the obsession of long range rests on any very reasonable foundation; and, next, whether the adequacy of the methods by which we have attempted to excel at it, can be satisfactorily defended.

14

It is not that both these problems are not already vigorously debated by many acute minds in the Service. I am well aware that opinions differ. But, speaking for myself, I find it impossible to defend my professed belief of ten years ago, viz., that to cultivate the art of hitting from the greatest possible distance is the beginning and end of tactical wisdom. Consistency is the acknowledged monopoly of the least eulogised of equines, and I am not concerned to defend my own lack of it; but I may be pardoned, perhaps, if I attempt to give my reasons for the weakening of my former faith.

15

First, and by way perhaps of unnecessary self-defence, may I say that my early hopes of long-range firing were based on obtaining that antecedent knowledge of the enemy's position and movements, which seemed necessary to make such fire effective? There are two, and apparently only two, ways in which long-range gunnery may be attempted. You may make a guess, open fire to correct your guess by spotting, and then use your gun as a rate finder – just as up to three years ago you at times used it as a range finder (and still must at great distances to get the gun range) – or you may find the rate by other means, before you begin to fire at all. It is the last method that I have always advocated. It is not the method that is actually adopted to-day. The plea that my condition precedent of 1901 has not materialised might, therefore, be put in to defend my change of mind.

16

The historic explanation of rate finding by guns, lies in the fact that the gun was used as a range finder before shooting at a moving target was begun. Spotting is the process by which the gun is so used; and is an effective process so long as the range is not excessive, and is either constant, or varies in only known amounts. It ceases, however, to be a warlike process when the rate is not known, not because it is ultimately ineffective for discovering the rate and range, though this must also be the case when the rate is high, but because the process is too dilatory a one for war. However, the habit was formed before the days of the moving target, and has been persisted in long after its inherent limitations have been passed. That these limitations were imminent, and hence a new basis for naval gunnery needed, was quite appreciated in 1905.

17

In that year a double experiment was tried in H.M.S. *Jupiter*. We had put forward a new invention that was at once a range and rate finding system. The range finding failed for mechanical reasons: the rate finding because it was man-worked or polyanthropic. This failure coincided with a vast improvement in the Barr and Stroud telemeter; and, abandoning our two-observer range finding, we designed gear that converted this instrument into an automatic rate finder. With the results obtained in the *Jupiter* fresh in their minds, the gunnery authorities, who quite appreciated the obvious truth that the distinction, between *knowing* the rate of change and *guessing* it, was vital, realised that, while manual methods must fail (because they were too slow), our method, being automatic (and, therefore, rapid), might possibly succeed; and a trial of our system was ordered.

18

But before this trial could take place, those who had recommended it were succeeded by others, whose attention had not been directed to the 1905–6 experiences in the *Jupiter*. They assumed that manual plotting was an untried novelty, and must succeed, and – still faithful to the doctrine that to know, and not to guess the rate, was essential – adopted it in lieu of the automatic system for the fleet. And, now, when after three years, its failure has been once more – and this time finally –

demonstrated, the search for a method of finding the rate of change *before* fire is opened, which for six years has been the cardinal point of fire control policy, has, for the first time, been apparently abandoned. Nevertheless, if, as seems to be the case, the production of an efficient and automatic rate finder is certain, it is most unlikely that guesswork will long survive as the foundation of action gunnery.

19

Method apart, however, to excel at the greatest possible distances, say, from 8,000 to 11,000 yards, does not seem to me now of one tithe of the importance of attaining such perfection as is possible at the lesser distances, that is, between 8,000 yds., and point blank. I am not laying it down that no shots are to be fired at 8,000 or beyond. But I do assert that to cultivate shooting at these ranges exclusively, and when the change of range is small, is to pursue an altogether false ideal.

20

The object of all preparation for war, whether it be in the creation of larger and faster ships, the provision of more formidable weapons, or the cultivation of skill in using these, is to obtain an advantage of some sort over our adversary. We all have a dim consciousness that, as movement has very prettily been called 'the soul of war', greater quickness of movement must make some form of superiority possible, even if no one has yet stated, how precisely that superiority is to show itself in action. Hence the huge public delight at hearing that a cruiser, larger, faster, and more expensive than any other in the world, is about to be laid down.

21

So, too, more powerful and longer ranging guns strike everyone as beneficial: though here the precise character of their superiority seems obvious enough. The mere possession of guns of such startling potentialities as the new 13.5's, warms the heart of the man in the street,[10] because the man in the street both takes hitting for granted, and is apt to over-emphasise the possibilities of the lucky first hit. That there are great possibilities in this, no thinking person will deny, and these possibilities must be given their due weight and value. There need be no fear that they will ever be ignored. But you cannot found a system of tactics on *possibilities*: you must build on the certainties. However, the fascination of the big shell, hitting from the very confines of the field, are undeniable. When we hear of things like this, we remind each other of Dewey at Manilla Bay, and then proceed to hug ourselves with the rousing reflection that, 'We can hit him before he can hit us.'

22

Now note that in defining the advantage of long-range hitting, we use the word 'before', an adverb of time, and not an adverb of space. We know, of course, that space, in a sense, is time. We talk of telegraphy in different forms annihilating space, because in our minds we connect space with the time taken to traverse it. Time is the essential obstacle to rate finding by guesswork and guns, because the process takes too long for the rate, when found, to be useful in action. If, when the rate is too difficult for this process, it can be got sufficiently rapidly by another process, we can say that we have *out-timed* our enemy; because we shall hit him before he can hit us, in exactly the same way as if we out-ranged him. If we had such a method at our command we could, by eliminating time, thus gain the whole tactical advantage of lengthening range; and, if we aim at developing such a process, we can find common ground in saying that our proper tactical aim should be *priority* in being both reasonably sure of the first hit, and absolutely sure of attaining that continuous hitting which alone can be decisive of an engagement. Let us see if we can illustrate the bearing of this by an example; possibly an extreme one.

23

Let us assume that the Battle Practices of 1908 and 1909 permit of the following fair deductions. First, with an unknown change of range of 200 yards a minute, the hitting efficiency of our fleet is 20 per cent of hits to rounds fired, between 6,000 and 9,000 yards. But in the 1907 Practice, before the change of range – i.e., the rate – was unknown (because then the target was moored and did not move on an unknown course), although the rapidity of the change of range was greater, the efficiency was higher by 50 per cent. So a further deduction must be allowed – viz., that, with every increase in the *unknown* change of range, there will be fewer hits made, until the vanishing point is reached, let us say, when the unknown speed and course of the target set up a change of range of 1,000 yards a minute. In such a case an enemy, 10,000 yards off at the opening of fire, would, for four minutes, be as immune from hurt at our hands as if our projectiles could not reach him.

24

Now let us suppose the existence of a perfect rate finder, which could be used in any sea, and whether under helm, or on a steady course indifferently. Such a rate finder would find the change of range in any conditions; and, if we had rate keeping gear as well, then whether the rate were 100 yards or 1,000 yards a minute, it would, theoretically, make no difference at all to hitting. The problem would be practically reduced to firing from a stationary gun at a stationary target; and, in the case we have been supposing, if one side had such gear and the other had not, the side possessing it would, at any range, have precisely the same advantage over his enemy as if he out-ranged him. Observe, I am not saying that to obtain priority by hitting from a greater distance is not a tactical advantage, or with better means at our disposal, should not be sometimes striven for. I am saying that with a change of range of 800 yards per minute, our present chance of hitting at such a range would be negligible. Similarly our chances of being hit would be negligible. The point is that there are other advantages to be sought.

25

The serious question, however, is whether these advantages are not far more important, far easier to obtain, and, for reasons that lie in the very nature of war, far more necessary?

And here we are faced by a consideration that meets us at the very threshold of the question. If a *Dreadnought* were anchored at 6,000 yards from a gun mounted ashore, assuming such a ship to be equal to a 30ft. target, and the range and the gun range were verified, the mere eccentricities of the gun would not permit even the latest 12in. ordnance, however admirably laid, to hit with every round; and, at 10,000 yards, in these exceedingly favourable conditions, not more than 40 per cent of hits could be expected.

26

If these are, as I understand they are, the brute facts of ballistics, it looks as if a long range action — when, even with ideal gear, it is irrational to expect the affair to be one of mere marksmanship — must exhaust the stock of ammunition and wear out the guns without any decisive result in the matter of hits. Material and economic considerations alone, quite apart from the danger of demoralising your fleet by a prolonged and ineffective bombardment, forbid the contemplation of very long range, either as the only, or even the normal, conditions of engagement. It may be wise, of this I cannot judge, to open fire during the approach — the mere chance of a hit may make it worth while. But it can never be wise to keep at a range at which you are not sure of continuous hitting — supposing, that is, you wish to fight.

27

Then there is the still graver question whether in point of fact — omitting the more obvious contingencies of bad light, thick weather, etc., you could always, or ever, have that choice of range which, at one time, figured so gloriously amongst the legend advantages of the all-big-gun ship. Supposing an enemy, knowing your pre-possession in favour of this form of fighting, elects to close you at 21 knots. You must turn away to keep your range, you must stop him with your guns, or you must resign yourself to a shorter range. With the means of getting tactical information now available, there seem to be great practical difficulties in carrying out either of the first two courses.

It is easy enough to talk of turning away, but it is, at least, as well to remember that the proceeding is not an automatic one. When tactical evolutions of this sort are proposed, one wonders what consideration is given to the elements that make up the momentum and the inertia of a fleet of ships in a line. Any turn of this kind, if it is to give the whole fleet the benefit of this choice of range, must, in most cases, be a movement which all must make together. The benefit of the chosen range will be very unequally distributed if the fleet is to follow the flagship. A concerted movement must be definite in character, a prescribed amount of turn and a prescribed moment for making it. The first step in the procedure, therefore, must be in the Admiral's mind. He must be precisely informed of, and quite comprehend, the movement that the enemy is making, and, having got the information and digested it, he must consider the different counter strokes open to him, and select the one that he is to act upon. Whatever his decision is, it must be signalled, his fleet must take in the signal, and, after it is taken in by every ship, the executive must be made. How long after the enemy's closing movement has begun, say at 10,000 yards, is it before the necessary knowledge is obtained, digested, and acted upon, so that a fleet of say, 16 ships, actually begins to turn away? If four minutes, the enemy will have already closed from 10,000 to under 8,000 yards, and it will become a serious question whether at this moment any turn can be made at all; and for the simple reason that the advantage of the superior ranging power has been lost before an effort has been made to regain it. To regain it now, a vast superiority in speed must be necessary. With a resolute enemy between 7,000 and 8,000 yards away, it will hardly be very inspiring for a fleet to be told to throw all guns out of action by what will look like precipitate flight; and, however well the manoeuvre might be defended (on scientific grounds) at the resultant court martial, it does not seem at first sight a thing that many seamen would be inclined to do, or be very proud of having done.

28

And now suppose, instead of turning, the Admiral, resolving to take advantage of having more guns bearing, holds his course, and tries to stop the enemy by his gun fire. The change of range will be 600 yards a minute, or more. If recent Battle Practices are a guide, we cannot expect a great number of hits to be made in these conditions; certainly not enough to stop an enemy whose determination it is to fight us at point blank range.

29

The moral of this example seems to be fairly simple. We may possibly be equipped with everything else we want in war, but we are demonstrably without the one thing which in war is most wanted.

War is not like a natural convulsion, a clash and conflict of blind forces. Man, when he exercises his strength in battle, is governed by his mind exercising itself upon his knowledge. His success will be proportioned to his possession of knowledge in two different forms. He must know how to use his weapon, and his use of it must depend on his knowledge of the enemy. If naval war is to be decided by artillery contests between hostile ships and fleets, it is evident that the side that can obtain the most accurate information about the other, can obtain it most rapidly, and has the means of turning that knowledge more instantly and more continuously to account, must have by far the greater assurance for success; and for the simple reason that it is upon these things that the efficiency of artillery, outside of point blank ranges, depends. It will hit sooner, more rapidly, more continuously than the other, and its tactics will be informed by instant knowledge of what its opponent is doing.

Hence it appears hardly disputable that a perfect means of information is the most obvious want of war; and, as we do not possess any such means, perfect or imperfect, it cannot be extravagant to say that we lack what is in war most wanted.

30

If we compare the actions off Ushant under Keppel, in the West Indies under Rodney, and of the 'Glorious First of June' under Lord Howe, with the three great battles that Nelson fought at the Nile, Copenhagen, and Trafalgar, the most obvious point of difference that we shall find is that the first were all inconclusive, because the opposing fleets were only intermittently, and then only partially, engaged, whereas the last were fought to a finish by fleets that were virtually stationary once the fighting had begun. Apart altogether from the hampering traditions of the line of battle, and of the merits of the tactical conceptions of Keppel, Rodney, and Howe, to have brought the fleets on these occasions into contact, and to have kept them long enough in contact to reach definite results, would have presented problems of almost insurmountable difficulty. It seems to me a most significant thing that Lord St. Vincent, an admirable judge, and certainly no grudging admirer of Nelson's genius, writing *after* Trafalgar, should have attributed all our naval victories from Rodney's to Nelson's, not to greater tactical genius, to greater courage, to better gunnery or discipline, but solely to superior seamanship. To bring about and maintain conditions in which gun-fire could be decisive, was clearly, to this enlightened critic, the single and sole test of leadership, both of the Commander of the Fleet and of the Captains of the individual ships. It was primarily by supreme seamanship alone that such conditions could be created and maintained at all; and, compared with such superiority, the whole of tactics, to use his own words, were 'only frippery and gimcrack'.[11]

31

There is something to be said for the view that the fundamental problem of sea fighting has changed since steam has revolutionised locomotion, and artillery has been developed which can be used at ranges of three or four miles. To-day nothing but the headlong flight of an enemy in the possession of superior speed can prevent ships being brought within effective ranges. There may be better and worse ways of closing, but closing *per se* presents no difficulties. The seamanship, then, of St. Vincent's admiration can no longer be appealed to as the most important of a sailor's acquirements.

32

But if the technique of propulsion and manoeuvring has become so straightforward an affair that no special genius is called for in a leader merely to keep his fleet in line and bring it somehow within range of the enemy, it undoubtedly is true that the technique of naval artillery, which was of very small account in the days of the sailing fleet, does now present most difficult and interesting problems. If, then, seamanship was the crux of naval war in the sailing days, it is probably no exaggeration to say that gunnery is the crux of naval war to-day.

And if superiority in technique, method, and skill in conquering the difficulties of gunnery is absent, no skill in tactics, and no superiority in number, speed, and protection of our ships, or ranging power or weight of our armament, can take its place. All else, this being wanting, is 'frippery and gimcrack', too.

33

The uninstructed public judge of the relative fighting strength of navies by the numbers, and tonnage, and cost of the ships composing them. They are impressed by superior thickness of armour, greater legend speed, larger and heavier armament. And these things are their only criterion. The verdict of history is against the setting up of so false a standard. I do not forget that a century ago a three-decker was considered the equal in action of two 74's; still less do I forget Nelson's 'Numbers only can annihilate.' But, good as these maxims were in their place, they did not deter Jervis engaging a fleet of 26 Spaniards with a fleet of only 15 British ships. Nor did they stop Nelson, in the *Captain*, taking two three-deckers single-handed at St. Vincent; nor frighten Cochrane when he set out to capture a 40 gun frigate with his cock-boat, the *Speedy*. Seamanship, discipline, and practical skill in the arts of war, inspired by the fighting spirit that takes the initiative and keeps it, and united with the determination to win, have redressed the balance of greater material power again and again. Would not superior skill with guns, joined to the same spirit, redress even more striking inequalities to-day? After all, with sailing ships at a decisive range, it was a certainty that both sides would hit. But, at between 5,000 and 7,000 yards, with a rapidly changing range, there could be no such certainty to-day. If one fleet never hit, and the other never missed, that twenty ships could not conquer two, but that two could conquer the twenty, would be a truism.

It is clear, then, that inferior material force, if used with better method, and greater knowledge, might often mean superior power. It is well, then, to beware lest, in being prodigal of pounds to the tune of millions, when it comes to the supplying of ships, we fall to being parsimonious of pence to the verge of miserdom, when it comes to providing means to fight them.

34

Nelson came to the problems of sea war as the inheritor of the lessons of a generation of sea-fighters, whose successes were hardly more instructive than their failures. By a supreme intellectual effort, he broke away from the hampering traditions of the line of battle, and availed himself of the incomparable seamanship of his Captains to bring his forces, with the highest possible speed and effect, to bear upon the enemy. The factor that ensured victory had been generated at sea. If the present generation of the British Navy is going to war, it can have no fighting traditions from the immediate past. What will be tested will be, first, its capacity to become adept in the use of weapons of untried possibilities, and, next, to apply the teachings of a remoter past to conditions of vastly altered implements of war. We may succeed, either because we have learnt both lessons well, and our opponents not at all, or we may fail because they have learnt them better than we have. But it is well to remember that the test on the day of battle will be very far from being only that of opposing leaders and opposing fleets. The battlefield is not only the trial of the forces, but of those who, perhaps long before, have been responsible for their preparation. If the technique of gunnery is truly the crux of war, the quality of the method is almost more vital than the quality of the practice. The greater the complexity, and the power of the weapons you use, the greater the influence your method of using them must have on the result. But the choice of method is not with the sea-going fleet: but with administration on shore. The things necessary to enable modern implements to be used in war do not defy analysis or anticipation. They have been made amply clear by argument, experiment, and three years' practice at a moving target by the sea-going fleet. If the common impression is correct that naval fighting is to be principally an affair of artillery, it is obvious that the requirements of artillery fighting must be supplied.

35

It is a curious coincidence, but a coincidence only, that the first experiment with the A.C. system, and the resolution to change the type of capital ship from one of moderate dimensions, moderate speed, and mixed armament, to far greater dimensions, far higher speed, and uniform armament, should have happened together. There seems clear ground for saying that the justification of the second policy must largely depend upon those necessities being successfully provided for, which until this year every effort has been made to meet; and, until those necessities are met, it can hardly be hoped, that the tactics on which the bigger and faster ship, and the bigger gun, have been based, can be verified by experiment to be practicable; possibly it is safe to say that there is no need of experiment to show that they are impracticable.

36

Not that the tactical objective known as the 'choice of range' — meaning thereby the effort to fight at a longer range than your enemy would choose — would be any sounder with a perfect rate finder than with none at all. It is essentially one of those advantages that looks less and less warlike the more often and the more closely it is examined. It seems to pre-suppose a very defensive sort of fighting, and appeals most strongly to those who, by keeping their enemy outside the physical limits of hitting, consciously or unconsciously have it in the back of their minds that they would like to make war without taking risks; and this, as Bonaparte had occasion to lament, is not the sort of Admiral that is very effective against those who act on more full-blooded principles.

37

There is no call to labour the value of *morale* in battle, nor how completely *morale* depends upon the vigour and decision with which the attack is made, nor how it grows by the contemplation of the efficacy of the attack upon the opposing forces. Possibly the spirit with which the attack is made is even more important than its efficiency when delivered. The judicious and mellifluous James, commenting on the battle of St. Vincent, explains the disproportion between the material strength and the fighting value of the Spanish Fleet thus:

On the other hand, the very front put on by the British Fleet was enough to sink the hearts of the Spaniards, for it is one of the characteristics of true valour to daunt by its intrepidity and to begin to subdue, ere it begins to combat.[12]

One can hardly picture to oneself the hearts of the High Sea Fleet sinking at the sight of our First Division of super-Dreadnoughts turning − with dauntless intrepidity − their backs upon the foe, and beginning to withdraw ere they began to combat!

38

The business of war is to destroy, and a leader may have to risk the wholesale destruction of his force to carry out this purpose. Hawke at Quiberon Bay, Jervis at St. Vincent, Nelson in his pursuit of Villeneuve, were each and all willing to chance the entire destruction of their own fleets so long as they could inflict still greater loss upon the enemy. The severest criticism of the tactics of Trafalgar was that the heads of the British columns might, and perhaps should, have been overwhelmed; but the taking of such risks is of the essence of war, and is its own justification. The comfort of security is not for the fighting man. You cannot often play for safety and for victory together. There are such things as defensive campaigns, and even in battle you may have to be on the defensive. But the wit of man has not yet devised a method by which you can take the tactical aggressive on any purely defensive lines.

39

The British Navy being ever a body of expert, active-minded, patriotic men, has of late been true to its history in being marked by the existence of what, for lack of a better word, one must call 'parties', that is groups of individuals distinguished by different ideals and following divergent trains of thought. One such group has been dubbed the historians, in that it seeks to restore the traditions of the fighting age by the study of its achievements; and by this revival, to eliminate the many false aims and standards that a long period of peace, coinciding with vast changes in the implements of combat, has, it alleges, brought into being. Thus, in the policy of an opposing group, it sees nothing but the gross worship of the material of war, and refers the recent increases in tonnage, armour, guns, and speed to nothing but this idolatry. The other group, so attacked, replies that industrial progress and invention have made engines of greater potential power possible, and, it being their business to prepare for war, they must see to it that the nation has the best that money and the arts can supply. The first retort that no greater warlike effect can be got with the larger and costlier units; that, in use, the paper potential of the new ships will not, and cannot, be realised. A *tertium quid* explains perhaps one of the more fundamental lines of cleavage between the two schools. Within the last few years the use of guns beyond point blank range has, for the first time, become an apparent possibility. The group the historians call the materialists have either, like our man in the street, taken the realisation of this possibility for granted, or counted, somewhat light-heartedly perhaps, on its early realisation, and so have built as if the legend power of the gun was a true index to its efficiency in action; and the historians, fully appreciating the intellectual weakness of this attitude, reply that, gunnery being as it is, the new departures are not only in a military sense valueless, but mischievous, partly because the money might be so much better spent, but more because the swagger of size has seduced both the Navy and the nation into living in a fool's paradise.

40

If it is not impertinent for a layman to pray a blessing 'on both your houses' (not a 'plague', for he has friends he greatly respects in either camp) it would be that reconciliation might be found in both parties merging in the endeavour that meanwhile has occupied the vast majority of the Navy. For the majority belongs to neither party, and contents itself with striving patiently, and with very inadequate means, to gain mastery of the weapons to its hands. And until this mastery is gained, neither the tactics of battle, nor the strategy of peace preparation for it, can be founded on anything but sand.

41

Should this mastery, by better means, now become attainable, then, if the 13.5, by its flatter trajectory, is an easier gun *to hit with* than the 12-inch, in that it can hit in more conditions, and not less rapidly; and if speed can be made of possibly supreme value in action, then the material school, who have given us these things, will not have been so far adrift in their recent works; and if, when the full use of guns and speed is made possible, the lessons of history can then, and only then, be embodied in the fighting tactics of modern fleets, the historians will have a full field for their labours.

42

It is largely with the hope that this may be the issue of the strenuous work and thought of the last few years that I venture on the conclusions drawn from the gunnery technique that now seems manifestly within our reach.

I suggest, then, that normally there can be only one tactic of battle — viz., attack.

And, just as quickness of brain in the leader is more than quickness of foot in the men, so superior speed in getting and using accurate knowledge of the enemy's movements is, tactically, of infinitely greater value than superior engine speed.

And high engine speed can be made of tactical value only if you have means of getting the knowledge necessary for using it.

Thus action speed is not engine speed, but Admiral speed. It cannot be measured by the volume of steam that causes the revolutions of the screws; but it can be measured by the swiftness of the knowledge that feeds the convolutions of the directing brains.

The only way to make your enemy keep any given distance from you is to make it too hot for him within that distance. You may wish to keep him away because you are in inferior force. It is a necessity of defence. It can be obtained only by establishing artillery conditions too difficult – *i.e.*, too great a change of range – for your enemy to overcome, but not too difficult for yourself. Flight is not a counterstroke.

If the enemy shows signs of wishing to keep at any given distance from you, the only tactical value of superior speed is to enable you to frustrate that desire.

In considering the mere distance at which the opening rounds should be fired, it is well to remember, that, if you go far in ensuring the demoralisation of your enemy by being the first to secure continuous hitting, you may go further in ensuring your own, by being in a hurry to begin continuous missing.

When you engage at long range, then, it can only be while the change of range is the highest that the skilful use of speed, exercised in the light of knowledge of the enemy's movements, enables you to set up.

Long-range shooting will thus always be at as rapidly diminishing a range as you can make it, for it will last only while you hasten to the ranges where there is a higher percentage of hits to rounds. This percentage depends, first, on the accuracy of the gun, but infinitely more on the suitability of our gunnery processes. Both should be of the best possible. That the method of using the guns should be the best, is vastly more important than that the gun itself should be the best.

And you will hasten to the shortest range that is for other reasons prudent, because your first and only instinct will be to make the business decisive, and as rapidly as you can.

Decisive conditions to-day are limited not only by the length of the range, but more by its unknown inconstancy. A scientific technique would eliminate change of range from the problem. For reasons already set out, this technique might not extend ranges – except as improvements in guns pushed out point blank limits – but it should extend almost infinitely the tactical conditions in which artillery could be made effective. Hence, whereas a constant range at a moderate distance would be sought by any Admiral who wanted decisive results to-day, with a better technique he would strive for the highest change of range possible within the distances he had marked out for action.

43

These seem to an outsider the axioms of tactical thought that arise inevitably from that flexibility and precision in the use of naval artillery that can only come, and will surely come, when we can command perfect knowledge of the enemy's movements. That they are in line with the military concepts of the heroes of old will not be denied by those who have enriched their leisure by the study of past glories. From Raleigh to Rodney, the fighting instructions insist that no shot shall be fired except at point blank. In every action of the French war, the signal to break the line, or 'to engage the enemy more closely', flew continuously. Broke waited to fire until the *Chesapeake* was within pistol shot of the *Shannon*. When fleet speed was three to five knots, Nelson said that 'five minutes make the difference between victory and defeat'.

But quotations are not needed to show that close action, of which the modern equivalent is not mere physical propinquity, but such mastery of conditions as result in continuous hitting, was ever the tradition of the British Navy; nor need the least historical be reminded that, from Alexander to Napoleon, and from Drake to Nelson, every great genius in war has been marked above all by the rapidity of his decisions, the swiftness of his movements, and the headlong impetuosity of his attack when he delivers it.

A.H.P.

POSTSCRIPTUM

After the final proofs of the foregoing had been corrected, I was privileged to peruse Sir Reginald Custance's War College Lectures on Lissa and the Yalu. To claim that the gallant Admiral's pellucid deductions from actual war, are in substantial agreement with the *a priori* arguments of my essay, must look like paying myself a gratuitous compliment. There is, of course, in a didactic sense, an element of disagreement. But the divergence is, I think, limited to my assumption that a scientific gunnery technique would give to naval artillery a certainty and flexibility it does not now possess; and that there must be such a divergence between my theses and the best practice in to-day's conditions, is anticipated in the last paragraph of section 42.

A.H.P.

January 1st, 1911.

XVIII

THE GUN IN BATTLE

(*FEBRUARY 1913*)

The mechanical indication of ranges that were changing at a rate that was itself varying required some means of driving a pointer across a dial at a speed that changed continuously. The motors of Pollen's day, however, could not be made to vary in speed continuously with the degree of accuracy that was sufficient for the purposes of fire control. In the Argo Clock Mark I this difficulty had been overcome by an arrangement in which a motor that ran at a constant speed produced a continuously variable speed result. This was achieved through a mechanism that was driven by an electric motor that ran at a constant speed, which after having been set with the courses and speeds of the firing ship and target and a range and a bearing, continuously reproduced the virtual course movement of the two ships – that is, the motion of the target relative to the firing ship, which was presumed to be stationary.[1] This action was, in turn – through mechanical linkages – used to control the movement of a pointer on a dial marked with ranges, and a pointer on a dial marked with bearings, whose indications of the range and bearing would reflect any continuous variation in the change of range and change of bearing rates, since such variation was inherent to the replicated motion of the two ships.[2]

The Argo Clock Mark I could not, however, compute ranges and bearings accurately while the firing ship was turning. This was because its mechanism was incapable of integrating the turning motion of the firing ship with the straight motion of the target to produce the virtual replication of relative movement that was required. Pollen and his design engineer, Harold Isherwood, thus turned to another means of using a constant speed motor to achieve a continuously variable speed result in order to produce the 'helm-free' clock that Captain Frederick Ogilvy had believed was essential.

The calculating mechanism of the Argo Clock Mark II consisted of a disc, a ball, and two closely spaced parallel rollers that were mounted on a slide which enabled them to be translated along their axes.[3] The path of rotation of the rollers ran perpendicular to a diameter of the disc, the roller axes were parallel with the disc surface, and the slide positioned so that the translation of the rollers along their axes took place directly above a diameter of the disc. The ball was positioned between the disc and the two rollers, and firm contact between the three units was maintained by a spring that pushed the disc towards the rollers. An electric motor controlled by a governor rotated the disc at a constant speed, the rotation of the disc caused the ball to spin, and the spin of the ball, in turn, caused the rollers to rotate. The speed and direction of the rotation of the rollers could be varied by moving the ball along a diameter of the disc. Friction between the ball and rollers during this movement along the diameter of the disc, that would have resulted in slippages had the position of the rollers been fixed, was eliminated by the movement of the rollers along the slide to match the travel of the ball. As the ball was moved from the rim of the disc to its centre its speed of spinning diminished from the maximum possible to nil, and as the ball was moved further along the diameter from the centre to the opposite rim its speed of spinning increased from nil to the maximum possible, but in the opposite direction. These motions were, in turn, imparted to the rollers.

The initial distance of the ball from the centre of the disc was determined by a change of range rate. The movement of the ball along the diameter of the disc was then determined by changes, if any, in the target bearing. If the courses and speeds of the firing ship and target were such that the range was constant, or changing at a constant rate, there would be no change in the target bearing. In the former case (constant range), the ball would be placed at the centre of the disc and would not, therefore, spin. In the latter case (change of range at a constant rate), the ball would be displaced from the centre of the disc, but its position would not alter, and the ball would thus spin at a constant speed. If the courses and speeds of the firing ship and target were such that the change of range rate was itself changing – which could occur when both courses were straight, or when the firing ship was turning – the target bearing changed. By moving the ball along the diameter of the disc at a speed that corresponded with the change in target bearing, the speed at which the ball was spun would vary in accordance with the change in the change of range rate.

The change of range rate that was required to set the initial position of the ball along the diameter of the disc could be obtained from a time and range plot. Alternatively, the Argo Clock Mark II was equipped with a trigonometric calculating mechanism – in essence a dumaresq[4] – that converted true and virtual course and speed settings into a change of range rate setting, e.g. which then determined the position of the ball. The bearings that determined the subsequent movement of the ball along the diameter of the disc were obtained by observation. The calculation of the target range was accomplished by the geared connection of one of the rollers of the disc-ball-roller mechanism to a pointer that rotated around a dial marked with a succession of ranges. After being set to an initial position that was determined by a range observation, the pointer rotated around the dial at a speed determined by the actions of the disc-ball-roller mechanism. The pointer thus indicated the range as marked on the dial face in accordance with the change of range rate, whether it was nil, constant, or varying.[5]

The disc-ball-roller mechanism of the Argo Clock Mark II was a development of the disc-ball-cylinder differential analysers that had been devised by Lord Kelvin in the 1870s, which in turn had been based upon ideas that had been formulated by his brother, James Thomson, a professor of engineering at Glasgow University.[6] In Kelvin's instruments, the motion of the ball was imparted to a cylinder whose position was fixed and whose length spanned the diameter of the disc. Kelvin may have suggested the disc-ball-cylinder scheme to Pollen in the spring of 1904 at which time, it will be recalled, the two had thoroughly discussed the fire control problem.[7] Alternatively, Pollen may have learned of the method from Charles Vernon Boys, who, it will also be remembered, had advised the inventor on gyroscopes in 1906[8] and who had a particular interest in the study of differential analysers.[9] But whatever the case, the initial attempt to apply the disc-ball-cylinder approach to the calculation of changes in the change of range rate failed because of the occurrence of slippages between the disc and ball that were caused by excessive friction between the ball and the cylinder as the ball was shifted along the diameter of the disc.[10] The design work to overcome this difficulty was not begun until early 1909,[11] even before the necessity of developing a 'helm-free' clock had been established by the discussions with Ogilvy. The solution, which took the form of the sliding double roller mechanism described above, was found in 1911 and allowed the design of the Argo Clock Mark II.[12]

A description and plans of the Argo Clock Mark II were submitted to the Ordnance Department on 8 May 1911.[13] Pollen then added a bearings calculator that embodied the virtual-course and speed-calculating mechanism of the Argo Clock Mark I.[14] When the firing ship steamed on a straight course, this device, when set with range and bearing rates and a bearing, or true or virtual courses and speeds, and a range and a bearing, generated bearings that were set on the disc-ball-roller mechanism automatically. When the firing ship turned, the automatic setting of calculated bearings was superseded by hand set bearings obtained by observation. Thus modified, the Argo Clock Mark II design became the Argo Clock Mark III design.[15] The Argo Clock Mark III design was in turn modified to become the Argo Clock Mark IV design by the addition of mechanisms that adjusted the change of range rate setting of the disc-ball-roller mechanism and the virtual course and speed setting of the bearings calculator, in accordance with any changes that might be made in the settings of the dials that indicated the computed ranges and bearings. Corrections in the settings of the range and bearing dials could be made whenever the computed data differed from observed ranges and bearings, which could be presumed to be accurate. Occasional individual range and bearing observations could, in this manner, be used not only to check the accuracy of the computed data, but to increase the accuracy of subsequent computations as well.[16]

A prototype instrument was constructed, and its operations demonstrated to several members of the Board of Admiralty and various naval officers in the London office of the Argo Company in February 1912.[17] The Argo Clock Mark IV was formally offered to the Admiralty for trials on 18 March 1912,[18] and arrangements were made to test the prototype in the new battleship *Orion* after further successful demonstrations were carried out in London in March, April, and May 1912.[19] Preliminary trials of the Argo Clock Mark IV in the *Orion* began in September and continued through October 1912.[20] In late September, Pollen and his engineers were allowed on board the *Orion* for tests of the clock's ability to keep the range accurately while the ship was turning. These trials were carried out in Bantry Bay off the coast of Ireland and were completely successful.[21] On 26 October 1912, the Admiralty – apparently impressed by the results achieved thus far – ordered five additional Argo Clock Mark IVs.[22]

On 27 September 1912, shortly after the preliminary trials had begun, Pollen had acted to obtain an independent scientific evaluation of his fire control instruments by asking Charles Vernon Boys to judge the practicability of the clock that was then under trial, an improved range finder that had been designed, built and tested by the Argo Company, and the drawings of an as yet unbuilt 'helm-free' true-course plotter.[23] Boys accepted the commission and, after examining the plans of the instruments and meeting the design engineers, submitted his report, which reached the Argo Company on 13 November 1912.[24] In his report, Boys concluded that all three instruments would produce the desired results, and further observed that the Argo Clock Mark IV was 'as near perfection as any mechanism which I have ever examined critically.'[25] Copies of the report were sent to Winston Churchill, the First Lord; Vice-Admiral Sir John Jellicoe, who had been placed in charge of the official trials of the Argo Clock Mark IV; and Rear-Admiral Richard Peirse, the late Inspector of Target Practice, in late November 1912.[26]

By 1912, however, an alternative mechanised system of fire control had been developed to rival that of the Argo Company. In 1910, Commander Frederic Dreyer had abandoned the manual plotting methods that he had previously insisted upon and had patented a mechanical system of obtaining a range and a bearing rate by plotting ranges and bearings separately against time.[27] Dreyer proposed that the two rates be set on a dumaresq, along with the known course and speed of the firing ship and the observed target bearing – a practice which was known as making a 'cross-cut' – in order to determine the target course and speed.[28] The dumaresq could then be reset with the target course and speed obtained from the 'cross-cut', the known course and speed of the firing ship and the target bearing, to obtain an indication of the change of range rate, which would change as the target bearing altered. The change of range rates indicated by the dumaresq could then be set on a Vickers Clock, whose pointer would, in turn, indicate ranges in accordance with the change of range rate settings. In mid-1911, the plans of the Argo Clock Mark II, which had been sent to the Admiralty by the Argo Company, were made available to Dreyer and his associate, Keith Elphinstone, of the instrument manufacturing company of Elliot Brothers Ltd, by Commander Joseph Henley, an assistant to Captain Archibald Moore, the Director of Naval Ordnance. Following this transfer of information, which was made without the knowledge or permission of the Argo Company, Dreyer introduced a mechanical connection between the dumaresq and the range clock that automatically translated the calculated rate into an indication of the range in the manner of the Argo Clock Mark II.[29]

When the courses and speeds of the firing ship and target were such that the range and bearing changed at a constant, or near constant rate, the time-and-range and time-and-bearing plots could give good results because their sequences of range and bearing markings ran in straight lines whose slopes, which each identified a rate, could be easily measured. The time-and-range and time-and-bearing plots did not produce good results, however, when the change of range and change of bearing rates were changing rapidly, which was likely to occur when the firing ship and target were converging or diverging at steep angles and at high speed, or when the firing ship was under helm. Under such circumstances, the sequence of range and bearing markings would take the form of curves embodying the continuous change in the change of rates, whose continuously varying slopes could not be measured to obtain the single rates that were required to set up the 'cross-cut' on the dumaresq. Any continuous change in the change of range rate indicated by the dumaresq after it had been reset with the courses and speeds of the firing ship and target, moreover, could only be represented by discontinuous resettings of the clock, whose motor speed could not be varied continuously. When the change of range rate was changing, therefore, the range indicated by the clock would become increasingly inaccurate, the degree of inaccuracy increasing as the rate of change in the change of range rate increased.

The deficiencies of the Dreyer Fire Control Table, as Dreyer's instruments were collectively known, were, however, of little account to the Admiralty. They believed that fleet actions would not involve convergence or divergence at steep angles and at high speed, or turning, that the change of range rate would thus be low, and that, as a consequence, any change in the change of range rate that did occur would be negligible.[30] During the official trials of the Argo Clock Mark IV, which were carried out on 19 and 20 November 1912 in the *Orion* under the supervision of Jellicoe, who commanded the Second Division of the Home Fleet, the problems of high or changing change of range rates were not, therefore, posed. And although the Argo Clock performed successfully, Jellicoe ruled in favour of the Dreyer Fire Control Table on the grounds that the latter was, given the existing view of tactics, comparable in capability to the Argo plotter and clock and far less costly.[31] Jellicoe's judgment was accepted by the Admiralty, and was explained to Pollen, who had been excluded from the official trials, in a meeting with the First Lord and the Director of Naval Ordnance on 6 December 1912.[32] On 19 December 1912, the Admiralty informed the Argo Company officially that no further orders for the Argo Clock Mark IV would be forthcoming.[33] In response to this decision, Pollen drafted a lengthy essay on the question of tactical requirements and fire control, which was entitled 'The Gun in Battle'. Copies of this work were printed and circulated privately from February 1913 onwards.[34]

1 INTRODUCTION

This pamphlet has been written to explain why a complete revolution in all preconceived notions of Naval Tactics is now certain. This certainly rests on the indubitable principle – incidentally that on which Admiral Sir Reginald Custance has based his exceedingly important new work – that weapons govern tactics.[35] It is an obvious expansion of this principle to say that the use of weapons is a vastly more important factor in results in war, than the weapons themselves. If, in certain conditions, a particular method enables a gun to hit, when without it hits could not be made, the weapon *plus* the method is not the same thing as the weapon without the method. Hence tactics that will prevail when the maximum hitting range is only a mile or two, cannot be the same tactics as those which prevail when the hitting range is extended to five or six miles.

2 CHANGING RANGES THE CRUX OF WAR

At the present moment it is not the shortness of hitting ranges that prescribes very rigid limits to Naval Tactics, but the inability of the present method to enable guns to hit beyond point blank ranges, when the opposing ships adopt even a very moderate departure from parallel courses in the same direction. If divergent courses are adopted, the guns will not hit, and the movements will be without result until the ranges are so short that battle will become a mere melée. It is confidently believed by many Naval authorities that it is the fixed intention of the Navy of one of the Great Powers to bring this melée about, and, as I shall show, there is no way of preventing this being done, except by running away, or adopting a system of Fire Control scientifically calculated to deal with a change of range of at least 1,000 yards a minute.

One does not have to be a seaman, or a technical expert, to understand that if opposing ships are free to adopt any speed and any direction of course, the distance between them must continually vary, and that the rate at which it varies must depend upon the angles of courses and the speeds adopted. To ascertain the direction and speed of the enemy's course, and to integrate this knowledge with the knowledge of your own movements – which you are supposed to possess – are the two functions of Fire Control. If the range is changing by 50 yards every three seconds, and if you cannot hit unless your sight is set within 50 yards of the right range, it is clear that you must know precisely what the rate is at which the range is changing and have means of keeping your sights set in its terms.

3 THE CONDITIONS IN WHICH FIRE CONTROL IS NEEDED

Now I submit that the experience of Naval officers, the world over, will bear me out in the following contentions:

First, that it is impossible while manoeuvring ships to keep them for any length of time on perfectly steady courses. It follows from this that a Fire Control system, to be available throughout the conditions of action, must not be affected by our ship turning, and consequently must be helm-free — both while keeping the enemy under observation, so as to track his movements, and while integrating those movements and our own to keep the sights set in the terms of the changing range.

Next, that with freely moving ships, all capable of very high speed, a very rapid change of range may occur; and consequently Fire Control system must be prepared to meet the highest that is likely to arise. If the tactics of closing as rapidly as possible to bring about a melée are used against us, the change of range, as I shall show, will be at least 1,000 yards a minute. This means altering the sights — if we are engaging at long range — at least once every 3 seconds. It is obvious, then, that a high rate must not prevent us either ascertaining the enemy's movements or keeping our sights set: and this means absolute instrumental accuracy in all the gear employed.

Thirdly, just as our being under helm ourselves must not prevent us tracking the target or keeping the range, so the enemy being under helm should be ascertainable by us as the turn takes place. Our system of ascertaining the enemy's movements must not, then, present greater difficulties in one set of circumstances than in another, if it is to be equal to them all.

And, finally, it is clear that, once fire has opened, it is not only probable, but certain, that the target we are firing at will be periodically obscured by its own smoke. This obscuration should not interfere with the processes of Fire Control. So long therefore as he does not change course while he is doing this, our system must be equal to keeping the sights set with perfect accuracy while he is obscured and we are turning.

In other words, because the greater includes the lesser, the ideal Fire Control system must be capable of dealing automatically with the last of the eight cases of rate-finding and the last of the eight cases of rate-keeping set out in Section 27 below.

4 WHY THE A.C. SYSTEM HAS SUCCEEDED

That no system except the A.C. System, is capable of dealing with these two cases will, I think, be generally admitted. And that the A.C. System can deal with these two cases is, after all, only what is to be expected.

The inventors of it stated categorically in 1904–5 exactly what the requirements of scientific Fire Control must be, and from that time to now they have been retained by the Admiralty, with all the benefits of sea experiments and Service co-operation, to bring it to perfection. They have spent over £80,000 on its development, and, in all, fourteen years on its study. Its final triumph is the fitting reward of their persistence and the natural outcome of the forethought of that large body of Naval experts who, from 1905 to to-day, have urged successive Boards to continue the policy of keeping the system secret, for the simple but sufficient reasons that it was bound to succeed, and that to be the sole user of it when it had succeeded was to secure sea supremacy.

I propose now to set out the accepted principles of artillery tactics – accepted in the absence of such a system as ours; to explain the aims of the A.C.System, and finally show how the attainment of those aims must affect the efficiency of Naval guns, and hence the tactics of the ships that only exist to use them.

5 THE TACTICS OF TO-DAY

When one reads the various studies of Naval Tactics that have appeared during the last ten years, one perceives a great variety in the points of view of the authors. And first among them all, one must rate the French writers, Darrieus[36] and Daveluy.[37] The acuteness of their historical criticism, the delicacy and sublety of their insight, and the charm of their respective styles, makes everything they write exceedingly pleasing to read; but it is impossible not to be struck, when it comes to their consideration of actual Tactics, with the fact that neither writer seems to treat of them in the light of any very defined reference to artillery technique: hence an extraordinary vagueness prevails in what they say, and one closes the book without any definite impression of what the actual scheme of tactics of either writer would be. Then there are other authors, like the writers of the various papers on Tactics that have appeared in the American Press, and the laborious studies that Austrian and Italian naval officers have produced. Of these, Bernotti is perhaps the most thorough student: his work is all characterised by an enormous amount of mathematical learning.[38] Of Naval Gunnery he has much to say, but more on its ballistic and mathematical, than on its technical side. So that in none of these do I find any definite guidance as to what the accepted principles of Naval Tactics are.

It is very much to be lamented from the point of view of the interest of the subject, though possibly not from the point of view of the interests of the nation, that the practice of publishing the accounts of the British Naval manoeuvres, together with the umpire's report, should have been discontinued. The last of these that I remember was the report on the 1906 manoeuvres, and I have unfortunately mislaid my copy, but from certain old notes I find the following material for comment.

6 MISLEADING ANALOGIES

It seemed a ship was ruled out of action, or at any rate, to have received a serious amount of damage, if it was for a prescribed period of time within a certain range of a more heavily gunned ship. Thus, if my memory serves me, when the *Drake* was attacked by three smaller cruisers, the fact that she was armed with 9.2 guns was held to put one or more of her opponents out of action, while she suffered little herself: although the range of the supposed engagement was between 9,000 and 11,000 yards, and all four ships were on independent courses and at very high speeds. It seemed meaningless to judge of manoeuvres by so curious a standard of gunnery. We find a general tendency to this same vagueness – in other words, taking gunnery effects for granted – both in many of the Austrian and some of the American writers; and I cannot avoid the reflection that this is largely owing to the authors being too little accustomed to analysing the technique of weapons, and too much accustomed to making sweeping deductions from peace experiences with towed targets.

As the American reports show, extraordinary percentages of hits have been made at 10,000 yards with big guns at a towed canvas screen, and indeed, equally good results – and possibly better – may have been obtained by our own Navy at battle practice. But surely it must be very unsafe to base any large deductions as to the value of gun-fire upon such slender premises. The limit speed at which a target can be towed is low – from 6 to 8 knots at the most – and if it is to be fired at it must be engaged by almost perpendicular fire. If the whole broadside is to be engaged it is not possible to do so at more than a very low rate of change of range. Proportions of hits, therefore, obtained at great ranges under these conditions are no guide whatever to the proportion of hits that would be got in action conditions, with a freely moving ship. Moreover, when the target is not firing back, fire can be corrected by observation with a success that it is not rational to expect in action.

Clearly, Naval manoeuvres, sham fights, etc., in which no fire takes place, must, unless the gunnery conditions are strictly limited to an exact replica of what the average performance of the fleet is actually known by experience to be, must be as unreal as would be a trial game of polo in which there was no ball employed. And if the rules of gunnery applied are anything less strict than ruthless deductions from accomplished facts they will become still more unreal through the application of a standard of criticism in itself faulty.

7 A TEST CASE

A little more than a year ago I was discussing with an eminent officer of flag rank a question then much debated, namely: Assuming an enemy's fleet to close at top speed with a view to turning to parallel when the range was close, say, 5,000 or 6,000 yards, what would be the right tactic to adopt to meet this form of onslaught?

I assumed the original *Dreadnought* doctrine that our speed would enable us to keep the enemy at the distance of our choice — that is, by turning away from him — would now have no adherents, and found that the generally accepted answer was that with our superior ranging guns and larger shells we ought to be able to destroy the enemy before he got to short range.

An examination of the actual change of range set up shows that this theory rests upon a very doubtful basis.

Suppose the 'A' fleet to be in line ahead: the 'B' fleet 13,000 yards away, formed in line abreast and steaming at right angles to the 'A' fleet. 'A's' speed is 18 knots; 'B's' 21. The bearing of 'B's' right of the line is 60° from 'A's' leading ship. 'B' is closing 'A'. The range at the end of each two minutes is approximately as follows:

0	13,000 yards
2	11,200 yards
4	9,450 yards
6	7,700 yards
8	6,000 yards

The rate has begun at nearly 1,000 yards a minute, and for the last minute has been over 800. What is the chance of hitting such fast targets, without a very highly developed system for controlling fire?

This is assuming one of the simplest and easiest of gunnery problems. If 'A' or 'B' changed course, the difficulties would be greatly multiplied.

It is clear that many deductions, wholly false, have been made from peace manoeuvres, and I remember, after reading one such study, asking myself this question: Supposing Nelson, in a peace exercise, had been able to take a fleet of twenty or thirty ships against a fleet of larger numbers, and had then and there produced the actual tactics carried out at Trafalgar, what would have been the verdict of the umpire?

There can be little doubt about the answer. Any application of rules as to the effects of gun fire would have resulted in the heads of both columns being ruled to have been annihilated, and the whole manoeuvre condemned as hare-brained, reckless, and a hopeless failure.

8 DISILLUSIONING ADMISSIONS – LOW RATE

But happily such vagueness in the discussion of tactics, though not rare, is by no means universal.

In Admiral Sir Reginald Custance, in England, and Rear-Admiral Bradley Fiske, in America, we have two writers who have certainly done their best to tear the veil of groundless optimism from the eyes of those who are blinded by it.

Sir Reginald's 'Ship of the Line in Battle' is quite uncompromising as to the gunnery conditions that must prevail in war.[39] He examines every Naval engagement from Trafalgar to Tsushima, and shows how in every case success or failure has followed upon either right or wrong gunnery conditions being brought about. His own summary of the right doctrine of gunnery will be found between pages 182 and 186 inclusive:

> When two fleets meet, and both intend to fight a decisive action, each will try to bring all his guns into action at a nearly constant range, and to keep them bearing. This means that they will sooner or later find themselves steering on parallel lines with their heads in the same direction. . . . Decisive action always means fleets, or parts of fleets, ultimately drawn up on parallel lines steering in the same direction. Any difference in the speeds forces curved courses instead of straight ones on the two opponents. . . . Whether the speeds be equal or unequal, neither side can take position against his opponent's will on the bow of the other with all guns in action. . . . So long as there is plenty of sea-room, any practicable difference in the speeds of the two fleets is of little importance. . . . On what does hitting depend? Firstly, on closing to ranges sufficiently short to make the fire decisive. On the 10th of August, and at Ulsan, the ranges were too long, but at Tsushima, Togo, profiting by experience, closed at once, and fought the battle at shorter ranges. Nevertheless, the lesson does not seem to have been taken to heart. The idea of engaging at very long ranges has been encouraged.
>
> Secondly, on counteracting the errors inseparable from firing at a moving object, either by firing a large number of rounds, which involves a more or less numerous battery, or by the use of accurate weapons and appliances. Two different principles are here seen at work. The principle of numbers is associated with the idea of 'decisive ranges'. . . . The principle of accuracy fosters the idea of long ranges, and in its present extreme development is a peace product. . . . The two principles, instead of being mutual aids, have become antagonistic. Guns have been reduced in numbers and increased in size to facilitate greater accuracy. The peace-tried principle of accuracy tends to undermine and destroy the war-tried principle of numbers. Is it wise and safe either

to encourage the tendency or to over-elaborate and over-centralise the control system? Are not the present gunnery ideals based on peace theory rather than on war practice?'

And he ends his book by an eloquent plea for larger numbers of ships and guns:

> An old and well-proved principle underlies the facts laid before you in these ... papers. That principle gave the English bowmen victory in the Middle Ages, ... and enabled Wellington's thin red line to defeat Napoleon's columns. The same principle made the three-decker of the past the most powerful instrument of war at sea. The principle is the development of fire effect to the fullest extent possible. The decline in the value of armour and its possible reduction, coupled with the increased range of modern guns, are the changed conditions which enable a return to be made to the old principle.

Admiral Fiske has treated the subject far less exhaustively, but is a veteran in the cause of Fire Control and a well-known inventor of ingenious Fire Control appliances. In a recent publication he speaks of a change of range of 160 yards a minute as a rate far higher than would be met with in battle.[40] It seems to me difficult to over-rate the importance of these two witnesses.

9 STEADY COURSES

But of course the actual change of range itself is not the only factor to be considered. It is almost more important that the factors that create it should remain constant, for any change of course by ourselves would change these factors, create a new rate, and so be fatal to the continued efficiency of the guns: not because it would throw off the gunlayer or trainer – for a gun can be trained much faster than a ship can be turned – but because it would set up a new formula for change of range, and hence valuable time would be consumed in getting this new formula rightly applied to the sights.

In a recent discussion with an officer highly distinguished for his gunnery services, I was informed that any Captain who changed course to keep station, while within fighting ranges of his enemy, would deserve to be hung at his own yard arm, and that any Admiral ordering such a manoeuvre was unfit to command. This principle is sometimes recognised in war games, when any considerable change of course is held to involve a loss of gun-fire for a certain period.

I conclude, then, from all the above authorities that the principal aim of an Admiral handling a fleet and bringing it into action, would be first of all to get into fighting ranges in such a formation as to have a concentration advantage over his enemy, next to begin, maintain, and finish the engagement in conditions of the minimum disadvantage to his gun-fire, by preserving a low rate and a constant course.

10 THE TORPEDO MENACE – ITS CHARACTER

I note that Sir Reginald Custance draws particular attention to the fact that the increased ranges of guns enable a longer line of ships to be brought into action, and to concentrate their fire more efficiently upon a smaller number, than was the case when ranges were shorter.

But the question of this tactical employment of fleets has, of course, recently been complicated, in the opinions of many, by the facts that the range of torpedoes is more than doubled; that their speed is very greatly increased; and that their efficiency (that is, the extent to which they can be relied upon to run well) has increased almost as much as their range and speed. This advance of the torpedo has followed very rapidly on the development of the submarine, and has led, quite naturally, to the suggestion that it should be employed on a considerable scale in a fleet action either from under-water craft or by squadrons of fast destroyers.

The torpedo menace has undoubtedly confused the problem of fleet action in a most bewildering manner, but with great respect to those who attach the most importance to this menace, there are, it seems to me, certain principles that should be borne in mind in estimating its probable influence.

Destroyers and submarines are not free of the seas in the sense that a big ship is. In heavy weather their mobility is cramped and their capacity to use their weapons very seriously diminished.

In any circumstances the torpedo, however highly developed, is not a weapon of the same kind as is the gun. It seems to belong to the same order of military ideas as the cutting-out expeditions and use of fire-ships in olden days and the employment of mines of more recent date. It is, of course, quite truly an element in fighting, and a most serious element; a means of offence far handier, and with a power of striking at a far greater distance than has been seen in any parallel mode of war hitherto, and yet I should be inclined to maintain that it and its employment remains more in the nature of a 'stratagem' than of a tactical weapon, truly so called.

Mines, torpedoes, a bomb dropped from an airship or aeroplane: these are all new perils of war. In the hands of a Cochrane, their employment might conceivably be decisive. But it would need the conjunction of an extraordinary man with extraordinary fortune.

11 THE TORPEDO MENACE – ITS TACTICAL EFFECT

Both Japanese and Russians lost ships by mines and torpedoes in 1904, and ships will be lost in future wars in the same way, but I find it hard to believe that, in spite of two apparent changes, the *essential* character of fleet actions or of naval war generally, can be affected by them. It seems indisputable that the future must be with the means of offence that has the longest reach, can deliver its blow with the greatest rapidity, and, above all, that is capable of being employed with the most exact precision. In these respects the gun is, and in the nature of things, must remain, unrivalled.

The two directions in which fleet-fighting seems likely to be most noticeably affected by the new weapon is in the formation of fleets and the maintenance of steady courses.

I think there are other reasons why the tactical ideals set out above — viz., that of using long lines of ships on approximately parallel courses at equal speed in the same direction — will be questioned; but even if there were not, that a mobile mine field can be made to traverse the line of an on-coming squadron, and that ships formed in line ahead offer between five and six times more favourable a target to perpendicular submarine attack than a line of ships abreast, will make it certain that sooner or later there will be a tendency in favour of smaller squadrons, and even with these, of large and frequent changes of course, so as to lessen the torpedo menace.

It may be said that these are mere counsels of caution, and that the danger is exaggerated, but the line of battle arose out of counsels of caution, because it was the strongest form of defensive formation for a fleet. I venture the opinion that it is a true principle of tactics to proceed on lines of caution as far as all danger is concerned that cannot be circumvented, prevented, or averted, by any other means; and that the true art of preparation for war lies in pushing to the furthest possible point the development of the art of the offensive; and for what it is worth I believe that the offensive in the future, as in the past, will lie with the gun. Nor do I for a moment believe that we have reached the end of art in our present preparations for attacking destroyers by gun-fire. It is, it seems to me, obvious that the development of scientific Fire Control in this field, both for day and night work, will bring about as striking a revolution in methods, as scientific control of the heavy artillery is clearly destined to effect in the larger field of battle tactics.

12 IS THE GUN DOOMED?

Assuming, then, that the tactical employment of fleets, looked at from an artillery point of view, calls for a low rate of change and steady courses, and that the recent developments of the torpedo make this ideal inordinately risky on account of the new weapon, we are driven to ask what will be the hopes of employing naval artillery efficiently when such radical changes of course are employed?

Undoubtedly some of those whose opinion on this question are entitled to the highest respect, are inclined frankly to throw the gun over. Admiral Reginald Bacon, for two and a-half years Director of Naval Ordnance, in a paper read before the Society of Naval Architects,[41] certainly gave it as his opinion that the battleship was defenceless against torpedoes, and seemed to look forward to a state of things when each fleet unit would have to be attended by a convoy of destroyers to chase away or sink the enemies from which the larger ship was unable to defend itself. According to this ideal, no fleet action could begin until the destroyer squadrons had destroyed each other. It would make the torpedo not only the most formidable enemy of the battleship, but also its most formidable weapon against other battleships. Nor does it seem to me to be credible that anyone could hold this belief unless he fully realised that in the kind of tactics which the torpedo had made necessary for fleets of big ships, the guns, in the generally accepted view of what is now attainable by Fire Control, would necessarily be reduced to impotence.

It is therefore worth examining into the practical application of a limited rate of change.

13 A LOW RATE AND THE CHOICE OF COURSE

I have quoted Admiral Bradley Fiske's opinion that 160 yards change of range is larger than would be expected in battle. The amount of change of range set up in action is within the control of those who fight it. If, then, an authority says that such and such is the limit, he means that such and such is the limit that he would allow to be set up if his plans could control the situation. It is obvious therefore that if we are to discuss tactics from the point of view of how much change of range can be permitted, we should start with some limit within which we may count upon our gunnery being efficient, and regard any rate outside of it as beyond our powers.

Probably no two people will be agreed as to the extent of change of range that a modern fleet could disregard as not affecting the efficiency of its gun-fire. In this connection it is interesting to note that by plotting out a range curve from the diagram of the 'Battle of Tsushima,' as given in Semenoff,[42] it appears that after the Russians opened fire at 1.49 until about 3.40, the range varied between 6,000 and 4,000 yards and the rate never exceeded 200 yards a minute; for the most part it was in the neighbourhood of less than 100 yards a minute. But, of course, this was before the days of modern Fire Control.

Taking 200 yards as the maximum change of range at Tsushima, I propose to double this — which is two and a-half times Admiral Fiske's limit — as the basis for what I have now to say.

14 THE 400 YARDS RATE DIAGRAM

The diagram shows the limitation in the tactical choice of courses that a restriction of change of range to 400 yards per minute for four minutes would impose.

O.S. stands for 'our ship', and T. for 'target'. Taking the bottom line on the left-hand side, T.12 and O.S.12 are two 12-knot ships. T.18 and O.S.18 are two 18-knot ships, and T.24 and O.S.24 are two 24-knot ships. The dotted portions of the disc O.S. represent the angular choice of course which a 12-knot ship has when 7,000 yards from the target and straight ahead of it. At any angle within the dotted section the rate of change for four minutes will be 400 yards per minute or less, and our ship is supposed therefore to have only that choice of courses open to it.

If we move O.S.18 so that its centre lies over O.S.12, and move T.18 so that its centre lies over T.12, we have the same conditions with regard to two 18-knot ships: but it will be observed that here the choice of courses to O.S.18 is smaller than in the case of O.S.12. Again, transferring O.S.24 so that its centre is superimposed over O.S.12 and T.24 so that its centre is superimposed over T.12, we get the case of two 24-knot ships, where a further restriction in the choice of courses takes place.

The above explanation is a key to the rest of the diagram. The arc of possible courses varies at each bearing. For instance, when the two ships are abeam, 7,000 yards apart, a 12-knot ship has the widest choice of courses, and a 24-knot ship the most restricted.

The diagram seems to make it perfectly clear that, if gunnery efficiency is a condition of any movement being tactically sound, every increase of speed must, for any given limit of negotiable change of range, bring about some corresponding restriction in the choice of courses. The converse is equally true; namely, that every increase in the negotiable change of range must make the tactical employment of higher speed in a less restricted choice of courses possible.

Note that 12-knot ships have at the most the choice of 27 points, and at the least of 15; 18-knot ships are limited to 16 points at most, and 10 at the least; and 24-knot ships never have a wider choice than 15 points, and at some bearings are limited to 8 only.

The radius of the discs, and the dotted lines, represent one minute's run of the ship.

It should here be observed, that if it were held to be tactically unsound to close an enemy to a shorter range than, say, 5,000 yards, a still greater restriction of courses would result.

FROM EACH TARGET TO THE CENTRE OF ITS CORRESPONDING CIRCLE IS 7000 YARDS.

15 THE DILEMMA OF TO-DAY

I therefore summarise the present situation as follows. Owing to the change of range difficulties, the ideal conditions for the tactical employment of artillery are a low rate, with its component factors constant. If we assume 400 yards a minute as the permissible limit of the rate, the tactical choice of courses to come within the desired conditions is exceedingly restricted.

The greater the speed, the greater the restriction, so that speed is thus an enemy to artillery efficiency.

But the development of the torpedo has made the maintenance of the desired conditions so risky as to be almost prohibitive. The tactician, therefore, is in this dilemma, that if he maintains courses favourable to his artillery he is exposing himself to conditions favourable to torpedo attack, and if he tries to diminish the torpedo risk he can only do so by the sacrifice of gun efficiency.

Before I pass to a short exposition on the origin and aims of the A.C. System, I draw attention to two writers, whose contributions to the problem seem to be prophetic.

16 THE PROPHECY

A writer in the 'Marine Rundschau', of a year or two ago, discussing the proper speed and course to adopt to ensure a favourable relative position at the opening of an action, goes somewhat deeply into the mathematics involved, but prefaces his study with the statement that his investigation can have no practical value until means exist of ascertaining the speed and course of the enemy, and this he declares is 'the unsolved mystery of naval tactics'.

Again, Mr John White, Naval Instructor at Greenwich, published about eighteen months ago a minute and most interesting investigation into relative velocity, a subject which, had he been treating of gunnery instead of tactics, would have been called 'Rate of Change of Range'. I cannot resist the temptation to quote the following from his introduction:

> Some of the problems here dealt with are well known and in common use. Many, however, can have no practical value till science and the arts have provided instruments by aid of which data may be found with a higher degree of accuracy and rapidity. Problems so closely allied to others that have a practical value, can hardly exist for no purpose whatever; and the more they are studied and discussed, the sooner may it be expected that they will enter the practical stage, and that the science and art of relative movement will reach their full development.[43]

It is clear that thinking people the world over fully realise the limitations of modern tactics, and as thoroughly realise that those limitations will be removed when the time Mr. White looks forward to arrives.

17 ITS FULFILMENT. THE ORIGIN

In the month of February, 1900, I was a guest on board the *Dido* at Malta, and, for the first time in my life, saw a naval gun fired. The range was 1,400 yards, the target was towed, and I was astonished at the excellence of the shooting. By a curious coincidence I had carried on board with me a copy of the 'Times' received by that morning's post. It contained the first published account of what the naval guns had done in the defence of Ladysmith. It stuck in my mind that the 4.7in. was said to have silenced the Boer 'Long Toms' at a range of five miles, and here at sea under my own eyes the identical gun was being used at less than one mile. I was told that no practices were carried out at any much greater ranges. The discrepancy between the sea and land use of the same gun was certainly startling, and, it appeared, the absence of an accurate range-finder accounted for its existence. Clearly, here was a wide chasm to be bridged.

It is to this accidental coincidence that I owe a fourteen years' sentence to the hard labour of producing the A.C. System.

18 FIRST PRINCIPLES

My first step was to have large scale plans made of two *Drake's* (then the fastest ship afloat) approaching each other at top speed on contrary courses, the first range being 10,000 yards. At that time my solitary scrap of ballistic learning was that the time of flight of a 6in. QF gun was somewhere in the neighbourhood of 30 seconds at this range. I was amazed to find that had the first *Drake* shot at the second the range would have altered nearly half-a-mile while the shell was in the air.

It seemed to me pretty clear that it was not only the absence of a range-finder that accounted for the naval limit of ranges, because it would clearly, in these conditions, be useless to fire, if all we knew was the target's range at the moment of firing; and equally useless to aim at the present position of the object that we wished to hit. The problem was clearly one of ascertaining the future position of the target, and being able to lay and train the gun accordingly.

19 FOUR AXIOMS

Further reflection made the four following inferences clear, viz.:

1   The only clue to the future position of the target must be found in its past movements.

Supposing that it doubled like a hare *after* you had fired, no earthly power could enable you to hit it, but if it continued on its old course, a knowledge of that course would make hitting easy.

2   Of its past movements, the only conceivable information that could be obtained must be a succession of synchronous bearings and ranges.
3   If these were plotted with due allowance for our own progress through the water, a plan of the paths of both ships would result, and
4   That from such a plan the forecasting of future ranges and the angle of deflection must be a mere matter of calculation.

20   EARLY EFFORTS

The absence of a range-finder turned me on to the problem of ascertaining the range by taking the target's bearings from each end of a long base and computing the triangle arrived at. Strangely enough the chief difficulty seemed to me the computation, and when a machine was designed for doing this I fondly thought that all serious difficulties in the way of realising my new ideal had been overcome. The certainty that the machine would work correctly and the cogency of the reasoning I have set out above, led to my being absolutely confident of my case, and I was dumbfounded, on presenting my proposals at the Admiralty, to find that they excited no other interest than a kindly concern for my mental condition. Now that I know the difficulties involved, I can realise that no crazier scheme was ever put forward; but it was only gradually that the difficulties came to be defined.

By 1904 a scheme for observing instruments had gradually been evolved. A second visit to Malta in 1903 had resulted in my finding many well-wishers to my scheme, and as soon as a not unhopeful set of instruments had been thought out and some were being designed for manufacture, I approached the Admiralty again.

By Christmas the scheme was complete, and after many vicissitudes an experiment with it was authorised in the following spring.

21   THE PROBLEM STATED

In July, 1904, and in the winter of 1905, I wrote and circulated two pamphlets, the first 'A Memorandum on a Proposed System for Finding Ranges at Sea and ascertaining the Speed and Course of any vessel in sight', and the other called 'Fire Control and Long-Range Firing'. In the first of these I projected a two-observer range-finding system and a method of plotting synchronously obtained ranges and bearings. I seem fully to have realised the importance of being able to obtain and plot these data under helm, for it was proposed that the plotting table should be adjustable to own ship's course. In the second pamphlet I set out not only the provisions of a plotting scheme, but described a change of range machine for automatically supplying a forecast of the ranges to the guns. This machine was not only to generate the future ranges and bearings of the target at the true rate of change of both, but, like the plotting table originally suggested, was to be corrected for any change of course by our own ship. I may add that in my first proposal to the Admiralty in January, 1901, I laid great stress on the necessity of ensuring ease and accuracy in the taking of the observations by controlling the observing instruments by gyroscope.

22 LORD KELVIN'S DICTUM

But, alas for the vanity of ambitious projects, the money and time available before the trials made it imperative to drop everything except the bare observing, receiving, and range-calculating instruments and a plotting table limited to steady-course work. Gyroscopic correction, a plotting table adjustable for the turn, and the change of range and bearing machine, all had to be put off to a future time.

When I first began the design of the observing instruments in the spring of 1904 I submitted my whole scheme to Lord Kelvin (who was my colleague on the Board of a company of which I was Managing Director). He was much interested, and, as always, extremely sympathetic with a new venture; and so exceedingly encouraging in urging me to go forward, that I hardly realised the significance of his warning when he added: 'It is a very big project. You have ten years' hard work before you.'

23 THE END FORESEEN

It has, I know, been thought by many, a singular thing that the writer, without ever having seen a shot fired at a greater range than a mile, a year before he was brought officially into contact with any gunnery officers or realised in the slightest degree what actual Service practice was – having indeed no knowledge of it except what could be gathered from chance conversations or from such casual mentions of Fire Control as he could gather from available publications – should have been the first to analyse the requirements of action gunnery to include the following things:

(*a*) Ascertaining with exact accuracy the speed and course of ships, so that data for the highest change of range possible should be available.
(*b*) Foreseeing that an automatic calculator to forecast ranges and bearings, by generating the true rates of both, was essential, and
(*c*) Realising that both the method of finding the data and for automatically converting them into future ranges and bearings

should be available whether our ship was on a steady course or under helm.

24 THE EXPLANATION

Every doctor knows that in diagnosing a case his chief difficulty is to marshall all the symptoms and, giving each its proper weight, to make the right deduction from all in their true relation to each other. The longer the case has been under his observation, the greater the danger that he will exaggerate the importance of one factor and overlook the significance of another. And it is because he knows only too well that such a preconception is his worst enemy, that he welcomes the detached, unprejudiced assistance of the independent specialist. The popular saying – 'The looker-on sees most of the game' – illustrates this truth. Well, I venture to say that this truth also explains why we diagnosed the gunnery disease rightly in 1904–5. We had no preconceptions, knew nothing of the difficulties, and, taking ultimate success for granted, let ourselves go in an analysis that, it seemed, committed us to no more for being logical and complete, than if it were tentative and partial. We had been bred to the belief that machines could be called, like spirits, from the vasty deep, and, unlike them, would come if only their requirements were once accurately defined. The meagreness of the funds at our disposal, the shortness of the time available for each experiment, and our ignorance of sea conditions, have made success a slow business, and if we owe success – as I think we do – to the fullness of our original programme, we paid a heavy price in failures for the optimism of our ignorance. And the difficulties began early.

25 ITS REALISATION

Our first experiment was a complete failure. The two-observer system was in every respect unsuitable to sea work. The absence of gyroscopic control was fatal, but for other reasons it would never, I think, have been practicable. With the development of the Barr and Stroud range-finder from a 4ft. to a 9ft. instrument, the observing part of the system seemed to me to become unnecessary. Its plotting and calculating side (i.e., what I shall herein call its 'Rate-finding' and 'Rate-keeping' character) was, after all, the essence of the undertaking, and we had no difficulty in applying the principles upon which we had worked to a different method of telemetry. But the collapse of our first experiment made it exceedingly difficult to persuade the Admiralty to give us another trial. It is no exaggeration to say that it was our success in getting this second start that alone made the ultimate completion of the system possible. This we owe solely to the enlightened policy that distinguished the D.N.O.'s Department in the eventful years of 1905–7, when undoubtedly the greatest progress, that has been made in recent years in gunnery practice and method, was achieved.

26 THE PRINCIPLES OF PROGRESS

I do not propose to narrate the various steps in our progress that has enabled us, out of the crude scheme tried in the *Jupiter,* to develop the highly organised gear of to-day — which seems to fulfil with absolute completeness the ideal of a helm-free rate-finding and rate-keeping system that we put so fully before the Service in 1904–5; I will instead pass to our detailed analysis of the problem of Fire Control, which for so many years has been the basis of our plans, and will preface this with a brief note as to the principles on which we have worked.

The reader will have noticed that my first introduction to the problem of change of range was the highest that could be set up by the then fastest ships in the world — namely, a rate of 50 knots. A somewhat long experience in manufacturing and business has impressed upon me the necessity, when any new project is put forward, of testing its character by ascertaining how it proposes to deal with the most difficult case that can arise. It is vital in all such cases to know what the breaking strain is. Consequently, in designing our gear, while we have not always gone for a complete solution of all the problems (witness the fact that our first two Automatic Plotters and our first Automatic Range and Bearing Rate Clock were none of them helm-free) we have, nevertheless, worked on the basis of their being able to deal with the most striking conditions of difficulty in the field they were designed to cover. Thus our plotters are intended to compete with the highest conceivable rate as rate-finders and our Clock to be equal to calculating the highest and most varying rates conceivable as a rate-keeper.

We have assumed from the first that unless a control system is to be judged by its employment in the most difficult conditions, there is danger of turning a blind eye to its limitations. A high and rapidly varying rate may have to be found, and may have to be applied at great distances where the danger space is small. If, to secure hits, the sight must be set within 50 yards, when the range is changing at 1,000 yards a minute, not only must the sights be re-set every three seconds, but the setting must coincide with each alteration in the range. Synchronism and accuracy, therefore, have from the first been standards from which we could not derogate. I am convinced that it was this fact that has impressed certain critics with the idea that we have adopted a quite unnecessary standard of complexity and accuracy, because in actual experience most excellent gunnery results are constantly obtained by much simpler methods. I mean mechanically simpler. I am not arguing at this moment that these simpler methods have succeeded only because the conditions themselves were simpler — I am merely stating the fact that we have worked on the lines of our system being judged by the most difficult case.

27 THE ANALYSIS OF THE CONDITIONS

Since 1909 we have been continuously at work on the application of these principles to the case contemplated in my first 1904–5 essays, to wit, that of our ship being under helm. This fact may excuse my setting out, in full, the list of conditions in which it seems to us Fire Control may have to be used. The list naturally falls into two categories – that of rate-finding, and that of rate-keeping.

It seems on analysis that there are the following variables in the conditions in which rate-finding may be necessary in action:

> Our ship might be on a steady course, or under helm;
> The target might be on a steady course, or under helm;
> The rate of relative movement might be high, or low.

Similarly, there were three variables in the problem of rate-keeping:

> Our ship might be on a steady course, or under helm;
> The rate of relative movement might be high, or low;
> The target might be visible, or invisible.*

Thus there are eight cases which a rate-finding system must be able to meet, viz.:

> 1 Our ship steady; target steady; rate low.
> 2 Our ship steady; target steady; rate high.
> 3 Our ship steady; target under helm; rate low.
> 4 Our ship steady; target under helm; rate high.
> 5 Our ship under helm; target steady; rate low.
> 6 Our ship under helm; target steady; rate high.
> 7 Our ship under helm; target under helm; rate low.
> 8 Our ship under helm; target under helm; rate high.

Similarly, there are the following eight cases of rate-keeping:

> 1 Our ship steady; target visible; rate low.
> 2 Our ship steady; target visible; rate high.

---

* This analysis is, of course, not exhaustive. I omit, for instance, weather which may be rough or smooth. The alternative of rough weather was forgotten when our Automatic Plotting System was rejected in 1908 in favour of manual processes. But the failure of this was so conspicuous that it has no partisans now.

3 Our ship steady; target invisible; rate low.
4 Our ship steady; target invisible; rate high.
5 Our ship under helm; target visible; rate low.
6 Our ship under helm; target visible; rate high.
7 Our ship under helm; target invisible; rate low.
8 Our ship under helm; target invisible; rate high.

Of the eight rate-finding cases, the A.C. rate-finder in the *Natal* could not deal with any of the four when our ship was under helm, but could deal with the other four; and the A.C. rate-keeper, while able to deal with cases in which the target was invisible and the rate high, could not deal with the four cases when our ship was under helm.

28 THE CASE FOR THE A.C. PLOTTER

The case for the A.C. Rate-finding device rests then upon:

1 The necessities of the case,
2 The experience obtained with previous rate-finders which, except for not being helm-free, were of similar type, and
3 The reasoned opinion of mechanical experts, that the 'helm-free' mechanism, as designed, should fulfil its purpose.

And to come to a useful conclusion as to whether or not, as a rate-finder, it is likely to be successful, these three subjects must each be investigated.

All the necessary material has recently been submitted to Professor Charles Vernon Boys, of the Royal Society, and his opinion is that the A.C. rate-finder ought, under helm, to get better plotting results than were ever obtained either in the *Ariadne* or in the *Natal*.

29 THE CASE FOR THE A.C. CLOCK

Fortunately, however, the case for the A.C. Clock rests upon the basis of ascertained fact, because practical experience has already established that it can do the following:

1 The Clock, when once set to the true range, speed, and course of the target, will, so long as the target maintains its course and speed, keep the sights set for range with *perfect* accuracy, no matter how high or low the rate may be, or how rapidly it changes.
2 Even for periods when the target may be obscured the sights will still be correctly set.
3 Rates will be accurately kept as above *while the ship is under helm*.

4 The range will be accurately *kept whilst turning,* at high speed and over large arcs, so that *hitting can be maintained uninterruptedly under helm* and continued when steady without any correction of the rate.

30 THE ENFRANCHISEMENT OF THE TACTICIAN

I now resume the discussion of the effect on Naval Tactics that the adoption of this system (or of any system that has an equivalent result in accuracy and the wideness of the conditions it can deal with) must have.

It seems clear from the development of the A.C. System that being under helm will not interfere with getting data or hitting when they are obtained, and that 400 yards a minute cannot remain the limit of change of range in action. Indeed, if Professor Boys's opinion is well founded, it is difficult to say what the limitation will be. A rate of 1,000 yards a minute would be ascertained without difficulty and with a rapidity hardly distinguishable from being instantaneous.

Again, the practical tests of the A.C. Clock seem to show that, far from it being necessary to maintain a steady course in action, the ship possessing it would be free to manoeuvre as her commander pleased. In present conditions the squadron commander keeps a steady course because otherwise he loses the rate; but this is just as favourable to the enemy as to himself. If gunnery is made 'helm-free', by the use of an efficient rate-keeper, the Admiral is relieved of this responsibility, may change his course as appears advisable, and by this very action will interfere with the 'rate-keeping' of his enemy.

By giving freedom of manoeuvring power without loss of fire efficiency – *i.e.,* by hitting being made possible while turning – station can be kept in action just as in peace manoeuvres; from which it follows that it would be possible to manoeuvre fleets with absolute precision in any way the fleet commander wished, and at a speed that would absolutely prevent any efficient gunnery without a rate-keeping device that was at once 'helm-free' and accurate.

It is common knowledge now that very fast ships cannot be used at their limit speed in action. Except on parallel courses in the same direction, high speed will nearly always mean a high rate of change and frequent changes of course – otherwise ships would come too close or get out of range – and all changes of course mean sudden changes in the rates. Hence high speed, by creating problems too difficult for present Fire Control, is incompatible with fire efficiency in action.

In other words, if it is a revolution in naval tactics to be able to close the enemy on any course, at any speed, irrespective of change of range, to change course and formation at will, to be able to keep station with the nice precision which would be necessary if action manoeuvres are to be complicated, and to do all these without loss of fire efficiency, then it seems obvious that, owing to the development of the A.C. methods of finding and keeping the rate, this revolution is virtually achieved.

31 PARALLEL AND A WARNING

It is a commonplace with the historians that, by the latter half of the eighteenth century, it was the respect for the doctrine of the 'line of battle' that created the rigidity in the tactical use of fleets, which explains the inconclusive character of practically all fleet actions (excepting Hawke's) up to Nelson's time. Nelson broke with this tradition, and was able to break with it because he could count upon the consummate seamanship of his Captains and a standard of gunnery excellence in his crews superior to that of his enemy's.

A steam fleet has such a flexibility of movement, such perfect command of variations of speed and course, that no particular evolution nowadays calls for any singular pre-eminence in seamanship to bring it about. But if the problem of handling ships has been made simpler by recent inventions, the complexity of the art of gunnery – if that art is to extend to be commensurate with the flexibility of ships – has grown so as to infinitely surpass all sea problems in the technical difficulties involved.

I have said that Nelson counted upon the relative superiority of the English gunnery over the French and Spanish; but the sea war which, one hundred years ago, was waged between England and America, should have been a final lesson as to the disadvantage of gunnery technique not being pushed to its full limit. Of fifteen engagements between single ships, we lost no less than twelve. The American gunnery was too much for us, and nothing can be more certain than this: that if, in any war, the technique of our gunnery is less scientific, and less complete, than that of an enemy's, the humiliating stories of 1812–13–14 must be repeated, only, with this difference: a ship of the line 100 years ago could hardly be put out of action under an hour. A modern ship exposed to a succession of hits from 12in. or 13.5 guns would not survive in action for five minutes. Obviously then, to risk being put under continuous fire before we can subject the enemy to it, is to imperil the very existence of the Navy, and with it, the existence of the Empire.

32 AN UNRECOGNISED REVOLUTION

There is surely no occasion to be so very sceptical about a new revolution in naval tactics being imminent. Have we not seen a complete revolution in naval tactics in the last ten years? In the year 1903, leaving isolated experiments on one side, the battle gunnery of the Navy (although then armed with 12in. and 13.5in. guns) could not have been assumed to extend to a greater range than 3,000 or 4,000 yards, whereas to-day it is certain that all tactics will be based upon fire in all probability commencing at 9,000, or even 10,000 yards, while 5,000 yards (a greater range than thought possible ten years ago) would be regarded as almost inside point blank conditions. But, so far, this revolution is only parallel to what has happened in land warfare. Sea tactics have changed, in this respect, just as the tactics of infantry, armed with a weapon of offence limited to one discharge a minute and to a hitting range of 100 yards, must be totally and absolutely different from the tactics of infantry armed with a weapon capable of a number of shots a minute and with a point blank range of 800 yards.

But is it not obvious that what fleets have gained by extended range they have lost in freedom of manœuvring power? Supposing that all guns were used inside, say, 3,000 yards, and the skill of the gunlayers to be what it is to-day, there would be no difficulty whatever in securing a reasonable number of hits even if ships were on contrary courses at the very highest speeds, because the danger space would be so large that perfect accuracy in estimating the change of range between shots would not be necessary for hitting. But remove these same ships from 3,000 to 10,000 yards apart, and the same manœuvres would mean absolute safety for both sides, if neither were equipped with adequate control systems.

33 CREATED BY RUDIMENTARY FIRE CONTROL

This revolution in Naval Tactics — already long accomplished — has been due to nothing but the simple fact that it has been discovered with improved range-finders and with the study of Fire Control, that under favourable conditions hits can be made at twice and three times the distance that ten or twelve years ago was thought the limit. With great respect, it seems to me that Sir Reginald Custance is confusing his arguments when he attributes the increase in range to the increased power of the gun. For all practical purposes the 13.5 gun of the *Royal Sovereign* class was just as effective for hitting purposes at 8,000 yards as any gun at present in existence, the difference in the danger space not being so great at this range as to change the problem materially.

The fact of the matter is, that it is impossible to ignore the technique of gunnery when we are discussing tactics. As the gallant and learned author says, 'The weapon governs tactics.' I should like to expand this phrase into 'The *method of using* the weapon governs tactics.'

It was not the number of bowmen, but their marksmanship that gave us victory at Agincourt; it was not the number of the *Shannon's* guns that gave her the victory over the *Chesapeake*, but the skill with which they were used; and it is not the increased power of the gun that has lengthened the range of actions, but the increased knowledge of how to use it.

34 THE IMPOSSIBILITY OF RETREAT

It follows from this that some method, some technique, some form of Fire Control, is inevitable. It is too late to object that Fire Control instruments are elaborate or costly or complicated. We are hopelessly committed to range-finders, telescopic sights, to machines for calculating rates, and devices for applying it to range. The choice does not lie between having Fire Control and having no Fire Control, but between having a complete, or an incomplete, system, having scientific or unscientific instruments; being able to control in all, or only in some conditions.

But the tradition that the accessories to naval gunnery should not be complicated dies hard. We find Sir Howard Douglas writing in the first quarter of last century deploring his inability to persuade naval officers that everything necessary to the ship's gunnery need not be 'rudely simple'.[44]

35 THE BOGEY OF COMPLEXITY

If the history for the last ten years of Fire Control is studied it will, I think, be seen that progress has been made, when all the conditions necessary to be dealt with have been resolutely faced, and that progress has been blocked when the adoption of some simpler expedient has been held to be good enough for the purpose. An unwillingness to face the complexities of the conditions, explains impatience with complex gear. Naval men are in this dilemma; nothing can alter the fact that with freely-moving ships there are the three variables in finding the rate, and again three variables in keeping it correctly applied to the sights. That the problem of being able to deal with — and for gunnery purposes to eliminate — all these variables by absorbing them in machinery, is a complicated one, no one who has given fourteen years' study to the question will be inclined to deny; but he will unhesitatingly deny that, when they have been absorbed in machinery, a complicated problem is left for the men who have to use it.

36 AN INDUSTRIAL PARALLEL

Few things in manufacture call for such a complex series of operations, or demand such accurate work, as the making of screws of exact dimensions to a standard pitch. But in every well-equipped works there are automatic machines turning these out by thousands, with a group of perhaps six or ten machines in the custody of one unskilled man. Has that man got a complicated task to perform?

Nothing in the A.C. Gear, either of its rate-finding or rate-keeping parts, excels in mechanical complexity the machine tools that are in everyday use in the workshops of England. Every such machine tool is the growth of − what might be called − ages (in the sense not of time, but the number of skilled contributors that have helped in its production). In its final form the Acme screw lathe is the product of an American firm, but the work and experience of tool designers the world over for one hundred years are embodied in it, as it stands to-day. For my own part, I have never been able to see that the elements of the first problem of naval fighting − namely, to get perfect flexibility in the efficiency of gunnery − are, properly considered, a problem essentially different from that which faces every manufacturer. It is a problem of providing the right means for getting a uniform product of perfect excellence at the highest speed and economy. The product of Fire Control is a continually-set sight; the raw materials are observations of the enemy; the factory, and the processes by which the last are converted into the first must be judged, as all commercial processes must be judged − by the simplicity of their demands on the personnel, and by the perfection of their results in getting an exact and regular product. Their intrinsic character must reflect the operations to be performed.

37 THE SIMPLICITY OF IGNORING DIFFICULTIES

But obviously the complexity of the means can be lessened by omitting some operations; if you are content to leave the harder part of the problem insoluble, it is not difficult to find simpler means of solving parts of it. In one or two of the eight cases of rate-finding − say, cases one and five − something other than the True Course Plotter can doubtless find the speed and course of the enemy; indeed, they could often be rightly guessed; and in one and five of the rate-keeping cases, possibly something short of the A.C. Clock can keep the sights rightly adjusted, if the range is not long, and perfect accuracy is not required.

I base my belief in the product of my life's work on this: that ours is the only system yet produced that, from the beginning, has been designed to deal with *all* the conditions, and after long years of work, at last can do so.

38 AN INSTANCE IN POINT

I will conclude my arguments with a demonstration of the conditions that seem about to be opened to Naval guns. In the *'Jupiter* Letters' (1906) and in 'An Apology for the A.C. System' (1907) I set out certain time and range curves under typical conditions, to illustrate my argument. The diagram opposite contains two figures by way of further illustration of what I have been contending for.

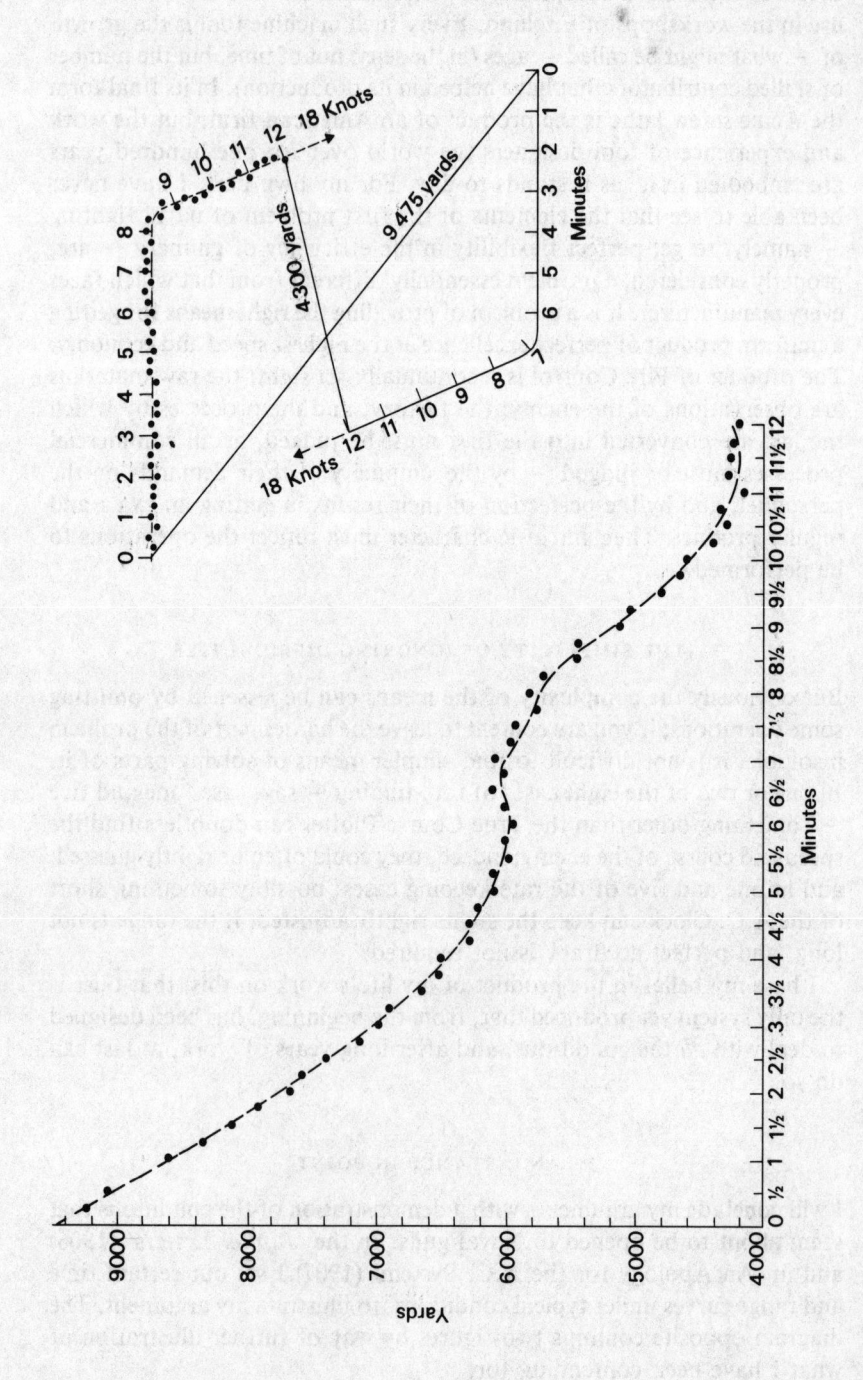

The right-hand figure represents the courses of two 18-knot ships, the initial range being 9,475 yards; a steady course is maintained by the lower ship for 6½ minutes, and by the upper ship for 8 minutes. The lower ship at 6½ minutes makes a turn of approximately 6 points, and at 8 minutes the upper ship makes a similar turn, the ships continuing on approximately parallel courses. The course of the upper ship is represented as if it had been plotted from the lower by means of the A.C. True Course Plotter. Nearly all the observations are supposed to be erroneous for range, the amount of each error having been determined by lot in making the figure.

The left-hand figure represents the time and range curve of the same conditions, similar to that shown in the *Jupiter* Letters' and the 'Apology', and in this the same observations used in the right-hand figure have been put down in the same sequence.

39 AS A RATE-FINDING PROBLEM

The gunnery conditions shown in these figures are clearly prohibitive, according to the principles put forward as fundamental by Sir Reginald Custance and other authorities, but assuming gyroscopic control of range-finders to be as perfect as it is generally acknowledged to be, and the range errors set out in the figures (which are considerable) not likely to be exceeded, there seems to be no reason why the A.C. True Course Plotter should not be able to make a plot exactly like the right-hand figure in the conditions shown. Many plots made in the *Ariadne* and *Natal* were not less exact or more exacting. In other words, so far as ascertaining the data is concerned, neither the relative speed nor the turns of the plotting or target ships introduce insoluble difficulty into the problem of ascertaining speed and course. And I submit it is clear that there is nothing in the *character of the record*, or in the nature either of the personal or mechanical operations, to make ascertaining the target's movements more difficult or doubtful in one set of conditions than in another.

40 AS A RATE-KEEPING PROBLEM

If we regard the left-hand figure as exemplifying a problem in rate-keeping, we shall notice that there is for the first three minutes — that is, until the range has come down to 7,000 yards — an approximately steady rate; from the third to the sixth minute there is very rapid variation; between the sixth and seventh minute there is a double change in the sign due to our own turn; at the eighth minute another complete change of curvature takes place; and from then on the rate is a rapidly varying one. Experiments with the A.C. Clock have definitely shown that the changes in the rate between the sixth and eighth minutes would be automatically accounted for by the Clock being adjusted to the ship's movements, as shown by a

Compass Receiver. Up to the eighth minute, therefore, if the data obtained in the first minute were set upon the Clock, the sights would be kept set in terms of the dotted line.

Between the eighth and the ninth minute there is a very large change of range and rate, owing to the turning of the target, and from the character of the plot it is not probable that the exact measure of the target's turn would have been taken until the ninth minute was reached; we should therefore have to regard the interval between the eighth and ninth minute as one in which fire would have to be suspended, but from the ninth minute the sights would continue to be correctly set to the end of the run. If we were further to assume that, owing to our fire, the target had been obscured between the fifth and eighth minutes it would, of course, not affect the sights having been correctly set during the whole of this period, in terms of the dotted line.

Assuming, then, that this was so, the two figures illustrate rate-finding conditions in which our ship is under helm, the target is under helm, and the rate is high; and as a rate-keeping problem, conditions in which our ship is under helm, the rate is high, and the target invisible.

There is, of course, no way of *keeping* the rate if the target turns, because its movements must be first ascertained – which is obviously a function of rate-finding, not of rate-keeping.

41 THE A.C. CLOCK AS A RATE-FINDER

But it should, for clearness, be added that as the A.C. Clock fulfils the function set out in 'Fire Control and Long-Range Firing' of 1905 – viz., that of generating and forecasting the bearing of the target independently of continuous observation – if we assume the target was visible between the eighth and ninth minutes, there would, very shortly after its turn had begun, have been a marked discrepancy between the bearing as forecasted upon the Clock and the bearing as shown by observation. The A.C. Clock not only generates the right rates of bearing and of range, but indicates them as well. Consequently, if the bearing were corrected on the Clock and a corresponding correction were made in the rate, the course and speed which such a new rate involves would appear on the Clock, and the range curve would continue in the new terms. If we suppose a similar correction of range and rate has been made, a further alteration in the target's presumed speed and course would be made; so that every information received would lead, with rapid approximation, to the Clock being set to the true conditions.

And in this particular, the A.C. Clock is not only an instrumentally perfect rate-keeper, but is so designed as to be of practical assistance in rate-finding – which will be valuable, if ranges cannot be obtained, but bearings can. But if we assume the range-finder is working satisfactorily the record of its performances, as shown in the right-hand figure, would clearly give a more rapid and more accurate indication of the change in the conditions, than could possibly be obtained by any corrections based upon altering the range or bearing, with corresponding corrections in rates.

42 THE COMPLICATIONS OF STATION-KEEPING

It has often been pointed out to me that with a line of ships the last vessels are seldom able to make as clean a turn as the leading ship. A turn for these ships, therefore, could not be supposed to be represented by the even plot shown in the right-hand figure. It is more likely to be a movement containing two or three, or even four, successive applications of helm, in opposite directions. No complication of this kind will in the least degree affect the accuracy with which the rates will be found or kept, so long as the movements of the ship's head and indications of her variations of speed were available to those in charge of the Clock. Nor is any previous information respecting the moment or amount of the turn required. In the plotter, the control for helm is automatic.

43 CONCLUSION OF THE ARGUMENT

It is, I submit, incontestable, that if the highly complicated problems of rate-finding and rate-keeping set out in these two figures can be solved instrumentally, by operations, as far as the personnel are concerned, already so familiar – namely, the keeping of the target in the field of the range-finder, adjusting the cut, and setting the Clock to clear indications of the angle of the ship's head and her speed – there can be no reason for supposing that tactics of the future need be so limited as Sir Reginald Custance and Admiral Bradley Fiske suggest; and, further, it is unreasonable to say that there is any doubt of Naval guns being able, in the near future, to keep an enemy under practically continuous fire, so long as the target is within the zone in which the gun can be trusted to shoot in the terms of the sight, irrespective of the relative velocities set up by the courses and speeds adopted, and unhampered by voluntary or involuntary changes of own course.

44 THE ESSENCE OF THE MATTER

If one could for a moment forget the last few years' experiments, with their partial failures and partial successes, the competition of rival systems, and the partisanship that has necessarily followed one or the other school of Fire Control being favoured – forget, too, not only the controversies that have arisen, but the intrinsic difficulties of the processes that both schools have put forward; and, having done so, look broadly at the problem of Naval Tactics, will not the following questions occur to the least imaginative?

> Is it thinkable that high speed in ships can be physically possible, but must remain tactically useless?
> That guns of such perfect inherent precision be produced, without the possibility of their being used with equally perfect precision in all conditions?
> Is it possible, as Mr White propounds, that all the subtle problems that arise out of the facts of the variations in relative velocities, arising from the wide choice of courses, bearings, and speeds that opposing ships and fleets can set up, can exist for no purpose, and without hope of their solution?

To me the answer to each is an emphatic negative. In any field which the human brain has invaded, obstacles only exist to be overcome. It is the lot of the man of routine to accept the limits he sees imposed as if the boundary were there by some divine dispensation that it were impious to question; but sooner or later some mutinous spirit rebels. He disputes accepted formulae, probes deeper into questions supposed to be finally answered, or pushes further into the uncharted and unknown. It is not always − or indeed often − given to the pioneer to reach the goal. For a hundred years intrepid explorers have left their frozen bodies in the Arctic ice − but at last the Pole is reached. All human history is a tale of failure, because it is a tale of effort; and victory grows out of defeat. Heaven knows the failures in the short period in which Naval guns have become five-mile instead of two-mile weapons, have been many and flagrant: the successes few and grudgingly admitted − but the substantial advance is there.

45 VICTORY INEVITABLE

Can it be doubted that complete success − absolute accuracy and universal flexibility will finally be won? There are many, and those not the least experienced in these baffling questions, who are resolutely confident that all these are won already. If they are right − and in the nature of things, the probabilities are on their side − the change in Naval Battle will be drastic, radical, revolutionary.

As in all similar activities, progress will come by an increase in the pace of the game. Just as the abolition of the offside rule has quickened polo, so will the freeing of the gun from the shackles of a limited rate and steady courses, quicken the tactics of Naval actions. The speed of polo is only pony speed if the players can hit and hit truly at top speed; and the speed of action can only be the legend speed of the ship when she can use her guns with effect at her highest rate through the water.

The conditions of this enfranchisement are obvious. The ship must be free to race, and free to turn, and free to hit while doing either − that is, no rate must be too high; no change in rate too sudden, to check her hitting.

Clearly, any less ambitious scheme than one that, at least, tries for this freedom, is not worth dallying with, if one so ambitious can possibly succeed.

XIX

THE NECESSITY OF FIRE CONTROL
(*SEPTEMBER 1913*)

From February to July 1913, Pollen attempted to persuade the Admiralty to reconsider their decision to reject the Argo Clock Mark IV and to give his new model range finder and redesigned plotter a trial. In July 1913, following the failure of these efforts, the Argo Company informed the Admiralty that the range finder, plotter, clock, and associated apparatus would be offered for foreign sale. The Admiralty responded by threatening to restrict such an action by invoking provisions of the Official Secrets Act. The Argo Company, for its part, upon being advised of the questionable legal validity of the Admiralty position, ignored the injunction.[1] The Argo Company's representations to foreign navies began in September 1913, and within four months, Pollen was able to inform Rear-Admiral Richard Peirse that, 'We are already negotiating orders to the value of nearly £120,000, and have already sold the total product of our factory for the next eighteen months.'[2] Pollen's views on the fire control were explained to foreign buyers in 'The Necessity of Fire Control', a pamphlet printed by the Argo Company in September 1913.[3] This brief and clear statement of the nature of the fire control problem undoubtedly contributed much to the success of the sales campaign.

(1) A glance at the range tables shows that for any gun to carry a specified distance, the gun itself must be elevated in relation to the line of sight. The elevation, for instance, that would take a projectile to a distance of 1,000 yards would be useless if you were shooting at a target 3,200 yards off. It is, therefore, axiomatic that a right range must be on the sight for there to be any hope of making a hit when the range is greater than about 1,500 yards. The accuracy with which the range must be set depends upon the distance at which it is hoped to make a hit, and the size of the target. Shooting, for instance, at a *Dreadnought*, at 5,000 yards with a modern 12in. gun, there might be an error of nearly 100 yards, and yet a hit be made, whereas at 10,000 yards a greater error than 30 yards would cause a miss, on account of the greater steepness of the angle of descent.

The first axiom therefore is:

(*a*) Within certain well-defined limits the range must be accurately set upon the gun if a hit is to be expected.

THE NECESSITY OF FIRE CONTROL

(2) The question then arises, how is this range to be ascertained and put upon the gun? We must dismiss at once the idea that the men at the guns can ascertain this range or form any judgment about it of the slightest value, when the range exceeds 1,500 yards. The moment we realise that some organisation of persons with instruments (other than the gun pointer himself) is necessary for hits, we have recognised the necessity of a Fire Control organisation. If, therefore, it is axiomatic that the right range must be on the gun if a hit is to be expected, and if the gun pointer cannot ascertain or set this range himself, then it follows that:

(*b*) Fire Control is necessary to effective naval gunnery.

(3) Now we have seen that the right range has to be on the sight at the moment of firing, but it is clear that in a naval engagement it will not be sufficient to fire one round from each gun with the range right. The engagement may last for several minutes or even hours. We shall therefore want to have the range often right, and it therefore becomes again axiomatic that:

(*c*) That form of Fire Control is the best that gives the range upon the guns most often correctly, and the *Ideal System* of Fire Control would be one that kept the correct range upon the sights *continuously*.

(4) Now let us consider what is involved in keeping a succession of ranges rightly upon the sights. If two ships, say, 5,000 yards apart, are travelling in the same direction and at the same speed, they will remain 5,000 yards apart, no matter how long they continue on their courses; but if either ship changes course or alters its speed, the range will at once begin to increase or to decrease, according to the character of the alteration of speed and course that is adopted. If we imagine two 20-knot ships engaging in a single ship action, the change of range will be nil if they get to parallel courses, equal speeds and identity of direction, and it may vary from anything from nil to the sum of the speeds of the two ships − i.e., 40 knots − if the ships are end on and going in contrary directions. We get, therefore, to our next axiom, that:

(*d*) The rate of change may vary between nil and the sum of the speed of the two ships.

(5) When ships are working at such speeds and on such courses that the lines of their courses, if continued, will intersect, and their speeds are such, that if the ships were progressing towards the point of intersection, a collision would take place at that point, in these conditions the rate of change of range will be constant. But in all other combinations of speeds and course the rate of change of range

will not be constant, but will be varying from moment to moment. We therefore arrive at our next axiom:

(e) With the exception of the case named, the rate of change is never constant, so that no Fire Control system can be of value unless it is capable of *automatically* altering the settings of the sight at a rate which is practically never constant.

(6) We have seen from the above that it is either a departure from identity of direction or inequality in speeds that causes the range to alter. It follows from this, that all changes in the range, and all changes in the rate of change, can be explained and foreseen, when we have knowledge of the speeds and courses of the ship and of the target, and from this it follows that:

(f) The efficiency of Fire Control depends upon its capacity to ascertain the speed and course of the target, and to keep a check upon our own speed and our own course.

(7) The foregoing axioms make it clear that the fundamental problem of Fire Control is not to ascertain a range, and then to put the ascertained range upon the gun, but to ascertain *the speed and course of the enemy*, to deduce from these both the *rate of change* and the *variation in the rate itself*, then to put the range at the desired moment on the sight, and then to apply the knowledge of the changing rate to the sight, so that once set it shall be varied in the terms of the ship's and the target's movements.

(8) That this is so, is apparent from the following: We have seen that the rate with 20-knot ships may vary from nil to 40 knots. With two of the fastest modern ships, excluding the extreme cases of ships running directly towards each other or directly away from each other, it may and will in action often vary as much as 1,200 and 1,500 yards a minute. If, therefore, we had an absolutely perfect range-finder, and if every big gun fired at every moment exactly as its sight indicated, and if it only took five seconds to ascertain the range and put it on the sights, we should still not succeed in hitting at 10,000 yards, because, in the five seconds that have elapsed between ascertaining the range and firing the gun the range will have altered more than four times the margin that is necessary for hitting at this range. We therefore see that the first step in Fire Control is not to ascertain the range, but the law by which it changes. Hence ascertaining the speed and course of the enemy, on which the change of range depends, and providing means of applying this change correctly, is the first step in Fire Control, and putting the right range upon the gun, which is to be continually varied once it has been set, is the second step.

THE NECESSITY OF FIRE CONTROL

(9) These necessities of Fire Control have only come to be understood very gradually. Until five years ago no navy in the world had attempted any regular practice at a moving target. When firing at a moving target was first attempted, the experiments were made exceedingly easy. The target was towed at a speed representing at most a quarter of the speed of a modern battleship. The firing ship adopted a course and a speed which set up a very low rate of change – never more than about 150 yards a minute. The conditions have been gradually made more difficult, and now a change of range of 200 yards and even 250 yards is sometimes attempted. But the results have not improved. They have become worse. Even the original small elements of difficulty brought about disastrous results to the shooting of the British fleet. In 1907 there was an average of 30 hits out of every 100 rounds fired in British battle practice. In 1908 this fell to little more than half – about 17 per cent of hits – the conditions having been made more difficult. And now the percentage is only slightly over 10 per cent. Every year large numbers of ships make practically no hits at all.

It is not that the men cannot aim. The Gunlayers Tests show that there is no falling off in skill. Indeed, it is curious that the ships which get the highest positions in gunlayers tests often get the lowest marks in battle practice. This can only be because the Fire Control organisation is inadequate. The reason so few hits are made in battle practice, therefore, is not because the gunlayers cannot aim the guns, but because the Fire Control system cannot keep the range accurately upon the sights, even when the conditions are so very much easier than they would be in war.

(10) It is now generally admitted by gunnery experts that the main defect of the Fire Control in use is that it consists almost entirely of operations performed by individuals. For instance, in a ship of the *Dreadnought* class, omitting the sight-setters, some 40 or 50 people are engaged in trying to ascertain the target's movements and to apply the laws of the change of range – the very small number of ships that shoot successfully points to the fact that it is impossible for these operations to be carried through with uniformity and success by such methods. The thing has been tried so long and so persistently, such immense trouble and pains are taken to train all the individuals in every ship, that no practical person now believes that the bad shooting is to be explained by any want of zeal or intelligence on the part either of officers or men. The explanation is only to be found in the fact that the method on which they are working is wrong.

(11) Fire Control, therefore, when carried out by a multitude of persons having proved itself a complete failure, the Argo Company have

produced a completely automatic system of (*a*) ascertaining the speed and course of the target in all conditions, and (*b*) of embodying the knowledge so obtained in a Clock which, again in all conditions, automatically generates the change of range and supplies the sights with the exact range to be kept upon the guns.

(12) The instruments required for this system are expensive if compared with the instruments required for the other system, but such a method of comparison is of course entirely fallacious. In the first place, the cheap instruments do not give the results that are required in war; this is obvious, because they cannot give the results required in peace, as the experience of all navies has shown. How then can they give it when the conditions are made infinitely more difficult by the target (*i*) travelling at 20 knots an hour instead of five; (*ii*) being free to adopt any course, no matter how great the change of range set up may be; and (*iii*) by the enemy's firing at us, and hence making spotting by observation of fire infinitely more difficult than it is in peace in exercises. The first and most important point is, therefore, that our instruments give a military result which is essential to war, and cannot be obtained in any other way.

(13) The question is, what is this military result? The answer is simple. These instruments enable the ship to do the only thing which it is built to do. A £3,000,000 battleship is not built because she is fast, because many light cruisers are much faster than the biggest battleships in the world. She is not built to fire and carry torpedoes. These can be better carried and fired to better advantage by smaller craft. The battleship is built solely to carry the largest number of the biggest guns into action. The guns are carried to destroy and sink the enemy's naval forces. If the battleship is not the material presentment of the greatest possible naval force, in other words, if she has not the means of bringing about the greatest possible destruction of the enemy's naval force, then she is a mere waste of money. But it must be admitted that she can only be a means of destruction if her guns hit.

(14) If, therefore, Fire Control is necessary to enable the guns to hit, the battleship must have some system of Fire Control, and that if there is a choice between one method of Fire Control and another, the proper way to compare the value of the best system over the second best is to compare the results which the ship obtains with one and the other respectively. Supposing, for instance, that in certain conditions a ship with the ordinary system of Fire Control were able to make 10 hits per hour in certain conditions, whereas with the Argo system she could make 100, it would be fallacious to say that latter system would be worth 10 times the former. The true comparison would be this. A £3,000,000 ship fitted with the

ordinary system is a 10 hit per hour ship — that is what is obtained for £3,000,000. But with the Argo system she becomes a 100 hit per hour ship. She therefore is, as an *engine of war*, 10 times as valuable. The relative value of the Argo system therefore must be measured in the value of the results which it enables the £3,000,000 ship to obtain. Viewed therefore simply as a commercial investment the value of a £3,000,000 ship could be enhanced at least 10 times by the addition of 1 per cent to its cost.

(15) But this comparison is based upon the supposition that in the conditions selected the ordinary system of Fire Control will enable the ship to make 10 hits per hour. It is, however, quite easy to demonstrate that the ordinary system of Fire Control could not enable a ship to make any hits at all, if the change of range were high or were continually being varied by the ship and target altering courses and speeds. But there are no conceivable conditions which would be too difficult for the Argo system to keep the sights almost continuously right. Hence, supposing the £3,000,000 ship fitted with the ordinary system, were engaging a ship fitted with the Argo system, the latter could so manoeuvre as to make all hitting by the first ship impossible, while able to make hits herself whenever she chose to do so. But if the first ship were fitted with the same system, equality would be restored. Thus the addition of the Argo system would not have multiplied the value of the first ship by ten, as in the previous case, but would have brought its value up *from nil to equality*. An increase of this kind cannot be measured by percentages — it is an *infinite increase in military value*.

(16) Not only does the Argo system confer a military value upon ships, that without it would possess none at all, but it is actually in practice a very economical investment. Leaving aside the enormous waste of ammunition which is involved in the employment of inferior systems — because with such systems it is always hoped to correct the inevitable errors by means of the observation of fire, the only chance of getting any hits at all lying in enough misses being made to afford a guide towards future hits — a waste which is almost entirely saved by our system, there is this fact to be borne in mind, that the working of our system can be carried out in its entirety with so small a number of men that very great economies in wages, clothing, feeding, etc., are made possible, so great indeed that the economies absolutely pay for the original installation of the instruments in a very few years.

(17) I therefore summarise the argument as follows. A battleship without a Fire Control system cannot expect to make any hits at all beyond 1,500 yards' range. The ordinary system of Fire Control, after being experimented with on the most gigantic scale, and regardless of

expense, for six years by the most efficient navy in the world, has proved itself unable to get more than an average of 10 per cent of hits with a very low change of range. The causes of this failure have been demonstrated to be the impossibility of a crowd of individuals making the necessary observations and calculations required for continuous hitting. But it is clear that these observations and calculations can be made automatically. The introduction of automatic Fire Control therefore is destined to bring about a complete revolution in naval war, so that ships fitted with it have an almost infinitely greater fighting value than those which are without it. Thus by an addition of 1 per cent of the cost of a ship it can be raised from utter inefficiency to the highest point of military value, and strangely enough by means which introduce such great economies in working that the actual cost of the instruments is saved in a very brief number of years.

XX

[UNTITLED PAPER]

(MAY 1916?)

The outbreak of war in August 1914 disrupted the Argo Company's negotiations with most of its foreign contacts, and only the Russian Navy was to receive delivery of a portion of its order in October 1914.[1] The Admiralty had in August 1913 responded to the Argo Company's decision to seek foreign buyers for its fire control equipment by severing all relations between the two parties,[2] and shortly after the start of hostilities a year later, refused Pollen's offer to supply the fleet with fire control gear at cost.[3] The production of fire control equipment by the Argo Company thus ceased, while Pollen turned to writing articles on naval aspects of the war for the journal *Land and Water*.[4] By the spring of 1916, the Royal Navy's war experience, which Pollen observed closely in his role as naval correspondent, had revealed the shortcomings of the service system of fire control. Pollen thus appears to have made preparations for yet another attempt to persuade the Admiralty to consider adopting his fire control system. In April or May of 1916, Pollen combined the texts of two letters that explained the principles of fire control, analysed several recent naval actions from the standpoint of gunnery, described the operation of his instruments, and gave an account of the rejection of the Argo Clock Mark IV in the summer of 1913.[5]

This work, which has survived only in the form of an untitled galley, appears not to have been circulated. On 31 May 1916, the British and German battle fleets fought the battle of Jutland. During the encounter between the opposing battle cruiser forces that opened this engagement, British fire control was hindered by change of range rates that were high and changing and targets that were either difficult to see or periodically obscured from view altogether. The unsatisfactory gunnery performance of the British battle cruisers at Jutland indicated that a reconsideration of the fire control question was necessary. Captain Frederic Dreyer, however, was placed in charge of the Jutland gunnery post-mortem in June 1916, and he was subsequently made the Director of Naval Ordnance in 1917. Pollen's views had little chance of serious consideration under these circumstances, and although he made some effort to bring about an inquiry into the fire control situation in light of the Jutland experience, he produced no further printed papers for private circulation on the subject of naval gunnery.

14, Buckingham Street,
Strand,
London, W.C.
April 17, 1916

My Dear ——,
The following notes will summarise our recent conversation.

The basic idea behind the Pollen Fire Control System is that only a mechanical solution can be found for the problems that arise from the fact that the firing ship and target will, in action, normally be manoeuvring at different speeds, and on divergent courses. It is necessary to ascertain and to integrate these, because it follows from them that the range will always be changing. And it is the inconstancy of range that is the main difficulty in the way of continuous hitting.

The changes in the range and in the speed (or the rate) at which it changes, are, of course, caused by the speeds and courses. But unless both ships are constantly under helm – an almost unthinkable condition – speeds and courses will remain, for certain periods, constant. This must be so where ships operate in squadrons. But I do not have to remind you that even when speeds and courses are constant the rate is not constant. In a very few conditions it will change at a constant rate; but only in very few; and never if either ship manoeuvres. If, then, we assume two freely manoeuvring ships, we can say that neither the range nor the rate can ever be constant.

If we wish to use the gun from a moving ship against a moving target with the ease and success with which we can use the gun from a stationary ship at a stationary target, then we must somehow eliminate the movement of the two ships, and with these movements the consequent variations in the range. If we can do this, *and do it always* – except, of course, when the target is making a turn – we can reduce naval gunnery to the simple problem, as far as range is concerned, of the Rufigi engagement. On that occasion two 6-inch guns in HMS *Severn* engaged the *Koenigsberg* at a range of about 10,000 yards, and, having found the range by spotting, made considerably over twenty hits in twenty minutes. It is probable that they made nearly forty.[6]

Contrast with this what happened when ships were steaming and occasionally under helm. At the Battle of the Falkland Islands,[7] sixteen 12-inch guns sank the *Scharnhorst* and *Gneisenau* with fifty hits, but they were spread over so long a period that the hitting efficiency of each gun was only about one hit per gun per 75 minutes! Why was this? Because at the Falkland Islands ships were travelling at high speed, and between one o'clock, when fire opened, and two minutes past six, when the *Gneisenau* sank, Admiral Sturdee made twelve turns, and Admiral von

Spee about eight turns. Twenty turns, then, reduced the efficiency of gunfire from a minimum of one hit per gun every two minutes to a maximum of one hit per gun every 75 minutes. The gun efficiency at the Falkland Islands, then, was one-thirty-seventh − or, say, 2½ per cent − of the gun efficiency at the Rufigi.

This is startling, because the targets at the Falkland Islands were nearly twice as big as the *Koenigsberg*, − the mean range was only 1,000, or, at most, 2,000 yards longer, and the angle of descent of the 12-inch guns was considerably less at 12,000 yards than that of the 6-inch guns at 10,000 yards. Had the ships been stationary, as they were at the Rufigi, the 12-inch guns should have got one hit per gun per minute, so that really the loss of efficiency is much higher than I have said. The loss was probably at least 98½ per cent.

How could the Falkland Islands conditions be reduced to Rufigi conditions? Three problems are involved:

(1) First, it would be necessary to find out the speed and course of the target ship;
(2) Second, changes of our own speed and course;
(3) Third, to integrate these factors − and so eliminate them.

(1) In this action the target's speed and course would have had to be ascertained, say, eight times − supposing, of course, that if they could be ascertained at all, and then rightly used, the action could possibly have lasted until von Spee's second turn. The speed and course of any target can be ascertained by a good range-finder and a properly designed plotting table − when the plotting table is controlled by a gyroscope − in from one-and-a-half to two minutes at the outside. The better the range-finder, the more quickly the plot giving the speed and course will be ascertained. It is entirely a question of definition, and definition is a question of light.

The conversion of a range-finder into an automatic speed and course plotter is a perfectly straightforward engineering problem, and one that we have solved perfectly. The fact that our ship is under helm does not affect the plotting gear's operation. It plots *both* courses as they are taken by the two ships.

(2) Let us assume that we have discovered the speed and course of the target. What is the next thing to do? We shall know the course of the target in terms of our own course. If our course is due north, and we discover that the target is travelling due west, this course in terms of ours will be 270°. We call our course then always zero. If we change course it will remain zero, and our turn will be measured in terms of an altered *relative* course by the target. A gyroscope will always indicate, and can be made to measure and give a continuous record of this change. We shall then always know the changes in our own course.

But we must have an indication of our speed, too. This is best got by a Forbes Speed Log.[8] The revolutions of our screws are not a reliable index, because if, when those revolutions show twenty knots, we put the ship under full helm our speed will gradually drop from twenty knots to fifteen or thirteen (according to the design of the ship) before a complete turn is executed, so that speed under helm is not reflected by the revolutions at all. We want an exact indication of every alteration in speed as it occurs, and, most particularly, we want it under helm.

With a speed indicator and a gyroscope we have now (1) a knowledge of the enemy's *future* speed and course, based upon our record of what he has done up to this moment, and (2) constant indication of changes in our own course and speed.

(3) If we can integrate all these quantities together, and do so by a continuously working automatic instrument, which reproduces *all* of these existing changing conditions; if we can control this instrument so that, as we change our own speed and course, the new conditions automatically combine with those that are constant; if we can, further, make this device control the sights of all the guns, then we shall have the sights of the guns moving for range just as the manoeuvring ships alter it by their movements. In other words, by ascertaining and integrating *all* the factors that make change of range, we should eliminate the change of range altogether, and so reduce the firing conditions, when ships are manoeuvring, to those of engaging a stationary target from a stationary ship. The Pollen Clock, of which some are already in use in the Royal and other Allied belligerent navies, is a device which carries out all these functions.

The Pollen Clock, then – if it had been in universal use – would have reduced the Falkland Islands conditions to Rufigi conditions, and have enabled Admiral Sturdee's 12-inch guns to begin hitting at the rate of at least one hit per minute per gun, as soon as

(a) Von Spee's speed and course had been ascertained, and
(b) A salvo or two had been fired so that the final range correction could be made.*

---

* It is, of course, obvious from the susceptibility of ammunition to temperature, from the variability of the atmosphere as a resisting medium owing to greater or less humidity, etc., and to other similar causes, that the range upon the sight which will hit – in other words, the actual gun range – will, in practical conditions, *never* coincide with the true geometrical range. The actual hitting range must then be discovered always by trial and error – that is, by the process known in all navies as *spotting*. At the Rufigi River, it is interesting to remember, the people in the *Koenigsberg* never saw the *Severn*, nor the people in the *Severn* the *Koenigsberg*. The fire of the *Severn* was corrected by an observer in an aeroplane that flew over the *Koenigsberg*. And the fire of the *Koenigsberg* was corrected by observers on a hill, from which a view over the stretch of river in which the *Severn* was moored could be obtained. It is an interesting demonstration of a truth on which I have insisted for years, that the range-finder *can never be reliable as a direct indicator of range to a naval gun*. I have always insisted that its chief – if not only – uses in action were to ascertain the enemy's speed and course, or to indicate his bearing.

When, in 1908, the *Dreadnought* battleships and battle cruisers were being experimented with, I developed my advocacy of helm proof gunnery along a different line. The theory of the *Dreadnought* was that a ship possessing both higher speed and guns of longer range than the enemy could compel that enemy to fight at a distance disadvantageous to himself. It appeared to me obvious that the weaker would not be helpless in these conditions, for if, at long range, he changed course with great rapidity, he would constitute himself a target so elusive that the longer-range ship would have a very poor chance of hitting him. If he closed the longer-range ship at full speed he would again set up so rapid a change of range that hitting would be highly improbable. In each case either a varying rate of change, or a rate so high that it could hardly be ascertained, could be put to a *defensive use* by the weaker enemy.

But it was also clear that if the slower, weaker enemy closed at full speed, the stronger, longer-ranging ship would either have to fight at a shorter range, which would go nearer to creating equal conditions, or she would have to retreat herself. In other words, she would have to manoeuvre – and manoeuvring would be just as fatal to continuous hitting as a high or varying change of range which the closing target would create. And, further, it seemed that the new type of ship inaugurated in 1906, viz., the lightly armoured battle cruiser, carrying the same armament as a *Dreadnought*, but without the *Dreadnought's* power to stand the fire of the enemy – would create almost a positive obligation to keep out of range of the inferior enemy.

Hence the introduction of the battle cruiser seemed to make it certain that in any future war there would be bound to be actions between battle cruisers of superior gun-power and slower and less well armed ships of perhaps superior resisting power. It would be the policy of the battle cruiser to keep a long range, and of the weaker ship to seek a short one. Conditions very unfavourable to continuous hitting would thus be set up, and sea actions would become interminable. If, then, the more powerful ship was to circumvent the defensive manoeuvring of the weaker, the bringing to perfection of helm-free gunnery was highly desirable.

By 1910 the long-range torpedo had been developed to a point of showing that it was bound in the future to take a dominating place in battle tactics. It would be used in shoals sent from large flotillas of destroyers. These shoals would be directed against the head of the enemy column, and the admiral in command would have to manoeuvre his ships to avoid what in effect would be moving mine-fields. If he was driven to manoeuvre, the turning of his ships would make hitting with his guns impossible.

Thus, in 1907, I argued for change-of-range-proof gunnery as a *luxury for tactics*; in 1909 as *highly desirable*, if a chase was to be successfully accomplished without undue waste of time; and in 1910 as a thing no longer a luxury or even desirable, but as a *plain necessity of the situation*.

The Falkland Islands affair proves my case of the chase conclusively. The Dogger Bank affair[9] equally demonstrates how long-range torpedoes can render an opponent's gunfire innocuous. For here Sir David Beatty, who was closing the range at well over 100 yards a minute at the beginning of the action, had no sooner brought the Germans under effective gunfire than they replied with destroyer attacks on our battle cruisers, and these attacks drove Sir David to manoeuvre, and doomed British gunnery to inefficiency. So far as I can see from published accounts, few, if any, hits can have been made after 9.38 in the morning, when the torpedo attacks began.

Thus, all the conditions anticipated have been reproduced in actual war.

You will thus see that our gunnery system was designed after analysing the closest anticipation of war conditions that it was possible to make. You will, I think, agree that the experiences of war has shown both the anticipation and the analysis to have been reasonably correct.*

It is interesting to add that apparently no other investigator seems to have attempted any other *complete* solution of the problem. Partial solutions are common enough. For instance, it is quite easy to produce a plotting (or rate-finding) and a rate-keeping system which will be perfectly effective *when the change of range is small* and the *rate of change regular*. It is quite a simple thing to equip a ship with fire control only capable of finding the target's speed and course, and keeping the rate when the *firing ship remains upon a steady course*, and the enemy's manoeuvres are not too divergent. It is easy, but not so very easy, to deal with the problem of the firing ship *being under helm*, if you assume, for the working of your fire control system, that you can *always keep an accurate bearing* of the target. But these, clearly enough, touch a comparatively small number of the conditions that must be dealt with.

To be *complete* you have to deal with the following variables in finding the necessary data:

(1) The firing ship may be steady or under helm.
(2) The target may be steady or under helm.
(3) The rate may be low or high.

---

* Had this system been employed at the Falkland Islands Admiral Sturdee could have opened fire on Admiral von Spee at 1.27 by English time. He could have bracketed for range by at latest 1.35, and he should have made the twenty-five hits necessary for destroying each of the German targets in a maximum of five minutes after the bracketing was completed. The action then should have been over by 1.40.

Three times in the course of the last eight years I have drawn up memoranda on the subject of the requirements of fire control. In 1907 I based my argument in favour of this method of eliminating change of range upon the great advantage which would be conferred upon those commanding fleets if they were enabled to manoeuvre their ships without any resulting loss in the efficiency of artillery. Freedom of manoeuvre, I argued, was a necessary preliminary to the evolution of any art of naval tactics.

When you have found the target's speed and course the variables change — for the target must be assumed to keep its course — but, against this, it may be invisible.

In keeping the rate there are thus, similarly, three variables:

(1) Our ship may be steady or under helm.
(2) The rate may be high or low.
(3) The target may be *visible* or *invisible*.

The different combinations of these variables gives us, then, eight cases in which we shall want to find the speed and course, and eight in which we shall want to keep the rate.

We can simplify the system, and hence the gear of which it is composed, by omitting any one or more of them. But can we agree that any single one is unessential? Manifestly, if our system only works when our ship is steady we are hopelessly handicapped both in getting tactical information and in using it when we turn. The same applies to a system that deals only with a target on a straight course, or with only a low rate. A possibly doubtful element is: Must we be able to keep the rate when the target is invisible?

I make a point of this because there is no doubt that a great deal of the elaboration of our system is due to the fact that we have felt compelled to produce an integrating device or clock which generates the future range of the target *by generating its bearing*. But the value of this feature can hardly be exaggerated from an action point of view.

The following case will exemplify what I mean:

(a) A target is obscured by its own or our smoke for two minutes. The moment it is seen on the smoke clearing its bearing is transmitted to the plotting-room by the range-finder, and this bearing is found to agree with the bearing automatically generated on the clock. We know at once that the change of range which the clock has generated in the meantime is correct, and we can reopen fire with the confidence that the range is right. By having generated the target's bearing *automatically*, we are able, within five seconds of obtaining a new view of the enemy, to open fire with the certainty of hitting.

(b) Let us suppose, after the interval of invisibility, the bearing is sent down, and we find that it *disagrees* with the bearing generated by the clock. We know that this discrepancy cannot be due to any change of our own, because these the clock takes care of automatically. It follows, then, that the enemy must have changed his course or his speed in an interval when we could not see him. If he has changed his course or his speed he cannot be at the right range. This is obvious.

Thus automatic generation of bearing saves us from wasting shots at a target that is not correctly ranged.

(c) But the clock does more than this. The clock has indicated the deflection angle continuously, and the deflection angle is dependent upon the bearing rate. If a target has changed speed or course so that the bearing has changed at a *different* rate from that which we anticipated, we can, by a rapid correction on the deflection dial, compensate for this error, and this correction will give the target the change of course *or* of speed, or of both, which has brought about the altered conditions. Thus, without waiting for a new plot to be made, we can, by the clock alone, by a mere bearing correction, and by, perhaps, one range correction, discover what the new movement is that the target has developed.

(d) And, finally, there are cases when the target is never visible. In the course of war it is almost inevitable that a battle fleet will, sooner or later, be compelled to bombard fixed defences. Such operations have taken place frequently during this war: for example, off the Belgian coast and at the Dardanelles. In this case the automatic generation of bearing is of vital consequence, because the target to be engaged will often be invisible to those on board the ship, so that it will have to be found both for range, *and for line of sight*, by means of spotting. That it can be so found, and with perfect accuracy, is proved by the Rufigi experience. But at the Rufigi, after the range was found, the firing ship never moved. But in such conditions a ship fitted with the Pollen Clock could now manoeuvre at will, retaining *unimpaired* the capacity to engage a target. For the clock would not only keep the range, in the sense of eliminating all the changes of range caused by the firing ship's manoeuvres, but it would also keep the *true bearing of the target*, which can be transmitted from the clock to the guns, so that an artificial line of sight can be provided by which the desired objective, though invisible from the ship itself, can be kept under accurate and continuous fire, while the ship manoeuvres freely.

Both off the Belgian coast and at the Dardanelles the utility of the bombarding ships was very greatly limited by their being unable to engage targets indirectly, or even directly, while they were manoeuvring. Their gunfire could only be effective while the ships were stationary. And it followed from this that the ships could only fire, with any hope of hitting, when they were easy and certain marks both for the enemy's long-range guns, and, above all, for his submarine attacks. For the experience in this war has shown it to be elementary that high speed and rapid manoeuvring are a ship's best defence against under-water attack. It was largely this inability to bombard *without thus risking capital ships* that involved us in expensive and embarrassing experiment of building monitors – ships that possess only the most limited utility.

The main, and the most costly feature, then, of the Pollen system of fire control arises from the fact that it was determined from the first to be satisfied with no integrating device that did not permit of maintaining both the range and the bearing of a target that was temporarily invisible.

And this leads me to another observation. It is impossible for anyone to see our fire control gear without being struck by its great elaboration, the extraordinary perfection of its manufacture, and its resultant high price.

It is curious that it should be so, but there are still some naval minds who have a prejudice against any form of instrument that is at all complex. When, nearly a hundred years ago, Sir Howard Douglas was urging the general sights of sights and firing locks to the muzzle-loaders of the day, he found that though no one disputed that gunfire could be made more accurate and more rapid by these rudimentary improvements, there was yet an inveterate prejudice against any additions to artillery that was not – as he put it – 'rudely simple'.[10] These objectors have their descendants today. Is it necessary to remind these gentlemen that experience in a thousand fields of activity has proved no machine is objectionable merely because it is complex? It is objectionable only when its complexities make it uncertain or irregular in its action. Nothing could be more complex than a ship's chronometer, or the machine tools in use in every factory in the world. When analysis shows that the only reason why these mechanisms are complex is because not otherwise can a complex problem be solved, when experience shows that these mechanisms can, in fact, be relied upon, all objections to their elaboration should vanish as a matter of course. The best defence – in this respect – of the gear composing the Pollen system is that while we have had over 100 units of our gear at work in different navies since 1911, we have not yet been asked to execute repairs to the value of a quarter of 1 per cent of the cost of these installations!

Now, as to the question of price. At the present moment, fighting ships all over the world are equipped with very cheap fire control gear. The moment the price of ours is mentioned we are put upon our defence. But the explanation of the high price is perfectly simple. We have maintained from the very first that there are three essentials of fire control. First, it must deal with all conditions as set out above. Secondly, it must deal with them with absolute exactitude. Thirdly, it must in every *essential* operation be automatic.

The phrase has been used that 'a fire control system is either perfectly universal and perfectly accurate, or it must be perfectly useless.' At any rate, this is the principle on which we have gone. As we have seen above, the system can be indefinitely simplified if certain difficulties – that are bound to arise – are ignored. It could be still further cheapened if its automatic character was abandoned.

The mistake that seems to have been made by some of our critics is the supposition that the difficulties we have provided against would be the exception. Experience has shown them to be the rule. It was thought we had contemplated an unnecessary standard of exactness. Engagements have to be fought at ranges at which absolute accuracy is essential to hitting. Ships have constantly to go under helm; the target is constantly invisible. If we could have found and kept the range at the Dardanelles with the ship under helm we should not have lost *Ocean*, *Irresistible*, *Triumph*, or *Majestic*, nor should we have had to spend millions on monitors.[11] If we could have kept the range under helm we should have sunk von Spee's battle cruisers in ten minutes after first engaging him; and *Emden* in twelve minutes instead of ninety. And *Derfflinger*, *Seydlitz* and *Moltke* would not have escaped at the Dogger Bank with their speed unimpaired and their armament unhurt.

These things show that the theory that fire control need not be complicated to deal with difficult conditions is a hopelessly erroneous theory, because the difficult conditions either inevitably are, or – what is much the same thing – can be made by the enemy, universal. The first element, then, in making the Pollen system costly is that it is designed to deal with *all* the circumstances that can arise.

The second reason is that it is, from beginning to end, mathematically and instrumentally perfect. The third is that it secures synchronism between the range changes on the sights with those changes being caused by the ships' movements. It is no use generating a change of range to go on to the sights unless the change which is generated is the right change. And clearly, in conditions like these, there are two elements of rightness. There is rightness in the *number* of yards, and there is rightness in the *time* at which this number is to be added or subtracted. It is this perfect standard of measurement accuracy which makes all this gear exceptionally costly to make. It is the need of *time* accuracy that causes the expense of making its system automatic.

It is just the same with the Cooke-Pollen range-finder. Range-finders of the Zeiss and Barr and Stroud type can be made for a very reasonable price, because they are of a design providing for an inferior illumination, inferior definition, and inferior resistance to temperature distortions.[12] To obtain a higher standard of visibility, of instrument accuracy, of constancy, a fineness of design, more costly in itself, and a standard of workmanship and accuracy of a higher grade are necessary.

The Pollen Plotting Table is similarly of a far more costly design than the methods of partially reading the desired result. But it is necessarily costly to give it three qualities which practical experience have shown to be indispensable. These are, first, that it should meet *all* the conditions; next, that it should be fool-proof and unbreakable; lastly, that the man who is operating it in action should always have the simplest, most straightforward and easiest functions at that critical juncture. It is therefore so arranged that the plot of the enemy's course is made at a fixed point of the machine, thus releasing the operator from ever moving from a single spot. The amount of gear is *multiplied* by four. The difficulties — and above all, the delays — of the operator's work are *divided* by four.

Throughout we have worked on the principle that it is far better to have a fire control system consisting of three complicated — but quite reliable — machines, than to leave the fortunes of fire control to the tender mercies of thirty complicated heads!

And this leads me to a final word upon the actual working of the system. The range-finder is placed under armour in the conning tower. We suggest a range-finder in every turret and one in each conning tower, fore and aft. The central system of control can be linked up with any one of these by change-over switches. Our range-finder is operated by two men. It is the function of one to keep the instrument trained on the target. The function of the other is to manipulate the split image so as to secure what is called the 'cut'. The range-finder is automatically connected to lamp indicators that have no moving parts. These indicators show continuously the *bearing* of the target and its *range* as shown by the range-finder. These are continuous indications. When the range operator has got the cut he is supposed to press a button which illuminates a tell-tale light. The indicators of the tell-tale light are down in the transmitting room. The moment the tell-tale light is shown, the plotting table is set to the range and to the bearing transmitted from the range-finder. It has previously been set to our own ship's speed. In the same interval the clock has been set to own ship's speed, to target's bearing and to the range. The clock operator will continue to vary these settings until he gets definite information from the plotting table as to what the enemy's course and speed are.

The first part, then, of the working of the Pollen system is this: The range-finder is brought to bear upon the target, and the enemy's range and bearing are found and transmitted below. The bearing will be accurately got in a very few seconds. The operator may take half-a-minute or more before he gets an accurate cut. While these transmissions are being attained the plotting table and the clock are being set to all the known elements — our speed, the bearing of the target, the range — as they are obtained.

The second process is as follows. So soon as the ranging operator records a *reliable* cut the plotting table is thrown into operation. From this moment a plan of the courses of both ships begins to be automatically generated by two pencils on the plotter, which not only mark the direction and movement of the two ships, but, by revolving at the end of each minute, record a time element from which the enemy's speed can be measured. If the conditions for seeing are good, if the range is within the capacity of the instrument,* if the skill of the R.F. operator is of

---

* In the case of our new 15 ft. instrument readings at 25,000 yards should vary so little that quite accurate plotting is to be expected.

a high order, at the end of the first minute's operation the course and speed of the target should be approximately known. The table operator will read off these elements from the plotting table and call them out to the clock operator.

The clock operator has in the meantime kept his instrument set to all the other elements. The speed and course of the enemy are set in a couple of seconds, the final correction is made to range and bearing, the clock is thrown into operation, and from that moment all the sights in the ship are recording the *geometrical range, as it changes* owing to the movements of the two ships.

The second and third operations are then, the second to obtain the speed and course of the target on the plotting table, the third to set these upon the clock from which the continuous setting of the gun sights is secured.

The fourth operation is as follows. There is now upon the sights the geometrical range as the meaned results of the range-finder give it. But this will not be the gun range. The gun range can only be ascertained by trial and error. The gunnery officer, if the range is within the capacity of his pieces, will now order a whole or half salvo, and will watch to mark the spot at which the impact of the projectiles upon the water is made. Judging from this impact, he will make the spotting correction so many hundred yards up or down. This process will be repeated until he has landed a salvo on the target. Supposing he proceeds by the principle of 'bracketing' the correction might be as follows: Up 600 yards; down 300; up 150; down 75; up 25; hit.

Let us assume it has taken five minutes to complete this series of salvoes. Let us next assume that the geometrical range was 12,000 yards when we began, that the range was *increasing* at 400 yards a minute. As the net results of these corrections we have found that the guns were shooting 325 yards *over* what we believed to be the geometrical range. We opened fire at 12,000 yards, but at the end of the spotting interval we have 14,325 yards on the sight, because *while the spotting process has been going on* the range on the sights has been *growing at the true rate*, quite irrespective of this process.

In other words, our spotter has dealt with the discovery of the *hitting range* by trial and error *exactly as if the firing ship and target were standing still*.

Once we have closed the bracket, and, indeed, while we are closing it, our ship is free to manoeuvre in any way it likes, change its course and change its speed *ad libitum* – and, of course, in doing so change the rate from 400, increasing to 600, decreasing, or in any other way – without introducing any range error whatever, so long as the target maintains the speed and course with which we began.

If the target changes speed and course obviously the whole process must begin *de novo*. But it does not necessarily follow that the whole process of plotting has to begin *de novo*. As we have seen in the case of a target which changed course while she was invisible, the fact that the clock generates future ranges and future bearings will enable us, if the target changes course suddenly, to find the angle of his turn with perhaps two or three range and deflection corrections. By so designing the clock that the settings of speed and course set deflection and rate of change of range, and the settings of deflection and rate of change of range, set speed and course, we are enabled to turn to immediate and accurate account *any* information about the target's range and bearing which is obtained either by range-finder or by spotting.

With the Pollen Clock, therefore, the spotter need never wonder how much the target has changed course, how much it may have changed speed. All he has to do is to tell the clock operator that the range is short or over, and the correlative readjustments of the clock will automatically translate themselves into the only changes of course and speed that can have brought about the new conditions.

I hope this description and account of the instruments and their origin will explain to you both their necessary elaboration, their unusually high cost, and the method of working them. I have never hesitated about elaboration or cost, if *a function of real war value can only be so obtained*, for the simple reason that it always seemed to me childish for people to build fighting ships costing from £300,000 for the lightest cruiser to £3,000,000 for the heaviest battleship, and to hesitate for five minutes about spending £8,000 on the first, and, say, £30,000 on the second, to enable them to give full efficiency to their guns.

If this war has shown nothing else, it has shown the amazing folly of being pound wise and penny foolish when it comes to the element on which the fighting efficiency of the Fleet depends. Can greater imbecility be conceived than to be prodigal of millions in building a Fleet, but parsimonious in pence when it comes to making it do the only work it is wanted for?

Yours very faithfully,
(Sgd.) A. H. POLLEN

Dear ———,

I understand from your letter that what you wish me to do is, first, to explain the theory of naval gunnery which I and my colleagues have sought to embody in the fire control system which goes by my name. Next, to give you a short sketch as to how it actually works in practice; and last, to account, if I can, for the British Admiralty having gone to so much expense and trouble to keep my system secret till 1912, and then deciding to allow it to become common knowledge to all other navies. In the following pages I deal, as far as I am able, with these three matters. [Editor's Note: Pollen, having dealt with the first two matters already in the previous letter, proceeded here to give only the account of the events of 1913, which received the heading 'Part 3'.]

PART 3

There is no public source of information as to the Admiralty's reasons for abandoning the monopoly of my system in 1912, except that given by Mr. Churchill, in answer to a question, June 30, 1913, which was reported as follows in the *Times* of the following day.

Mr. Churchill (Dundee): I am informed that some portions of Mr. Pollen's apparatus were used, but they were not used in accordance with the Pollen system. The good shooting was not attributable either to the Pollen apparatus or to any employment of the Pollen system. It is not intended to adopt the Pollen system, but to rely on a more satisfactory one which has been developed by Service experts. In reply to the last part of the question, the Admiralty has given Mr. Pollen considerable assistance in the hope of obtaining a valuable system of fire control for the Navy. The results obtained with his system and the principles underlying it are such that it is not proposed to take any steps to prevent his making public use of all its essential features. This he can do without divulging official secrets connected with the Service system, and he will, of course, be precluded from disclosing Service secrets of which his connection with the Admiralty has given him knowledge, or from any infringement of the Official Secrets Act, which I have every confidence will be respected.

I should like to add by way of caution that all these questions connected with fire control are of a highly technical character, and their discussion in the newspapers could not lead to any intelligible conclusion or be attended by any public advantage. So far as the relations between the Admiralty and Mr. Pollen are concerned, I shall be quite prepared on a suitable occasion, if desired, to explain them fully to the House. So far as the technical aspects are concerned, I must decline on behalf of the Admiralty to take any part in their public discussion. I can only say that in coming to the decisions which I have stated to the House, and for which I accept full responsibility, I have been guided by the representations of my naval colleagues and the advice of the experts on whom the Admiralty must rely.

I became aware of the character of Mr. Churchill's answer on the afternoon on which it was given, and wrote the letter, which appeared in the same issue of the *Times*, before waiting to see Mr. Churchill's answer in print.

FIRE CONTROL IN THE NAVY
To the Editor of the 'Times'.

SIR, – I understand that Mr. Churchill, in replying to Mr. Harcourt to-day, said that, after assisting me for some years to produce the A.C. system, the Admiralty, in view of the results obtained with it, had adopted a more satisfactory fire control system developed by Service experts. A meaning, which, I am sure, Mr. Churchill did not intend, might be conveyed by these expressions – namely, that two fire control systems had been in competition with each other, and the more satisfactory one accepted. As this is an inference which might prove damaging to an enterprise in which a large amount of capital has been invested, I ask to be allowed, without waiting to see the text of Mr. Churchill's reply, to make the following statement in your columns:

The A.C. Fire Control system, invented by Mr. Isherwood and me, consists principally of –

(1) A group of instruments for ascertaining the target's range and movements, generally known amongst experts as the plotting or 'rate-finding' unit; and
(2) of a device known as the Argo 'Clock', which uses the information gained by the first group to control the sights of the gun automatically.

Both processes are essential to fire control.

The first group has never been supplied for trial; the clock alone was tried for the first time about two months after monopoly had been abandoned.

Yours faithfully,
A. H. POLLEN

The Argo Company, Limited,
14, Buckingham Street, Strand, W.C., June 30.

To understand the purport of Mr. Harcourt's question, the following extracts from the *Times* that appeared before this question was answered, should be read:

THE FIRING OF H.M.S. *ORION*
Successful Battle Practice

The battleship *Orion*, which was at the top of the list of ships armed with the 13.5-in. gun in last year's battle practice, has just completed this year's test with continued success. The actual figures of her practice are, of course, confidential, but it may be said that she has broken all 'records', and that unless the firing of other ships is very exceptional, her performance cannot be eclipsed.

This is the third occasion on which the *Orion* has recently carried out special gunnery tests. She took part in a competition with the *Thunderer* when the latter was fitted with Sir Percy Scott's director, that test being confined to a comparison of the two systems of aiming. In the one ship all the guns were laid by an officer at the director, and in the other by individual layers at each gun. In the circumstances, the system of simultaneous firing exhibited manifest advantages, but it was felt by many naval officers that in a trial under more difficult conditions the *Orion* might have made a better showing. After the director trials the *Orion* carried out experiments to test certain parts of the Pollen automatic fire control system, an account of which appears for the first time in the current issue of the *Naval Annual*. These tests were particularly interesting, not only as indicating the advantages to be obtained by using the Pollen automatic change-of-range and bearing clock, but because of the unusual conditions under which they were carried out. The *Orion* ran at a far higher speed than is customary at battle practice, which with the course chosen for the target resulted in a very high rate of change, and the tests included an attempt to keep the target under fire while the firing ship was under full helm. The success of the *Orion* in maintaining continuous hitting under these conditions is held by many naval officers to suggest that the almost universally held conception of the tactics which must prevail in battle requires modification.

Writers of every nation who have attempted an analysis of the conditions which must prevail in action have maintained that in order to hit the enemy continuously the ship must be kept on a steady course, the fleets moving in the same direction on approximately parallel lines, so as not to interfere with the fire control. The experience of the *Orion*, however, when only fitted with a portion of a scientific fire control system, indicates that continuous hitting can be ensured in spite of changes in the course. Possibly, when the rest of the Pollen system is tried, the highest changes of range and direction will be tackled without difficulty, and the necessity for keeping on a course parallel to that of the enemy's line during action will no longer exist. It does not appear that on the third occasion the innovations made in the December tests were repeated, yet it is likely that, if they had been, the *Orion*, using the Pollen Clock, would have recorded an equally good percentage of hits to that which has been obtained under the easier conditions of Service battle practice.

NAVAL AND MILITARY INTELLIGENCE
The Firing of the Orion

Our Portsmouth Correspondent writes:—The interest in the shooting of the *Orion* deepens as the circumstances come to be better known. The full significance of any number of battle practice performances can never be satisfactorily established until the record that is taken from observations on various ships of the fall of every shot is fully analysed by the Inspector of Target Practice's staff. Until this is done the precise result cannot be arrived at, but although this is so, certain broad facts with regard to the *Orion's* practice appear to have been clearly established. In the first place, she is credited with between 40 and 45 per cent of hits to rounds fired. Next, it is stated that her number of hits would have been 50 per cent higher but for certain deflection errors that crept in.

It is possible that mere aiming errors by the gun-layer may account for other misses. When the shortness of time available for finding the object and course of the target is considered, the rapidity at which the guns have to be fired to get in the prescribed number of rounds, and the new elements of actuality which have been introduced into this year's battle practice, it is remarkable that, deflection errors excepted, the *Orion* should have obtained almost the full number of hits that could be expected. That so many hits should have been lost through erroneous deflection raises the question as to whether it will not be advantageous to apply the principles of automatic calculation to deflection as well as to range. The *Orion* is the only ship so far fitted with the Pollen Clock, and it is said that all the battle cruisers are to have it, as well as the Scott Director. The director complements the clock. While one finds the spot at which to aim, the other centres the fire of all the guns on that object. There seems to be a widespread feeling here that as soon as these appliances are in more general use the conditions of battle practice should be made more exacting, so as to ascertain exactly what extension of gunnery possibilities the Pollen system throws open.

Two days after Mr. Churchill's reply, the following letter appeared in the *Times*:

FIRE CONTROL IN THE NAVY
Mr. Churchill and the Pollen System
To the Editor of the 'Times'.

SIR, – The First Lord of the Admiralty, in reply to Mr. Harcourt's question in the House on June 30, stated that it was not intended to adopt the Pollen system, but to rely on a more satisfactory system which has been developed by Admiralty experts.

The inference to be drawn from this pronouncement is surely that a series of tests has been made at sea between two ships, respectively fitted with the Admiralty system and the Pollen system in its present up-to-date form. Such a procedure would certainly recommend itself to most people as an indispensable preliminary to the Board's decision. But it appears from the inventor's letter in your issue of July 1 that Mr. Churchill, in speaking of the results obtained by the A.C. system, could merely have been alluding to the opinion of the Admiralty experts in question, for as it happened no trials had taken place. The very explicitness of the First Lord's statement makes one wonder whether he himself realised that fact.

It must always be remembered that the problem Mr. Pollen set himself to solve so far back as 1900 was that of automatically obtaining range and bearing of an enemy's ship when in motion, so as to enable the attacking ship (being also in motion) to maintain a continuous fire, whatever the change in bearing and speed of the respective ships might be. Until Mr. Pollen had actually solved this problem no naval officer had recognised that such a problem even existed. And if the Service ever acquires a satisfactory system on this basis it will be in consequence of his having led naval thought in that direction.

In 1908 his system in its then form was tried on board H.M.S. *Ariadne* in competition with the Service system, when the latter was held to be the superior. It was subsequently found that the Service invention was merely a revival of one given up by Mr. Pollen some years before. It would be interesting to see whether history would repeat itself in further trials.

I note Mr. Churchill says he has been guided by the advice of experts on whom the Admiralty 'must' rely. A curious, and apparently a defensive, phrase. Had his predecessors felt this compulsion, secrecy would have been abandoned finally in March, 1908, again in February, 1909 – although by that time the failure of the Service system was manifest – and again in the spring of 1910, when there was no Service system in the field at all. The secrecy of the A.C. system was saved by the fact of the disagreement of the experts being brought to the notice of the First Lord on each occasion. And each time it was realised that it was unsafe to give up secrecy till conclusive tests had been made.

For fourteen years Mr. Pollen has worked at his invention, and successfully kept it secret in the interests of the country, but to his own commercial detriment. In these circumstances it seems less than just, and much less than generous, that the First Lord should in his speech have included a covert threat of using the Official Secrets Act against him.

That there is a strong feeling in the Navy in favour of the A.C. system is undeniable. With the experts divided, and no definite authoritative results by which the controversy can finally be settled, it is surely premature to part with a secret of which so high an opinion is held, for the believers in the Pollen system are quite as eminent in professional attainments as the official experts who are once more in opposition to it.

Lastly, the wisdom of giving up a fire control system acknowledged to be the foundation of our own, developed entirely within the Navy and largely at public expense, must be very questionable, especially in the absence of any evidence that foreigners possess any system as scientifically thought out or as thoroughly developed.

'EMERITUS'[13]

July 1

None of the foregoing statements has ever been disputed, and the following are material to answering your question.

The abandonment of the monopoly of the Pollen system seems to have been decided upon some months before the trial of the Pollen Clock in H.M.S. *Orion*, which trials took place in the autumn of 1912. The rangefinder and the plotting table were never tried at all. In the trials of the clock which finally did take place the *Orion* is stated to have made a succession of hits at long range, when the angle of the target's course with herself made the rate of change high, and when the firing ship was actually under helm, so that the rate, in addition to being high, must also have been varying in a very exceptional manner. This entirely unprecedented performance is said to have evoked amongst those naval officers who could appreciate its portentous character the conviction that the naval tactics of the future must be revolutionised by so drastic an advance in the art of using guns. It is not surprising – as we learn from the letter of 'Emeritus' – that there arose a sharp division of opinion between the experts in the Service and those who advised Mr. Churchill. So that in refusing to reconsider the abandonment of the monopoly after the efficiency of the Pollen Clock to effect this tactical revolution was proved, Mr. Churchill must have accepted the decision of the Admiralty expert or experts blindly, and have ignored the diametrically opposite opinion of the experts to whom 'Emeritus' alludes.

Were these official experts the same as those who refused my urgent requests that the South Coast ports should be defended by my scheme?

'Emeritus' further states that an exactly similar position was brought about in the course of a previous trial of Pollen gear, and that had the opinion of the official experts been followed secrecy would have been abandoned at a far earlier date. But it was saved on this occasion by Mr. Churchill's predecessor being made aware of the disagreement of the experts, and by his decision, in face of it, to maintain secrecy until a competition between the Pollen system and the Service system could show which was really superior.

But no such trial, as appears from the inventor's letter, ever in fact took place before monopoly was abandoned. That Mr. Churchill lightheartedly decided that no such trial was necessary may perhaps be explained in his naive admission, in his reply to Mr. Harcourt, that the subject of fire control was an unintelligible one. The admission is interesting, and throws a curious light upon the success of Mr. Churchill's advisers in explaining so simple and elementary a subject as fire control to so brilliant and appreciative a listener as the then First Lord. The inference is surely irresistible that the officer who essayed this task failed not through any defect of Mr. Churchill's understanding, but through an incapacity to understand the subject himself.

Nor is this altogether impossible, for I find a naval officer writing to the *Times* of June 7, 1914, in the course of the debate on the submarine threat. He argues in this letter that the theory that submarines had superseded battleships arose from the fact that naval officers are aware that they have not adequate means for using guns accurately. And he sees in Sir Percy Scott's defection an admission of one of the protagonists in the newer gunnery, that the effort to put the use of naval guns on a scientific basis has failed. He continues:

'Your readers will miss the entire point of this controversy unless they realise the fundamental principle that ships of war can be efficiently protected only by the destruction of those who seek to hurt them, and that the main means of destruction is the armament of guns. So that by far the most urgent and vital of all naval problems is to discover the best means for using them. They will never be discovered, or if discovered adopted, until an organisation exists by which the necessary steps are left entirely to the gunnery experts, and naval policy in this matter is decided by them and by them alone. So long as the *whole* of gunnery policy – methods as well as material – is in the hands of a single officer there is very little prospect of progress being made. Of recent Directors of Naval Ordnance one was not a gunnery man, another had hardly been identified with gunnery in a single particular since the Scott renaissance and the institution of long-range firing. They were, doubtless officers of great administrative capacity, and their want of special knowledge would have mattered little if the choice of methods had been left to those who understood them. The office and staff of the Inspector of Target Practice – in constant touch with the experts at sea, witnessing their performances and knowing their needs – afforded just the authority that was wanting. But when it was no longer wanting the authority was no longer wanted. Far from the N.O.D. following its guidance, it is an open secret that from 1908 to 1912 the policies of the two Departments were in such hopeless conflict that at last the post and Department of Inspector of Target Practice were abolished. It is worth remarking that the failure here was not inability to organise the brains of the Service, but inability to use them when already organised. Nor is there anything very encouraging in this precedent.

The answer to the submarine and the torpedo scare is to be found not in flying machines or any hastily extemporised panacea, but by the quite commonplace bringing to perfection of the use of guns, whether great or small, whether in battleships, cruisers, or destroyers, whether used by day, or night. It is the failure of gunnery that gives the scare the only reality it has.'

If this officer was right in saying that 'of the recent Directors of Naval Ordnance, one was not a gunnery man, another had hardly been identified with gunnery in a single particular since the Scott renaissance and the institution of long-range firing', then there was nothing surprising in an officer so poorly qualified having been unable to explain the elementary principles of fire control to Mr. Churchill.

And if he could not explain these principles, it is equally intelligible that he was at any rate sincere in telling Mr. Churchill to make the astonishing misstatements which he delivered in the House of Commons. Note, for instance, that while Mr. Churchill denies that the Pollen gear was used according to the Pollen system in *Orion* battle practices, he does not state that in the December tests it was only by the use of the Pollen Clock that this ship was able to make a succession of hits under full helm at top speed.

Let us follow the statement a little further: 'It is not intended to adopt the Pollen system, but to rely on a more satisfactory one which has been developed by Service experts.'

He did not say what one gathers from the 'Emeritus' letter to be the case, that there was no Service system in the field when the Pollen system was abandoned, and that when produced it was again an adaptation of the Pollen invention, and likely to fail, as previous plagiarisms had already failed.

The next statement is, as is obvious from the 'Emeritus' letter, a flat untruth. 'The Admiralty has given Mr. Pollen considerable assistance in the hope of obtaining a valuable system of fire control for the Navy. The results obtained with his system and the principles underlying it are such that it is not proposed to take any steps to prevent his making public use of all its essential features.'

The statement here is quite simple, viz., that the system had been tried and had not given the desired results. The system, of course, as a whole, had never been tried. The only part of it tested was the clock, and that was only tested after monopoly had been abandoned.

The rest of the paragraph is, as a Press commentator puts it, 'a veiled menace' to the inventor of the Pollen system. In fact, however, the menace was anything but veiled. For, six weeks after the agreement for secrecy between my company and the Admiralty had expired, and we had been given complete liberty to take *all* our gear abroad, I was threatened with proceedings under the Official Secrets Act if I ventured to make the character and design of my clock public!

This, of course, threw an entirely new light upon the policy of giving up monopoly. Had those who advised it supposed that, while foregoing the advantages of the system for the British Navy, they would be able, by intimidating me, to prevent any other navy profiting by them? It was known that I had worked for fourteen years profitless and unpaid, and that the keeping of my invention secret was entirely due to the pressure that I had personally been instrumental in bringing upon successive First Lords of the Admiralty. Was it now proposed to reward me for this not only by refusing to adopt my system when its success was proved, but by preventing my offering it elsewhere? This is not a proceeding which it is necessary to qualify by adjectives.

Leaving its moral character for others to decide, my own concern was that it raised a far more important matter either than my own commercial interests, or the obligation I was under to my shareholders, to see that they suffered no injustice at the hands of the Admiralty. It was a direct challenge to my honour and integrity. On the many previous occasions on which the Admiralty had abandoned secrecy, in spite of strong inducements to put my work on a purely commercial basis, I had, at my own risk and expense, carried on because I was convinced that the new fire control would revolutionise naval war. And it seemed to me monstrous that if any effort of mine could preserve it as a secret to my own country, that that effort should be wanting.

But the Admiralty menace of 1913 left me powerless in the matter. To have yielded would have been tantamount to acknowledging that the accusation that I had stolen a Service invention and called it my own was true. I had, then, no alternative but to insist upon the rights and wrongs of this matter being settled by the Royal Courts of Justice, which alone were capable of deciding finally in such grave matters as the vital interests of the nation on the one hand and the primary rights of an individual on the other.

I accordingly put myself into the hands of the most eminent legal authorities, and in the action I took carried out literally the instructions of the only barrister in private practice who had held the high office of Attorney-General to the Crown.[14] His advice was that I should inform the Admiralty that I intended on a certain date to apply for patents and otherwise take steps for the publication of my invention, and to say that I gave this notice expressly to give the Admiralty an opportunity of proceeding against me, either by warrant or by injunction, so that the title to the invention could be legally settled before its publication had become irrevocable. He thought it my duty, as a loyal citizen, to do this, because the Admiralty claimed the right to forbid the publication of an essential part of my system on the ground that it 'embodied a *vital* secret' of the British Service. If there was a tittle of evidence to support this view the Department would have to defend the interests of the Service.

The rest of the story can be guessed. Mr. Churchill and his advisers took no steps whatever to carry their threats into execution. It had long been obvious, both to my advisers and myself, that they had acted in this matter without consulting the law officers, and that no law officers, in view of the overwhelming character of the proof of my claim to the invention, would encourage them, when it came to the point, to take the action that they threatened. Their doing so throws a sinister light on the good faith of the effort to intimidate me, and inflicted a great injustice, because to this day the accusation that I stole and misused a Service secret lives in the traditions of a great Department.

It still remains to be explained, however, how it was that the Board of Admiralty considered that my system included a secret that was 'vital' to the interests of the British Navy, and took no steps, beyond a manifestly empty threat, to prevent that secret being communicated to others.

I need hardly add that the abandonment of the monopoly of my system was a very bitter disappointment. The only consolation I have is that I exhausted every effort to make its abandonment impossible, and the final publication of its character abroad was in the end forced upon me by an act that can hardly be defended.

The disappointment did not arise from any fear that my financial future had been clouded by this precipitate and foolish action. I had not studied this problem for fourteen years without understanding it. I had not discussed it with all the most accomplished artillerists in the British Service without realising the advance in gunnery technique that my system had made certain. I had not been in almost constant conference with the acknowledged authorities in naval strategy and tactics without fully appreciating the influence our instruments must have in naval war. It seemed to me certain therefore, that the abandonment of monopoly, far from being a commercial injury, would, in fact, be a commercial favour of unexampled liberality, had it been possible to regard my freedom from any such point of view. After all, I have had unique opportunities for studying the problem at sea, and unique and most generous financial assistance in bringing each instrument to perfection. First and last, I can hardly have benefited by less than £80,000 when the Admiralty in effect abandoned all claims on the work that they had so signally subsidised. To have complained of a *material* grievance would have been absurd. Nor need I add, what is now well known, that during the brief twelve months during which I was able to explain my system to the different naval administrations of the world, before war terminated all possibilities of commercial development, I succeeded in getting the promises of more orders than the factories at my disposal could execute in the next three years.

No, my grievance was very different. I knew the value of my work, and, as it was undertaken and carried through with the single object of helping my own country, I felt its being renounced by those who acted for the country as an acute, and almost unbearable, sorrow.

APPENDICES

I BIOGRAPHICAL NOTES

Persons with entries in the *Dictionary of National Biography*, *Who's Who of British Members of Parliament*, or *Who Was Who* are noted as follows: full name and highest title attained; birth and death dates; naval offices, assignments, and promotions; and major title awards from the time of first mention in the text. Assignments through 1914 and promotions, which were culled from the *Navy List*, are given for naval officers who are without entries in the *Dictionary of National Biography* or *Who Was Who*.

Bacon, Sir Reginald Hugh Spencer (1863–1947): Director of Naval Ordnance, 1907–9; Colonel Second Commandant Royal Marines, 1915; Senior Naval Officer, Dover, 1915–17; Rear-Admiral, 1909; Vice-Admiral, 1915; Admiral, 1918; knighthood, 1916.

Barr, Mark: biographical information on Barr from standard sources does not exist.

Barry, Sir Henry Deacon (1849–1908): Director of Naval Ordnance, 1903–5; Admiral-Superintendent, Portsmouth Dockyard, 1905; Commander-in-Chief, Third Cruiser Squadron, 1906–8; Rear-Admiral, 1905; Vice-Admiral, 1908; knighthood, 1906.

Battenberg, Prince Louis Alexander of: *see* MOUNTBATTEN, Louis Alexander.

Bayly, Sir Lewis (1857–1938): President, Royal Naval War College, Portsmouth, 1908–11; Commander-in-Chief, First Battle Cruiser Squadron, 1911–12/Third Battle Squadron, 1913–14/First Battle Squadron, 1914/Western Approaches, 1915–19; Vice-Admiral, 1911; Admiral, 1917; knighthood, 1914.

Beresford, Lord Charles William De La Poer, Baron Beresford (1846–1919): Second-in-Command of the Mediterranean Fleet, 1900–2; MP, Woolwich, 1902–3; Commander-in-Chief, Channel Squadron, 1903–5/Atlantic Fleet, 1905/Mediterranean Fleet, 1905–7/Channel Fleet, 1907–9; MP, Portsmouth, 1910–16; Vice-Admiral, 1902; Admiral, 1906; created Baron Beresford, of Metemmeh and of Curraghmore, 1916.

Boys, Sir Charles Vernon (1855–1944): physicist; knighthood, 1935.

Carr, Henry C.: *Jupiter*, 1905–8; *Nile*, 1908–9; *Tenedos*, 1909–10; Royal Naval Barracks, 1910–11; *Psyche*, 1911–14.

Churchill, Rt Hon. Sir Winston Leonard Spencer (1874–1965): First Lord, 1911–15, 1939–40; Conservative and Liberal politician and statesman; knighthood, 1953.

Colville, Hon. Sir Stanley Cecil James (1861–1939): Commander-in-Chief, Nore Division, Home Fleet, 1908–9/First Cruiser Squadron, 1909–11/First Squadron, First Fleet, Home Fleet, 1912–13; Special Service and with Grand Fleet, 1914–16; Commander-in-Chief, Portsmouth, 1916–19; Vice-Admiral, 1911; Admiral, 1914; knighthood, 1912.

Craig Waller, Arthur: *see* WALLER, Arthur Craig.

Custance, Sir Reginald Neville (1847–1935): Second-in-Command, Channel Fleet, 1907–8; Admiral, 1908; knighthood, 1904.

Dreyer, Sir Frederic Charles (1878–1956): assistant to the Director of Naval Ordnance, 1907–9; *Vanguard*, 1909–10; *Prince of Wales*, 1910–12; War Staff, 1912; *Amphion*, 1913; *Orion*, 1913–14; *Iron Duke*, 1915–16; Assistant Director, anti-Submarine Division, Admiralty Naval Staff, 1916–17; Director of Naval Ordnance, 1917–18; Director of Naval Artillery and Torpedoes, Admiralty Naval Staff, 1918–19; many subsequent appointments through the Second World War; Commander, 1907; Captain, 1913; Rear-Admiral, 1923; Vice-Admiral, 1929; Admiral, 1932; knighthood, 1932.

Dumaresq, John Saumarez (1873–1922): *Canopus*, 1900–2; *Magnificent*, 1903–4; Hydrographic Department, 1904; Torpedo Boat 116, 1906–7; *Nith*, 1908–10; *Prince of Wales*, 1912–13; *Shannon*, 1913–14; Commander, 1904; Captain, 1910; Rear-Admiral, 1921.

Elphinstone, Sir (George) Keith (Buller) (1865–1941): engineer; partner in the firm of Elliott Brothers, 1893–1941; knighthood, 1920.

Fisher, John Arbuthnot, 1st Baron Fisher, of Kilverstone (1841–1920): First Sea Lord, 1904–10, 1914–15; Admiral of the Fleet, 1905; Baron, 1909.

Gipps, George: *Excellent*, 1905–7; *Hindustan*, 1908–10; *Newcastle*, 1912–14; (Pollen Papers: employee of the Argo Company, 1911–12; killed in action at the Dardanelles, 1915).

Goodenough, Sir William Edmund (1867–1945): various appointments, 1900–1; Commodore, First Light Cruiser Squadron, 1913–16; Commander-in-Chief, Second Battle Squadron, 1916–19; various appointments, 1919–30; Commander, 1901; Captain, 1905; Rear-Admiral, 1916; Vice-Admiral, 1920; Admiral, 1925; knighthood, 1919.

Harcourt, Robert Vernon (1878–1962): Liberal MP, Montrose Burghs, 1908–18.

Harding, Edward W.: Naval Ordnance Department, 1904–8; *Bellerophon*, 1909–12; Major, 1914.

Henley, Joseph C. W.: Naval Ordnance Department, 1909–11; *King Edward VII*, 1911–13; *Marlborough*, 1914; Captain, 1916; Rear-Admiral, 1927.

Hughes-Onslow, Constantine H.: Royal Naval War College, Portsmouth, 1908–9; *Grafton*, 1910–12; *Revenge*, 1914.

Isherwood, Harold: biographical information on Isherwood from standard sources does not exist.

Jackson, Sir Henry Bradwardine (1855–1929): Controller, 1905–8; Commander-in-Chief, Third Cruiser Squadron, 1908–9/Sixth Cruiser Squadron, 1909–10; President, Royal Naval War College, Portsmouth, 1911–13; Chief of the War Staff, Admiralty, 1913–15; First Sea Lord, 1915–16; President, Royal Naval College, Greenwich, 1916–19; Rear-Admiral, 1906; Vice-Admiral, 1911; Admiral, 1914; Admiral of the Fleet, 1919; knighthood, 1906.

Jellicoe, John Rushworth, 1st Earl Jellicoe (1859–1935): Director of Naval Ordnance, 1905–7; Second-in-Command, Atlantic Fleet, 1907–8; Controller, 1908–10; Commander-in-Chief, Atlantic Fleet, 1910–11/Second Division of the Home Fleet, 1911–12; Second Sea Lord, 1912–14; Second-in-Command, Home Fleet, 1914; Commander-in-Chief, Grand Fleet, 1914–16; First Sea Lord, 1916–18; Rear-Admiral, 1907; Vice-Admiral, 1910; Admiral, 1915; Admiral of the Fleet, 1919; knighthood, 1907; Viscount, 1919; Earl, 1925.

Kelvin, Lord: *see* THOMSON, Sir William.

Kerr, Lord Walter Talbot (1839–1927): First Naval Lord, 1899–1904; Admiral of the Fleet, 1904.

Lafone, Albert S.: *Ariadne*, 1907–8.

Landstad, D. H.: biographical information on Landstad from standard sources does not exist.

BIOGRAPHICAL NOTES

Lawrence, Sir Joseph, 1st Baronet (1848–1919): businessman and Conservative politician; Baronet, 1918.
Lock, William Henry: biographical information on Lock from standard sources does not exist.
McKenna, Reginald (1863–1943): First Lord, 1908–11; Liberal politician and banker.
Marjoribanks, Edward, 2nd Baron Tweedmouth (1849–1909): First Lord, 1905–8; Liberal politician.
May, Sir William Henry (1849–1930): Controller, 1901–5; Commander-in-Chief, Atlantic Fleet, 1905–7; Second Sea Lord, 1907–9; Commander-in-Chief, Home Fleet, 1909–11/Devonport, 1911–13; Rear-Admiral, 1901; Vice-Admiral, 1905; Admiral, 1908; Admiral of the Fleet, 1913; knighthood, 1906.
Moore, Sir Archibald Gordon Henry Wilson (1862–1934): Director of Naval Ordnance, 1910–12; Controller, 1912–14; Commander-in-Chief, Second Battle Cruiser Squadron, 1914–15/Ninth Cruiser Squadron, 1915–19; Rear-Admiral, 1911; Vice-Admiral, 1916; Admiral, 1919; knighthood, 1914.
Mountbatten, Louis Alexander, 1st Marquess of Milford Haven, formerly styled Prince Louis Alexander of Battenberg (1854–1921): Commander-in-Chief, Atlantic Fleet, 1908–10/Third Division, Home Fleet, 1911; Second Sea Lord, 1911–12; First Sea Lord, 1912–14; Vice-Admiral, 1910; Admiral, 1919; Admiral of the Fleet, 1921; Marquess, 1917.
Ogilvy, Frederick Charles Ashley (1866–1909): *Natal*, 1908–9.
Palmer, William Waldegrave, 2nd Earl of Selborne (1850–1942): First Lord, 1900–5; Liberal-Unionist/Conservative politician and statesman.
Parr, Alfred Arthur Chase (1849–1914): President of committee 'Finding and Keeping Ranges at Sea', 1905–6; Admiral, 1908.
Peirse, Sir Richard Henry (1860–1940): Inspector of Target Practice, 1909–11; Commander-in-Chief, First Battle Squadron, Home Fleet, 1911–12/East Indies Station, 1913–16; Board of Invention and Research, 1916–18; Vice-Admiral, 1914; Admiral, 1918; knighthood, 1914.
Scott, Sir Percy Moreton, 1st Baronet (1853–1924): *Terrible*, 1899–1902; *Excellent*, 1903–5; Inspector of Target Practice, 1905–7; Commander-in-Chief, Second Cruiser Squadron, 1907–9; Rear-Admiral, 1905; Vice-Admiral, 1908; Admiral, 1913; Baronet, 1913.
Selborne: *see* PALMER, William Waldegrave.
Slade, Sir Edmond John Warre, (1859–1928): Commandant, War Course College, Portsmouth, 1905–7 (Royal Naval War College from 1907 to 1914); Director of Naval Intelligence, 1908–9; Commander-in-Chief, East Indies, 1909–12; Rear-Admiral, 1908; Vice-Admiral, 1914; Admiral, 1917; knighthood, 1911.
Smith, Bertram H.: *Excellent*, 1905–7; *Duke of Edinburgh*, 1907–8; *Black Prince*, 1908–10; Royal Naval War College, Portsmouth, 1910–14; *Vengeance*, 1914; Captain, 1912.
Thomas, Sir Charles Inigo (1846–1929): Admiralty Permanent Secretary, 1907–11; knighthood, 1907.
Thomson, James (1822–1892). professor of engineering, and older brother of Lord Kelvin (William Thomson).
Thomson, Sir William, 1st Baron Kelvin of Largs (1824–1907): physicist and inventor; Baron, 1892.
Tweedmouth, Lord: *see* MARJORIBANKS, Edward.
Waller, Arthur Craig (1872–1943): *Excellent*, 1905–7; *Pelorus*, 1908–9; Assistant Director of Naval Ordnance, 1909–11; *Orion*, 1911–13; *Albemarle*, 1913–14; Captain, 1908; Rear-Admiral, 1919.

Warren, Herbert A.: *Jupiter*, 1905–6; study at Portsmouth, 1906–7; Rear-Admiral, 1907.

Wilson, Sir Arthur Knyvet, 3rd Baronet (1842–1921): Commander-in-Chief, Channel Squadron, 1901–3/Home Fleet, 1903–4/Channel Fleet, 1905–7; First Sea Lord, 1910–11; Admiral, 1905; Admiral of the Fleet, 1907; knighthood, 1902; Baronet, 1919.

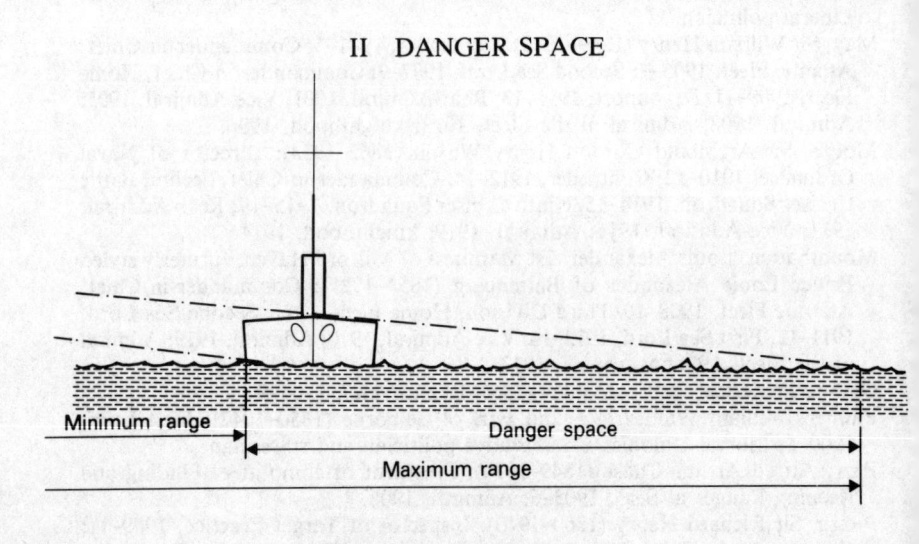

II DANGER SPACE

Figure 1

Figure 2

Source: 'The Elements of Change of Range', a print submitted to the Royal Commission on Awards to Inventors by the Argo Company and Arthur Hungerford Pollen (1925), in the Pollen Papers.

III RANGE FINDING BY TRIANGULATION

Optical range finding is a matter of triangulation. By trigonometry, a knowledge of the base length and the two base angles of a triangle is sufficient to determine the lengths of the remaining two sides. In range finding, lines of sight to the distant object are taken from two separate points, the known distance between the points being the base, with the angles formed by the intersections of the base and the lines of sight comprising the base angles (see Figure 1). The determination of the distance to the object along one of the lines of sight is then simply a matter of measuring the base angles and solving the trigonometry.

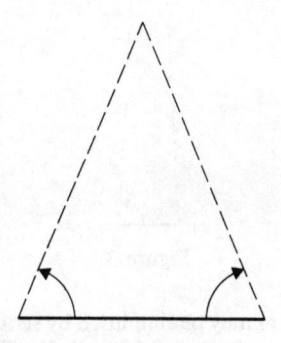

Figure 1

When a triangle has a base that is long relative to the lengths of its sides, there is the possibility that the sides will differ considerably in length (see Figure 2). In range finding problems at sea, however, where the maximum base possible is

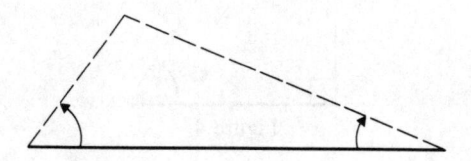

Figure 2

the length of a ship, and is usually much less, the base is very small relative to the distances along the two lines of sight, and thus these distances are virtually equivalent (see Figure 3). The choice of which side to measure in practice is therefore immaterial.

362   THE POLLEN PAPERS

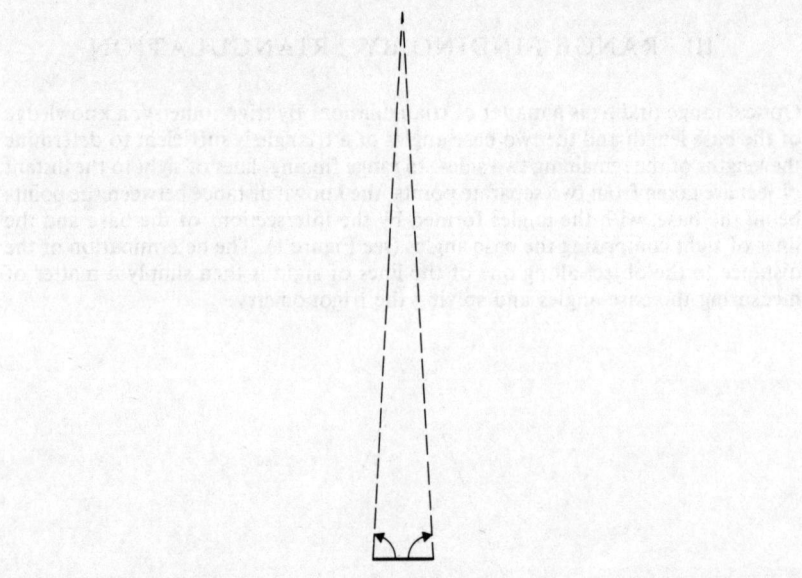

Figure 3

The range finding problem may be simplified by setting up the triangle in such a way that one of the base angles is a right angle (see Figure 4). In this case, the same trigonometric rules apply, but only one base angle has to be measured since the other is known from the start.

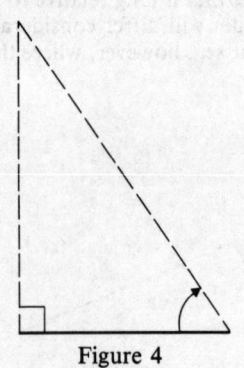

Figure 4

As the range to be measured increases, the amount of increase in the oblique base angle for every equal increase in the distance decreases (see Figure 5: x = x'; angle α is larger than angle β). The measuring limits of a range finder are thus reached when changes in the range result in changes in the oblique angle that are too small to be measured. The range can then only be determined by increasing the length of the base to the point that the amount of change in the oblique base angle for each significant change in the range is large enough to be measured (see Figure 6: x″ = x' = [from Figure 5] x = x'; angle α' is smaller than [from Figure 5] angle α; angle β' is *larger* than [from Figure 5] angle β).

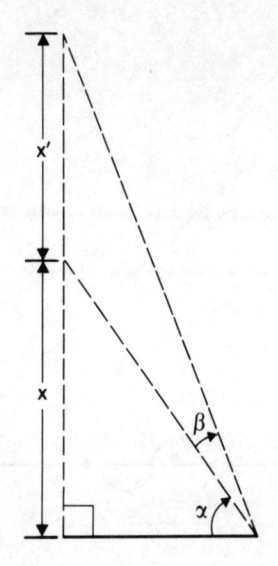

Figure 5

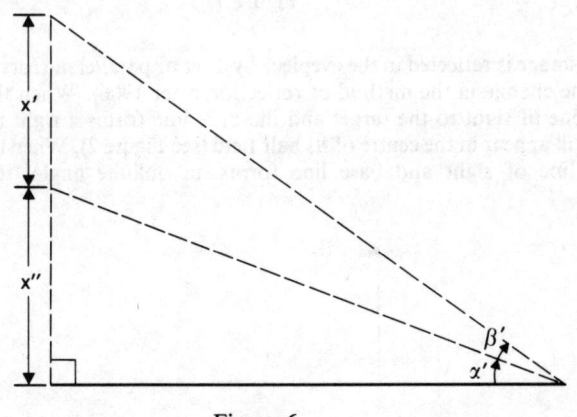

Figure 6

IV TRIANGULATION BY COINCIDENCE

The field of view of the single eyepiece of a coincidence range finder is divided in half by a fine line, the upper field being formed by the light reflected from one end of the instrument, the lower field being formed by the light from the other. When the range finder is trained to view a vertical object such as a mast or funnel, the upper portion of the object appears in one field, the lower portion in the other, the complete image of the object being formed as a result from two different sources (see Figure 1).

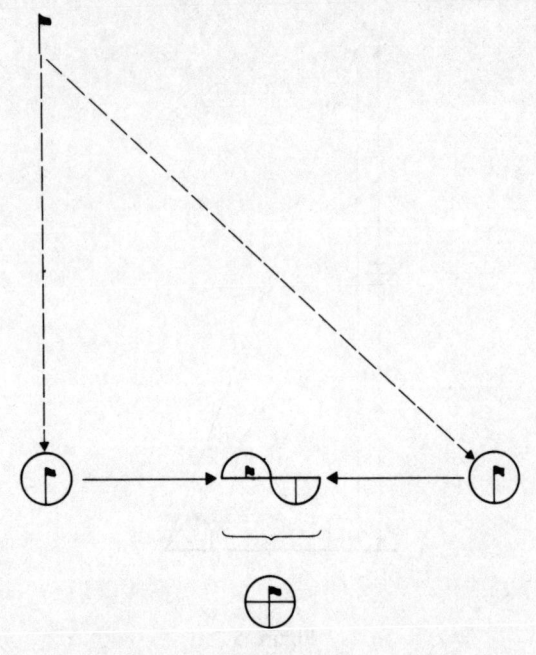

Figure 1

Each image is reflected in the eyepiece by a set of parallel mirrors (see Appendix V for the change in the method of reflection after 1906). When the intersection of the line of sight to the target and the base line forms a right angle, the half image will appear in the centre of its half field (see Figure 2). When the intersection of the line of sight and base line forms an oblique angle, the half image

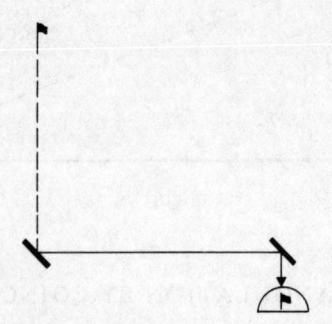

Figure 2

produced will be displaced from the centre by an amount corresponding to the extent of the deviation from the normal (see Figure 3).

To set up the range finding triangle, a right angle is first formed at one end of the instrument by centring the half image from that end in the half field of the eyepiece. An oblique angle will thus be established at the other end, resulting in a displacement of the half image from that source at the eyepiece (see Figure 4).

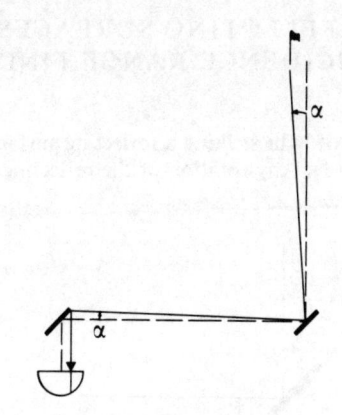

Figure 3

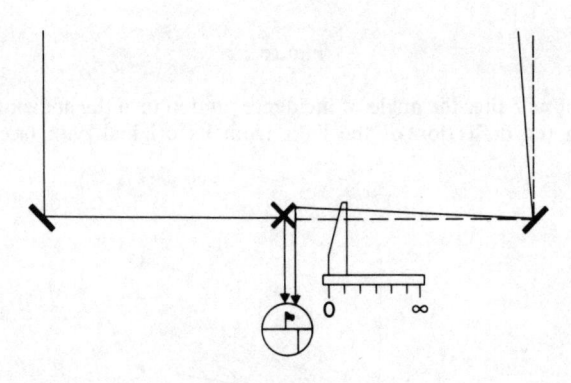

Figure 4

To obtain the range, the displaced half image is made to coincide with the centred half image – hence the term 'coincidence' – by the deflection of the path of light by the movement of a low power prism, the position of the prism on a calibrated scale after coincidence has been obtained giving the range (see Figure 5).

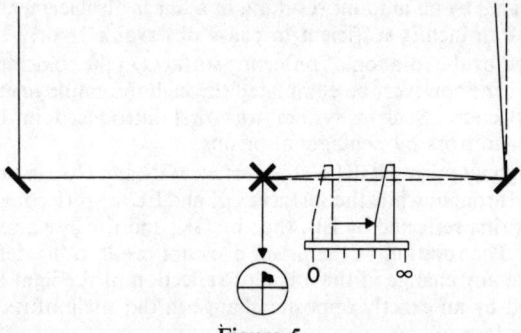

Figure 5

V REFLECTING SURFACES IN COINCIDENCE RANGE FINDERS

The angle of reflection of light striking a reflecting surface is equal to the angle of incidence (see Figure 1). Any rotation of the reflecting surface relative to the

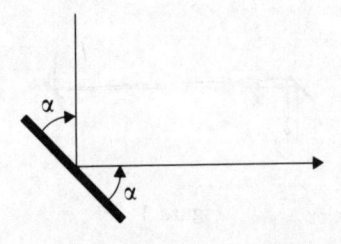

Figure 1

path of light will alter the angle of incidence, and in turn the angle of reflection, resulting in the deflection of the light from its original path (see Figure 2).

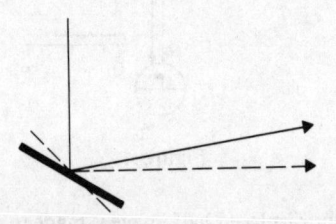

Figure 2

In coincidence range finders using mirrors, the rotation of either of the two reflecting surfaces by an amount resulting in a linear displacement of as little as 1/200,000th of an inch is sufficient to cause observable error.

Errors caused by the rotation of reflecting surfaces in the coincidence plane (see Appendix IV) can, however, be eliminated through the employment of a system of double reflection. Such a system was first introduced in 1906 with the replacement of mirrors by pentagonal prisms.

In the pentagonal prism NGKLE, the surfaces GN and NE are polished so that light may pass through, while the surfaces GK and EL are reflecting: light passing through GN is first reflected by EL, then by GK, and finally passes through NE (see Figure 3). The rotation of the prism does not result in the deflection of the light path since any change in the angle of reflection of the light by the surface EL is cancelled by an exactly opposite change in the angle of reflection of the light by the surface GK.

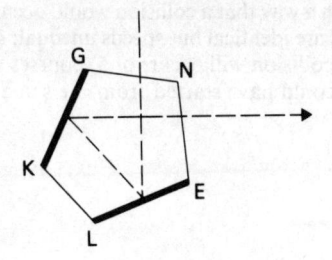

Figure 3

Source: Great Britain, Admiralty, Gunnery Branch, *Handbook for Naval Range-Finders and Mountings: Book I*, G. 0303 (November 1921), N.L.M.D.

VI CHANGE IN RANGE

The range is constant when 1) both ships are stationary; 2) courses are straight and parallel, and speeds equal; 3) courses are identical and speeds equal; or 4) courses are concentric and speeds proportioned to traverse equal arcs in equal times; (see Figure 1).

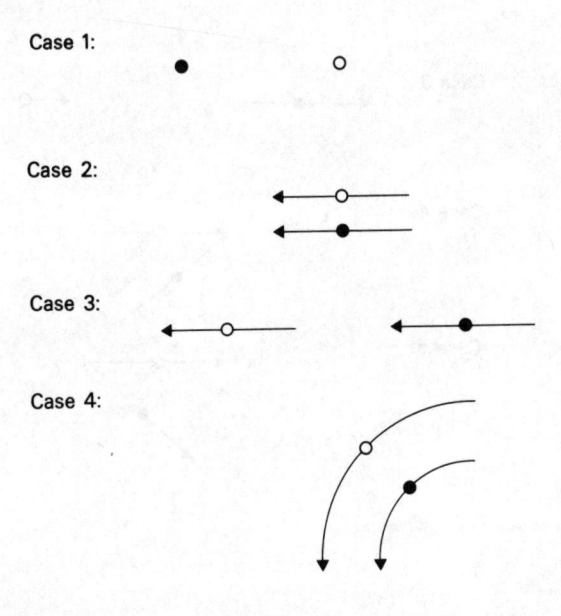

Case 1:

Case 2:

Case 3:

Case 4:

Figure 1

The range changes at a constant rate when 1) one ship is stationary while the other approaches in such a way that a collision would occur; 2) courses are identical but opposite; 3) courses are identical but speeds unequal; 4) courses are converging and speeds such that a collision will occur; or 5) courses are diverging and speeds such that both vessels could have started from the same point at the same time; (see Figure 2).

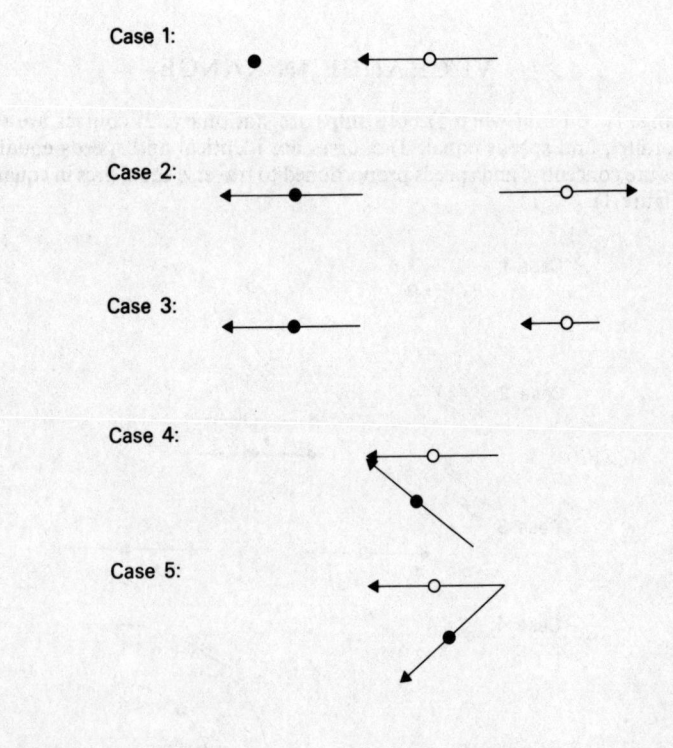

Case 1:

Case 2:

Case 3:

Case 4:

Case 5:

Figure 2

CHANGE IN RANGE

The change of range rate changes at a constant rate when 1) one ship is stationary while the other approaches in such a way that a collision will not occur; 2) courses are parallel and speeds unequal; 3) courses are parallel and opposite; or 4) courses and speeds are such that if converging a collision would not occur, or if diverging that both vessels could not have started from the same point at the same time; (see Figure 3).

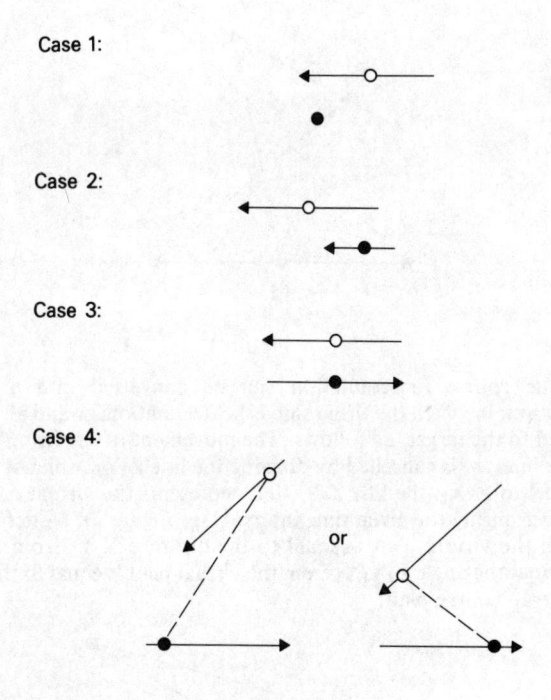

Case 1:

Case 2:

Case 3:

Case 4:

Figure 3

VII VIRTUAL COURSE

Given two ships, with courses and speeds such that if converging a collision would not occur, or if diverging that both vessels could not have started from the same point at the same time (as in Technical Appendix III, Case 4). During a given time interval, the speed of the firing ship is such that it moves the distance X_1X_2; during this same time interval, the speed of the target is such that it moves the distance Y_1Y_2: thus at the start of the time interval, the distance between the two ships is X_1Y_1, and at the end of the time interval, the distance is X_2Y_2 (see Figure 1).

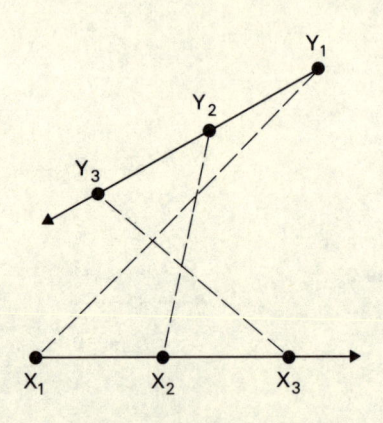

Figure 1

This true course representation can be converted into a virtual course representation, in which the firing ship is held as stationary and all relative motion is imparted to the target, as follows. The movement of the firing ship during the given time interval is cancelled by drawing the line Z_2Y_2, which is equal in length and parallel to X_1X_2; the line Z_2Y_1 thus represents the virtual course and speed of the target during the given time interval (see Figure 2). Note that the distance X_1Z_2 from the virtual plan is equal to the distance X_2Y_2 from the true course plan, and that the angle $Z_2X_1X_2$ from the virtual plan is equal to the angle $Y_2X_2X_3$ from the true course plan.

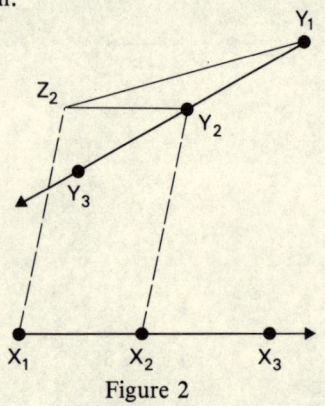

Figure 2

VIII TRIGONOMETRIC COMPUTATION OF CHANGE OF RANGE AND BEARING RATES, AND DEFLECTION

Given two ships, with courses and speeds such that if converging a collision would not occur, or if diverging that both vessels could not have started from the same point at the same time (see Figure 1: ● = Firing Ship; O = Target).

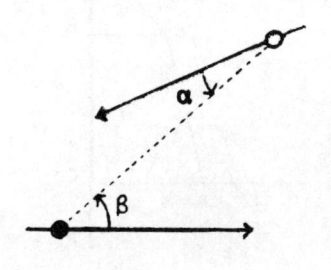

Figure 1

Thus, if:

t = Target speed,
s = Firing Ship speed,
R = Range,
β = Bearing (Angle of Firing Ship's course to line of sight),
α = Angle of Target's course to line of sight;

then:

$s(\cos \beta) + t(\cos \alpha)$ = the Change of Range Rate,
$s(\sin \beta) + t(\sin \alpha)$ = the Deflection, and
$\dfrac{s(\sin \beta) + t(\sin \alpha)}{R}$ = the Change of Bearing Rate

Note that changes in the change of range rate are determined by changes in the angles α and β. Given the conditions of course and speed in the above example (and also in any of the other cases illustrated in Appendix VI, Figure 3), the angles α and β change by the same amount over time, resulting in a constant change in the change of range rate. If the conditions of course and speed are such that the angles α and β do not change (as in the cases illustrated in Appendix VI, Figure 2), the change of range rate remains constant.

A trigonometric calculator (dumaresq) indicates the change of range and deflection when set with the firing ship and target courses and speeds, and the target bearing (see Figure 2: long arrow AR rotates around point A; adjustable length cA pivots on A; adjustable length ac pivots on c; and length QC slides along length AR).

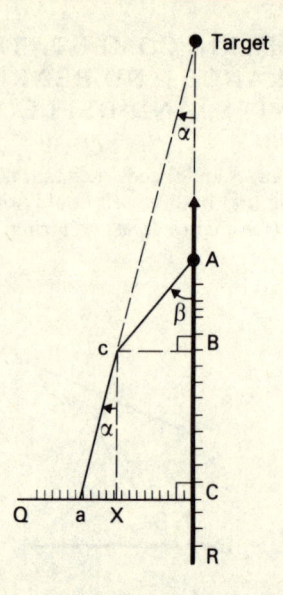

Figure 2

The instrument is operated as follows. The target bearing is set by pointing the long arrow AR at the target. The firing ship course is set by fixing the adjustable length cA at the angle of the firing ship's course to the line of sight to the target along AR (angle β from Figure 1). The firing ship's speed is set by adjusting the length cA. The target's course is set by fixing the adjustable length ac at an angle relative to cA such that the intersection of the line of sight along ac to the target, and the line of sight along AR, forms an angle equal to the angle of the target's course to the line of sight to the firing ship (angle α from Figure 1). The target's speed is set by adjusting the length ac.

In the right triangle cBA, the length of the side aB is equal, by trigonometry, to $s(\cos \beta)$. In the right triangle aXc — where the length Xc is parallel to the long arrow AR and the angle acX thus is equal to the angle α — the length Xc is equal, by trigonometry, to $t(\cos \alpha)$. Since the length Xc is equal to the length BC, BC is equal to $t(\cos \alpha)$. Thus, AC = AB + BC = $s(\cos \beta) + t(\cos \alpha)$ = the Change of Range Rate.

In the right triangle aXc, the length aX is equal, by trigonometry, to $t(\sin \alpha)$. In the right triangle cBA, the length cB is equal, by trigonometry, to $s(\sin \beta)$. Since the length XC is equal to the length cB, the length XC is equal to $s(\sin \beta)$. Thus, aC = aX + XC = $s(\sin \beta) + t(\sin \alpha)$ = the Deflection.

The change of range rate is indicated by the position of the slider QC along the graduated scale AR (giving the distance AC) as determined by the intersection of ac and QC at a. The deflection is indicated along the graduated scale QC by the intersection of ac and QC at a (giving the distance aC) as determined by the lengths ac and cA and the angles α and β. The division of the deflection, $s(\sin \beta) + t(\sin \alpha)$, by the range, R, by means of a simple mechanical linkage not shown, results in an indication of the change of bearing rate.

As the arrow AR is rotated to account for changes in target bearing (as would occur in the case illustrated in Figure 1, and the other cases illustrated in Appendix VI, Figure 3), the position of the angled length acA would remain fixed, resulting in changes in the angles α and β, and thus changing indications of the change of range rate and deflection along QC and AR, and the change of bearing rate. A constant target bearing (as would occur in the cases illustrated in Appendix VI, Figure 2), would result in no change in the orientation of the arrow AR, no change in turn in the angles α and β, and thus no change in the indication of the change of range rate and deflection, or the change of bearing rate.

Note that the instrument can be worked in reverse, with settings of the change of range rate (the distance AC), the deflection (the distance aC) or the change of bearing rate, the target bearing (the direction of the arrow AR), and the firing ship's course and speed (the angle β and the length cA) causing it to indicate the target's course and speed (the angle α and the length ac), a procedure known as making a 'cross-cut.'

Source: 'The Elements of Change of Range', a paper submitted to the Royal Commission on Awards to Inventors by the Argo Company and Arthur Hungerford Pollen (1925), in the Pollen Papers.

IX DISC-BALL-ROLLER MECHANISM

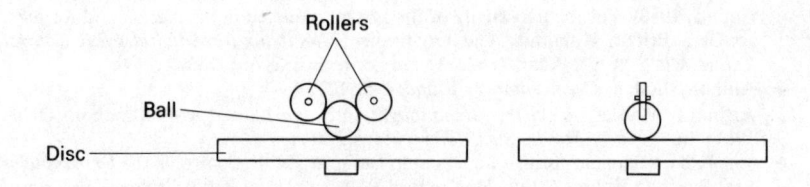

Figure 1: Elevation

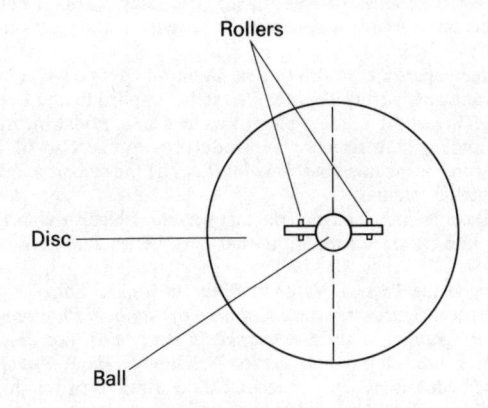

Figure 2: Plan

Source: British Patent 17,441/1912 (submitted April 4, 1912; accepted, October 24, 1912).

NOTES

NOTES TO PREFACE

1 See editorial introduction to 'The Gun in Battle', and Appendix IX, 'Disc-Ball-Roller Mechanism'.
2 'Report of Proceedings by Commander Richard T. Down, R.N., during visit to Washington – 6th May to 27 [*sic*] June' (1917), Adm. 137/1621, Public Record Office, Kew, and Herman H. Goldstine, *The Computer from Pascal to von Neuman* (Princeton: Princeton University Press, 1972).
3 For Pollen's influence on the development of the Royal Navy's fire control instruments, see Anthony Pollen, *The Great Gunnery Scandal: The Mystery of Jutland* (London: Collins, 1980). For the particulars of the system mounted in the *Nelson* and *Rodney*, see Great Britain, Admiralty, Gunnery Branch, *Handbook for Admiralty Fire Control Tables Mk. I* (n.d.), Adm. 186/273, Public Record Office, Kew.
4 Anthony Pollen, *Great Gunnery Scandal*, p. 105.
5 Arthur J. Marder, *From the Dreadnought to Scapa Flow*, 5 vols. (London: OUP, 1961–70), I, *The Road to War, 1904–1914*, p. 404n.
6 Jon Tetsuro Sumida, 'British Capital Ship Design and Fire Control in the *Dreadnought* Era: Sir John Fisher, Arthur Hungerford Pollen, and the Battle Cruiser', *Journal of Modern History*, 51 (June 1979), pp. 205–30.
7 Copies of a few of Pollen's privately circulated printed works are held by the Naval Library of the Ministry of Defence, the National Maritime Museum, and the wardroom library of the HMS *Excellent* (Whale Island). The complete set of Pollen papers, along with most of his surviving correspondence, is currently in the possession of Mr Anthony Pollen.
8 The two omitted papers are 'Pollen to Rear-Admiral Lewis Bayly' (January 1909), and 'The Pollen Manual Charting Table as Privately Supplied to and Used in Certain HM Ships' (1909). The text of the first paper was to a large extent incorporated in a later one, 'Memoranda and Instructions Introductory to the Use of Pollen's Tactical Instrument', which is included in this volume, while the second paper is nothing more than an instruction manual.
9 The abridged sections contained material that was either highly technical and of relatively little general interest or information that was better summarised in an editorial introduction.
10 'Some Aspects of the Tactical Value of Speed in Capital Ships'.
11 For naval historical figures, see the *Dictionary of National Biography*, the *Dictionary of American Biography*, or the *Encyclopaedia Britannica*. For descriptions of naval battles, see R. Ernest Dupuy and Trevor N. Dupuy, *The Encyclopedia of Military History*, revised edition (1977). For technical descriptions of warships, see *Conway's All the World's Fighting Ships, 1860–1905*, and Antony Preston, *Battleships of World War I*.
12 Jon Tetsuro Sumida, 'Financial Limitation, Technological Innovation, and British Naval Policy, 1904–1910' (Ph.D. Dissertation, University of Chicago, 1982).

NOTES TO GENERAL INTRODUCTION

1. See Appendix II, 'Danger Space'.
2. See Appendix III, 'Range Finding by Triangulation'.
3. See Appendix III, 'Range Finding by Triangulation'.
4. See Appendix IV, 'Triangulation by Coincidence'.
5. See Appendix III, 'Range Finding by Triangulation'.
6. See Appendix V, 'Reflecting Surfaces in Coincidence Range Finders'.
7. See Appendix VIII, 'Trigonometric Computation of Change of Range and Bearing Rates, and Deflection'.
8. See Appendix VI, 'Change in Range'.
9. For the sources of the account of the development of naval gunnery in the late nineteenth and early twentieth centuries given in the General Introduction, see Jon Tetsuro Sumida, 'Financial Limitation, Technological Innovation, and British Naval Policy, 1904–1910' (Ph.D. Dissertation, University of Chicago, 1982).

NOTES TO 'THE POLLEN SYSTEM OF TELEMETRY'

1. For Pollen's account of the origins of his interest in fire control, see the text of 'The Gun in Battle', included in this volume.
2. For extracts of Pollen to Kerr, 26 January 1901, see the appendix to 'Notes, Etc., on the *Ariadne* Trials', included in this volume.
3. Pollen to Selborne, 4 February 1901.
4. The Admiralty Permanent Secretary, not to be confused with the Admiralty Parliamentary and Financial Secretary, received all official communications to the Admiralty and signed and sent all official communications from the Admiralty. In February 1901, the Admiralty Permanent Secretary was Sir Evan MacGregor.
5. Pollen to the Admiralty Permanent Secretary, 25 February 1901.
6. Admiralty Permanent Secretary to Pollen, 7 February 1901.
7. Beresford to Pollen, 17 March 1901.
8. For a more detailed explanation of the two-observer system of range finding, see the two later prints (included in this volume): 'Memorandum on a Proposed System for Finding Ranges at Sea and Ascertaining the Speed and Course of any Vessel in Sight' and, especially, 'Fire Control and Long-Range Firing', which provides a diagram.
9. Pollen did not provide a specific proposal with regard to the plotting of ranges on a chart until he wrote his 'Memorandum on a Proposed System for Finding Ranges at Sea and Ascertaining the Speed and Course of any Vessel in Sight' of 1904, included in this volume.
10. The 'Barr and Stroud machine' must refer to the Barr and Stroud 4½-foot, single-observer, self-contained, coincidence range finder (see General Introduction).
11. By 'ships of the new first-class cruiser type', Pollen evidently meant armoured cruisers, which replaced the protected cruiser type. The Royal Navy's first armoured cruisers were the six vessels of the *Cressy* class, which were built under the estimates of 1897–8 and which entered service in 1901 and 1902.

NOTES TO A 'MEMORANDUM ON A PROPOSED SYSTEM FOR FINDING RANGES AT SEA AND ASCERTAINING THE SPEED AND COURSE OF ANY VESSEL IN SIGHT'

1. British Patent 6838/1902 (submitted 20 March 1902; accepted 19 March 1903). This appears to have been a later version of British Patent 12,952/1900, which, according to a Patent Office index of inventors and their patents, was submitted to the Patent Office on 18 July 1900, but which was later withdrawn. The index mention of Archibald Barr of the instrument firm of Barr and Stroud as a co-inventor of the device described in the 1900 patent was probably an error, the intended reference being to Mark Barr.

376   NOTES TO PAGES 14–22

2   British Patent 11,535/1904 (submitted 19 May 1904; accepted 6 July 1905).
3   Pollen memorandum on two-observer range-finding, 9 May 1904, in Adm. 1/7733, Public Record Office, Kew, and Great Britain, Admiralty, *Pollen Aim Correction System: General Grounds of Admiralty Policy and Historical Record of Business Negotiations* (February 1913), p. 21, Naval Library of the Ministry of Defence, London.
4   Lawrence to Selborne, 24 May, and Lawrence to Selborne, 27 May 1904, in Adm. 1/7733.
5   Pollen to Vincent Wilberforce Baddeley, secretary to the First Lord, 5 July 1904, in Adm. 1/7733.
6   Pollen to Selborne, 14 November 1904, in Adm. 1/7733.
7   Date of writing given in 'The Gun in Battle', included in this volume.
8   Pollen to the Admiralty Permanent Secretary, 21 December 1904.
9   Commander Wilfrid R. Nicholson, secretary to the First Sea Lord, to Pollen, 19 November 1904 [in the Pollen Papers], and Pollen to Baddeley, 24 November 1904, in Adm. 1/7733.
10 Pollen to the Admiralty Permanent Secretary, 21 December 1904. The Committee members were: Captain Henry F. Oliver and Lieutenant Thomas B. Crease, Royal Navy; Captain Edward W. Harding, Royal Marine Artillery; and Captain B. J. W. Locke from the War Office; for the preceding information, see 'Proof of Evidence of Colonel E. W. Harding, RMA' (January 1920), paragraph 5, in the Pollen Papers.
11 Pollen to the Admiralty Permanent Secretary, 21 December 1904, and Captain Edward W. Harding, RMA, to the Director of Naval Ordnance, 'Report on Pollen System of Range Finder' (3 April 1905), in the Pollen Papers.
12 See General Introduction.
13 See 'Fire Control and Long-Range Firing', included in this volume, Figures 9 and 10.
14 Gun calibration — the process of determining how far a given gun will fire a projectile under given circumstances and the adjusting of gunsights accordingly — was necessary if guns were to be aimed accurately at long ranges because of the slight consistent variation in performance between guns of the same model. The practice of calibrating guns was not standard in the Royal Navy at the time that Pollen wrote his pamphlet. Captain Percy Scott, the well-known expert in naval gunnery and most likely one of the 'eminent authorities' referred to by Pollen, had advocated gun calibration as early as December 1903 in a privately circulated pamphlet and in proposals to the Admiralty. Scott's proposals were rejected by the Admiralty in March 1904, but under the new Board of Admiralty that took office in October 1904 under Admiral Sir John Fisher, a Gunnery Calibration Committee was appointed and Scott made its chairman. The Gunnery Calibration Committee conducted trials in the battleship HMS *Commonwealth* off the southern coast of Ireland in Bantry Bay in mid-1905, and as a result of its recommendations, gun calibration was made mandatory for all Royal Navy warships upon commissioning and recommissioning. See Percy Scott, 'Remarks on Long-Range Hitting' (14 December 1903), in Percy Scott, *Gunnery* (London: by the author, 1905), p. 26, in the Arnold White Papers, WH1/65, National Maritime Museum, Greenwich; 'Calibration of Guns: Report of Committee &c.' (20 July 1905), Adm. 1/7835, Public Record Office, Kew, Admiral Sir Percy Scott, *Fifty Years in the Royal Navy* (New York: George H. Doran, 1919), pp. 180–2, 189; and Admiral Sir Frederic Dreyer, *The Sea Heritage: a Study of Maritime Warfare* (London: Museum Press, 1955), p. 45.

NOTES TO 'FIRE CONTROL AND LONG-RANGE FIRING: AN ESSAY TO DEFINE CERTAIN PRINCIPIA OF GUNNERY, AND TO SUGGEST MEANS FOR THEIR APPLICATION'

1   Date of composition marked in Pollen's hand on the copy in the Pollen Papers.
2   Date of distribution and the number of recipients marked in Pollen's hand on the copy in the Pollen Papers. According to Pollen's testimony to the Royal Commission on Awards to Inventors in 1925, Lieutenant Frederic Dreyer received a copy of this print in May 1905; see Royal Commission on Awards to Inventors, *Minutes*, vol. 9 (1 August 1925), p. 80.

3   Great Britain, Admiralty, *Pollen Aim Correction System: General Grounds of Admiralty Policy and Historical Record of Business Negotiations* (February 1913), p. 21, Naval Library of the Ministry of Defence, London.
4   'Proof of Evidence of Colonel E. W. Harding, RMA' (January 1920), in the Pollen Papers.
5   Captain Edward W. Harding, RMA, to the Director of Naval Ordnance, 'Report on Pollen System of Range Finder' (3 April 1905), in the Pollen Papers.
6   Captain Edward W. Harding, 'Report on Pollen System of Range Finder'.
7   Admiralty Permanent Secretary to Pollen, 3 May 1905, CP 7491/9793.
8   In Pollen's next paper, 'A.C.: A Postscript', included in this volume, the factors that determined the 'True Range or Actual Range' were called the 'geometrical elements'. In a letter written by Pollen to an unnamed admiral in the Royal Navy in January 1906, which became the third of six letters that Pollen printed in the '*Jupiter* Letters', included in this volume, the terms 'True Range or Actual Range' were replaced by the term 'geometric range'.
9   The bore of a gun was measured in inches, while its length from breech to muzzle was measured in terms of a multiple of the bore diameter, that multiple being the calibre of the gun. Thus a '30 cal. 12-in. gun' was one with a bore of 12 inches and a length of 30 times 12 inches, or 30 feet. Higher muzzle velocities could be obtained by lengthening the barrel, and thus a 46-calibre 12-inch gun would be more accurate ballistically at extreme ranges – all other things being equal – than a 12-inch gun of 30 calibres. Pollen's example was purely illustrative, since the Royal Navy was never equipped with 12-inch guns of either 30 or 46 calibres.
10 For Pollen's revised view on the question of 'the percentage error of the day or hour', see 'A.C.: A Postscript', included in this volume.
11 Pollen must here be referring to the combined rate of fire of the several guns of a warship's entire armament, because even the fastest firing quick-firers could not do much better than 12 aimed rounds per minute, while the larger pieces that made up the main and secondary batteries of a battleship fired even more slowly.
12 See General Introduction.
13 The instruments referred to in this section are the dumaresq, the Vickers Clock, and the follow-the-pointer sight-setting system. For the first two instruments, see General Introduction. For the third, see Vickers Sons & Maxim Ltd, *Vickers' Fire Control System, 1910 (The Vickers Mark III Follow-the-Pointer Instruments)* (1910), in the Naval Library of the Ministry of Defence, London.
14 On 4 November 1904, Pollen and William Henry Lock had applied to the Patent Office for what was to become British Patent 23,872/1904, which covered an improved mechanism for taking bearing observations and a plotting mechanism. At the time that Pollen wrote his paper – December 1904 – he had yet to apply for patents covering either the 'Change of Range Machine' or the 'Deflection Machine' that he described as being part of his proposed system.
15 These were the battleships *Kashima* and *Katori*, which had been laid down in February 1904 in British yards.

NOTES TO 'A.C.: A POSTSCRIPT'

1   The copy of this text in the Pollen Papers provides no indication of the precise date of composition and circulation. The text states that the Admiralty had authorised the trial of Pollen's instruments, which would put its date of composition no earlier than May 1905, while the lack of any reference to the HMS *Jupiter* suggests that it was written before September 1905.
2   Wilfrid Arthur Greene, brief: 'The Claim of the Argo Company and Mr Pollen before the Royal Commission' (1925), part 2: 14, in the Pollen Papers. For gun calibration, see Note 14 of the 'Memorandum on a Proposed System for Finding Ranges at Sea and Ascertaining the Speed and Course of any Vessel in Sight' given in this volume. For Percy Scott's views on using spotting corrections to account for the change of range,

see Percy Scott, 'Remarks on Long-Range Hitting' (15 December 1903), in Percy Scott, *Gunnery* (London: by the author, 1905), p. 24, in WH 1/65, Arnold White Papers, National Maritime Museum, Greenwich.
3   Jellicoe to Pollen, 14 April 1905.
4   For the origins of this use of the term 'fire control', see Captain Edward W. Harding, RMA, writing under the pseudonym 'Rapidan', in *The Tactical Employment of Naval Artillery* (London: Offices of *Engineering*, 1903).
5   The design of the battleship HMS *Dreadnought* had only just been approved by the Board of Admiralty on 17 March 1905 (Board of Admiralty Minutes, 1905, Adm. 167/39, Public Record Office, Kew,) and construction did not begin until October of that year. The 'rumoured' disposition of the new battleship's armament, which was in fact correct, probably came from the London technical journal *Engineering*, which published a roughly accurate description of the *Dreadnought* in May 1905.
6   The *Duke of Edinburgh* and *Natal* classes of first-class armoured cruisers were armed with main batteries of six 9.2-inch guns, of which only two were mounted on the centre line while the remaining four were placed on the beam, two to each side.
7   T. A. Brassey, ed., *The Naval Annual, 1904* (Portsmouth: J. Griffin, 1904), p. 355.
8   Pollen was incorrect to claim that the antiquated 13.5-inch guns of the *Royal Sovereign* class would have been 'overwhelmingly superior' to the new model 10-inch guns of the *Triumph* and *Swiftsure* at 10,000 yards range. The 10-inch gun could not only fire much more rapidly than the 13.5-inch gun, but more accurately at long range as well because its longer and more strongly built barrel enabled it to exploit the advantages of cordite charges, which were much more powerful than the brown powder charges for which the 13.5-inch gun had been designed. For the comparative performances of the old model 13.5-inch and new model 10-inch guns at 10,000 yards, see Fred T. Jane, editor, *Fighting Ships*, 8th edition (London: Sampson Low, Marston, 1905), p. 34.

NOTES TO 'THE *JUPITER* LETTERS'

1   Wilfrid Arthur Greene in Royal Commission on Awards to Inventors, *Minutes*, vol. 2 (29 June 1925), p. 37, and testimony of Harold Isherwood in Royal Commission on Awards to Inventors, *Minutes*, vol. 10 (3 August 1925), 104–5, both in the Pollen Papers.
2   Arthur Hungerford Pollen, 'The Gun in Battle', included in this volume; Admiralty Permanent Secretary to Pollen, 3 May 1905, C.P. 7491/9793, in the Pollen Papers; Pollen to Lord Tweedmouth, 14 February 1906, in the Pollen Papers; and Great Britain, Admiralty, *Pollen Aim Correction System: General Grounds of Admiralty Policy and Historical Record of Business Negotiations* (February 1913), pp. 6, 21, Naval Library of the Ministry of Defence, London.
3   Members of the 'Finding and Keeping Ranges at Sea' committee: Vice-Admiral Alfred Arthur Chase Parr, Captain Frederick Tower Hamilton, Captain Herbert A. Warren, Captain Bernard Currey, Lieutenant Ralph Eliot, and Commander Francis H. Mitchell, who acted as secretary; see 'List of Committees, 1906', Admiralty paper (31 December 1906), in Adm. 1/7782, Public Record Office, Kew.
4   Typescript memorandum: 'Instructions for the Committee Appointed to Carry out Trials of "Aim Corrector" System of Range Finding &c. in HMS *Jupiter*' (25 September 1905), DRYR 2/1, in the Dreyer Papers, Churchill College, Cambridge.
5   Log book of HMS *Jupiter* (15 August 1905–20 September 1906), Adm. 53/22479, Public Record Office, Kew.
6   Pollen to Chase Parr, 2 February 1906, in 'The *Jupiter* Letters'.
7   Wilfrid Arthur Greene, brief: 'The Claim of the Argo Company and Mr Pollen before the Royal Commission' (1925), part 2, pp. 18–19, in the Pollen Papers.
8   Pollen to Vice-Admiral Lord Charles Beresford, 21 June 1906.
9   Pollen had proposed the use of gyroscopes to correct bearings against yaw as early as 1901; and in the spring of 1905, Pollen had warned the Ordnance Department that gyroscopes would be required, but was denied the funds that were necessary to develop a suitable mechanism; see Greene, brief, part 2, pp. 13–14, in the Pollen Papers, and the appendix to 'Notes, Etc. on the Ariadne Trials' included in this volume.

10   Testimony of Arthur Hungerford Pollen in Royal Commission on Awards to Inventors, *Minutes*, vol. 9 (1 August 1925), 45, in the Pollen Papers.
11   By the beginning of 1906, a 9-foot base instrument from the firm of Barr and Stroud of Glasgow, and a 10-foot base instrument from the firm of Thomas Cooke and Sons of York, had been developed in response to an Admiralty advertisement of 1904 for a range finder of greater accuracy at long range than the existing 4½-foot base service model; see Wilfrid Arthur Greene, brief, part 2, p. 17, and 'Notes of a Meeting held at the Admiralty in the Board Room on 9 August 1906', C.P. 13313/1906, in the Pollen Papers; and 'Summary', statement of the Dreyer case for the Royal Commission on Awards to Inventors, p. 5, courtesy of Sir Desmond Dreyer to the editor, 1979, and General Introduction.
12   Greene, brief, part 2, p. 19, in the Pollen Papers. For the Pollen-Isherwood Gyroscope, see British Patent 11,795/1909 (submitted 19 May 1909; accepted 7 July 1909).
13   Greene, brief, part 2, p. 19; British Patent 11, 795/1909.
14   Pollen to the Admiralty Permanent Secretary, 18 June 1906, C.P. 10542/1906.
15   On the basis of correspondence contained in the Pollen Papers, the following persons are known to have received copies of 'The *Jupiter* Letters' in 1906:

Lieutenant Oliver Backhouse, HMS *Excellent*
Lieutenant Roger R. C. Backhouse, HMS *Excellent*
Captain Reginald Bacon, HMS *Dreadnought*
Vice-Admiral Lord Charles Beresford, Commander-in-Chief, Mediterranean Fleet
Lieutenant John Evelyn Bray, Sheerness Gunnery School
Commander H. C. Carr, HMS *Jupiter*
Captain Bernard Currey, inactive
Lieutenant Frederic Dreyer, HMS *Dreadnought*
Lieutenant Ralph Eliot, HMS *Jupiter*
Admiral Sir John Fisher, First Sea Lord (acknowledged by Captain Charles E. Madden, Naval Assistant to the First Sea Lord)
Captain William Goodenough, Royal Naval College, Dartmouth
Captain Frederick Tower Hamilton, HMS *Excellent*
Captain Sir John Jellicoe, Director of Naval Ordnance
Admiral of the Fleet Lord Walter Kerr, inactive
Vice-Admiral Alfred Arthur Chase Parr, retired
Rear-Admiral Sir Percy Scott, Inspector of Target Practice
Lord Tweedmouth, First Lord
Admiral Sir Arthur Knyvet Wilson, Commander-in-Chief, Channel Fleet

There were, undoubtedly, others who received copies of the paper for whom there is no record, such as Captain Edward W. Harding, RMA, a gunnery expert and close associate. Vice-Admiral Sir Reginald Custance received a copy in 1908 (see Pollen to Custance, 10 July 1908).
16   Copy of the second printing of 'The *Jupiter* Letters', which was bound together with 'Notes on a Proposed Method of Studying Naval Tactics', included in this volume, in the Pollen Papers.
17   'Captain Savage' and 'Mr Falcon' were characters in Captain Frederick Marryat's novel, *Peter Simple*, which first appeared in the *Metropolitan Magazine* in three instalments in 1832 and 1833. Mr Falcon 'never forgot what he called *zeal*' (chapter 28), but it was 'Gentleman (William) Chucks', the boatswain, whose 'zeal for the service' obliged him to use strong language 'to prove in the end that I am in earnest' (chapter 14).
18   Between 1884 and 1904, the gunnery of warships in the Royal Navy was tested by the Annual Prize Firing, in which a ship steaming at low speed on a set course fired at a stationary target at ranges of from 1,400 to 1,600 yards. Under the direction of Percy Scott, who became the first Inspector of Target Practice in February 1905, the Annual Prize Firings were replaced in 1905 by the Gunlayer's Test, and Battle Practice. In the Gunlayer's Test, gunners fired from a stationary ship at a stationary target at a range of 2,000 yards, while in Battle Practice, gunners were required to shoot at a stationary target from a ship steaming at low speed on a set course at ranges of from 5,000 to 7,000 yards. Firing from a moving ship at a moving target was not introduced as standard

until the Battle Practice of 1908. For the Annual Prize Firings, the Gunlayer's Test, and Battle Practice, see Viscount Hythe, *The Naval Annual, 1913* (Portsmouth: J. Griffin, 1913), pp. 311–14, and Admiral Sir Percy Scott, *Fifty Years in the Royal Navy* (New York: George H. Doran, 1919), pp. 189–92.

19 'An extraordinary miscalculation' probably refers to the failure to anticipate that heavy seas would produce movement in the ship which was such that the training arcs of the telescopes were inadequate; see Pollen to Admiral Chase Parr, 2 February 1906, below.
20 For the Vickers Clock, see General Introduction.
21 Experience of the Japanese Navy in actions against the Russian Navy in the Russo-Japanese War, 1904–5.
22 'Rapidan' (Captain Edward W. Harding, RMA) *The Tactical Employment of Naval Artillery* (London: Offices of *Engineering*, 1903).
23 Charles Lamb (1775–1834), essayist and humorist, 'A Dissertation upon Roast Pig', one of twenty-five articles published between August 1820 and December 1822 in the *London Magazine*, and in 1823 published together as *The Essays of Elia*.
24 Ignacy (Jan) Paderewski (1860–1941), Polish pianist, composer, and statesman.
25 Battle of the Yellow Sea, 10 August 1904.
26 *The Globe*, a London evening paper that was published from 1803 to 1921, was in Pollen's day a strong proponent of increased naval strength.
27 'Maga' (Vice-Admiral Sir Reginald Neville Custance), 'The Growth of the Capital Ship', *Blackwood's Magazine*, 179 (May 1906), 577–96.
28 Alfred Thayer Mahan, 'Some Reflections upon the Far-Eastern War' (renamed 'Retrospect upon the War between Japan and Russia'), *National Review*, 157 (May 1906), 383–405; also in *Living Age*, 32 (July 14, 1906), 67–81.
29 See Notes 7 and 8, 'A.C.: A Postscript'.
30 The larger figure represented the approximate cost of an all-big-gun battleship such as the *Dreadnought*, while the smaller figure represented the approximate cost of a first-class battleship built from the Naval Defence Act of 1889 to the estimates of 1900–1. The *King Edward VII* and *Lord Nelson* classes, which were built under the estimates of 1901–2 to 1904–5, cost approximately £1,500,000.
31 Probably George Louis Palmella Busson Du Maurier (1834–96), artist and novelist, reference unknown.

NOTES TO 'NOTE ON THE POSSIBILITY OF DEMONSTRATING THE PRINCIPLE OF AIM CORRECTION WITHOUT THE USE OF INSTRUMENTS DESIGNED FOR THE PURPOSE'

1 Jellicoe to Pollen, 16 May 1906; Wilfrid Arthur Greeene in Royal Commission on Awards to Inventors, *Minutes*, vol. 2 (29 June 1925), p. 25; and Pollen to the Admiralty Permanent Secretary, 18 June 1906, C.P. 10542/1906; all in the Pollen Papers. No copy of the Boys report of 1906 (not to be confused with the Boys report on the Pollen system of 1912 that is contained in the Pollen Papers) has been found.
2 The record of Edward W. Harding before the Royal Commission on Awards to Inventors gives the names of Jellicoe, a Captain Percy Smith, and a Commander Crook, as the members of the committee; See Royal Commission on Awards to Inventors, *Minutes*, vol. 9 (1 August 1925), 14, in the Pollen Papers. Commander Henry Ralph Crook was then assigned to the Ordnance Department. The Captain Percy Smith mentioned was probably meant to be Commander Bertram H. Smith, a participant in the *Jupiter* trials; see list of officers associated with the recommendations to the Board of Admiralty in regard to the Pollen system from 1905 to 1907, made up by Pollen in 1908, an untitled and undated typescript (henceforward cited as 'list of officers, 1905–7'), in the Pollen Papers.
3 Wilfrid Arthur Greene, brief: 'The Claim of the Argo Company and Mr Pollen before the Royal Commission' (1925), part 2, p. 22, and 'list of officers, 1905–7', in the Pollen Papers.

4   Pollen had rejected the magnetic compass for fire control purposes as early as in 1900 on the advice of Lord Kelvin, who warned that the firing of heavy guns in close proximity would disrupt the magnetic field of a magnetic compass. During the *Jupiter* trials, Pollen had attempted to obtain bearings with a magnetic compass when no guns were firing simply to prove the practicability of plotting, but had discovered that the compass needle pivoted rapidly enough to be affected by yaw. For Lord Kelvin's advice in 1900, see Wilfrid Arthur Greene, brief, part 2, p. 6, in the Pollen Papers. For the experiments with a magnetic compass during the *Jupiter* trials, see Wilfrid Arthur Greene, brief, part 2, p. 16, and Wilfrid Arthur Greene in Royal Commission on Awards to Inventors, *Minutes*, vol. 1 (22 June 1925), 43, both in the Pollen Papers.
5   Wilfrid Arthur Greene, brief, part 2, pp. 24–5, in the Pollen Papers.

NOTES TO 'SOME ASPECTS OF THE TACTICAL VALUE OF SPEED IN CAPITAL SHIPS'

1   Wilfrid Arthur Greene, brief: 'The Claim of the Argo Company and Mr Pollen before the Royal Commission' (1925), part 2, pp. 25–30; Edward W. Harding, 'Proof of Evidence of Colonel E. W. Harding, RMA' (January 1920); and testimony of Harding in Royal Commission on Awards to Inventors, *Minutes*, vol. 9 (1 August 1925), 15–16; all in the Pollen Papers.
2   Board of Admiralty Minutes, 1906, Adm. 167/40, Public Record Office, Kew. The Board members present on 7 August 1906 were: Lord Tweedmouth, the First Lord; Rear-Admiral Henry Jackson, Third Sea Lord and Controller; Captain Frederick S. Inglefield, Fourth Sea Lord; Edmund Robertson, Parliamentary and Financial Secretary; and C. Inigo Thomas, acting for Sir Evan MacGregor, the Admiralty Permanent Secretary.
3   For the disagreement between Pollen and the Admiralty in August 1906, see Pollen to Tweedmouth, 27 August 1906, in 'Notes, Etc., on the *Ariadne* Trials' of April 1909, included in this volume.
4   Fisher to Tweedmouth, 10 September 1906, in Arthur J. Marder, ed., *Fear God and Dreadnought: The Correspondence of Admiral of the Fleet Lord Fisher of Kilverstone*, Vol. II: *Years of Power, 1904–1914* (London: Jonathan Cape, 1956), 87. For Fisher's intervention on Pollen's behalf, and the influence of the Harding report, see Jon Tetsuro Sumida, 'British Capital Ship Design and Fire Control in the *Dreadnought* Era: Sir John Fisher, Arthur Hungerford Pollen, and the Battle Cruiser', *Journal of Modern History*, 51 (June 1979), pp. 205–30.
5   Admiralty Permanent Secretary to Pollen, 21 September 1906, C.P. 14079.
6   For the special Admiralty Committee, see editor's introduction to 'Memorandum on a Proposed System for Finding Ranges at Sea and Ascertaining the Speed and Course of any Vessel in Sight', and 'Fire Control and Long-Range Firing'.
7   Pollen to the Admiralty Permanent Secretary, 14 May 1906, C.P. 5745/10654.
8   Admiralty Permanent Secretary to Pollen, 3 May 1905, C.P. 7491/9793.
9   British Patent 13,082/1906 (submitted 6 June 1906; accepted 25 April 1907).
10   'Maga' (Vice-Admiral Sir Reginald Custance), 'The Speed of the Capital Ship', *Blackwood's Magazine*, 180 (October 1906), pp. 435–51, and 'Black Joke', 'Tactical Speed', *The United Service Magazine*, 155 (October 1906), pp. 1–15.
11   Pollen to Jellicoe, 3 November 1906.
12   Date of composition marked in Pollen's hand on the copy in the Pollen Papers.
13   Introductory remarks to 'Some Aspects of the Tactical Value of Speed in Capital Ships'.
14   These are the diagrams, presumably, that are to be found in 'Notes on a Proposed Method of Studying Naval Tactics', included in this volume.
15   Battle of Tsushima, 27 May 1905.
16   No diagrams were provided in the copy of the 'Notes' in the Pollen Papers. For diagrams reconstructed from the text by the editor, see Editorial Appendix immediately following the text.
17   Togo's report on the Battle of Tsushima was widely reported, and thus could have been available to Pollen, in whole or in part, from a number of sources; see the annotated

bibliography in Edwin A. Falk, *Togo and the Rise of Japanese Sea Power* (New York: Longmans, Green, 1936).

18 Pollen appears to have known that the three 25-knot first-class armoured cruisers of the 1905–6 estimates, the *Invincible* class, were scheduled to be armed with a uniform calibre armament of 12-inch guns, although this was then a tightly guarded secret.

NOTES TO 'NOTES ON A PROPOSED METHOD OF STUDYING NAVAL TACTICS'

1 Wilfrid Arthur Greene, brief: 'The Claim of the Argo Company and Mr Pollen before the Royal Commission' (1925), part 2, p. 40, and copy of the formal contract signed on 18 February 1908, both in the Pollen Papers.
2 Beresford to Pollen, 18 May 1907, acknowledging the receipt of a copy. Admiral of the Fleet Sir Arthur Wilson to Pollen, 20 December 1907, noting that he possessed a copy.
3 Untitled, undated, typescript memorandum written in 1909, apparently by Pollen, in the Pollen Papers.
4 At the Battle of Trafalgar on 21 October 1805 the flagship of the British fleet, the HMS *Victory*, led an attack that was to result in the breaking of the French and Spanish line of battle, which began with the raking of the *Bucentaur*, the flagship of the French fleet.
5 At the Battle of Lissa, on 20 July 1866, the ramming and sinking of the Italian broadside ironclad *Re D'Italia* by the Austrian broadside ironclad *Erzherzog Ferdinand Max* convinced many that ramming tactics were capable of producing decisive results.
6 The torpedo, a self-propelled device that delivered an explosive charge underwater, was introduced in 1860 by Robert Whitehead. Subsequent improvements in design persuaded many that the large armoured warship had been made obsolete by torpedo-bearing fast surface craft or submarines.
7 'Barfleur' (Vice-Admiral Sir Reginald Custance), *Naval Policy: a Plea for the Study of War* (London: Blackwood, 1907).
8 Tactical exercises were known as 'P.Z. exercises' after the flag signal 'P.Z.', which indicated that they were to be performed; see Lord Chatfield, *The Navy and Defence: The Autobiography of Admiral of the Fleet Lord Chatfield* (London: William Heinemann, 1942), p. 106.

NOTES TO 'AN APOLOGY FOR THE A.C. BATTLE SYSTEM: BEING NOTES FOR A LECTURE TO THE WAR COURSE COLLEGE, PORTSMOUTH'

1 Admiralty Permanent Secretary to Pollen, 21 May 1907, C.P. 8110/11374; 'Proof of Evidence of Colonel E. W. Harding' (January 1920), p. 12; and testimony of Arthur Hungerford Pollen, in Royal Commission on Awards to Inventors, *Minutes*, vol. 9 (1 August 1925), pp. 82–3; all in the Pollen Papers.
2 'Proof of Evidence of Colonel E. W. Harding, RMA' (January 1920), p. 12, and testimony of Arthur Hungerford Pollen, in Royal Commission on Awards to Inventors, *Minutes*, vol. 9 (1 August 1925), pp. 82–3; both in the Pollen Papers.
3 See 'Extracts from a Letter to Capt. Reginald H. S. Bacon, CVO, DSO, Royal Navy. Dated 27 February 1908' included in this volume.
4 For plans and descriptions of Dreyer's devices, see the Dreyer Papers, DRYR 2/1, Churchill College, Cambridge.
5 Bacon to Pollen, 19 November 1907, and Wilson to Pollen, 4 December 1907.
6 Pollen to Wilson, 28 May 1906.
7 Pollen to Dreyer, 18 December 1907; see also testimony of Arthur Hungerford Pollen, in Royal Commission on Awards to Inventors, *Minutes*, vol. 9 (1 August 1925), p. 84, both in the Pollen Papers.

NOTES TO PAGES 131–146

8   Pollen to Lieutenant George Gipps, 26 August 1911.
9   See the editor's introduction to 'Some Aspects of the Tactical Value of Speed in Capital Ships'.
10 Anthony Pollen, *The Great Gunnery Scandal: The Mystery of Jutland* (London: Collins, 1980), pp. 56–7. For the date of the inspection, see Bacon to Pollen, 19 November 1907.
11 Wilson to Pollen, 8 December 1907.
12 Pollen to Wilson, 9 December 1907.
13 See Foreword to 'An Apology for the A.C. Battle System', and Pollen to Wilson, 9 December 1907.
14 See Foreword to 'An Apology for the A.C. Battle System', and Pollen to Wilson, 9 December 1907.
15 See Foreword to 'An Apology for the A.C. Battle System'.
16 For the copies sent to Bacon and Dreyer, see Wilfrid Arthur Greene, in Royal Commission on Awards to Inventors, *Minutes*, vol. 2 (29 June 1925), p. 108, in the Pollen Papers. For the copy sent to Jellicoe, see Jellicoe to Pollen, December 1907. For the copy sent to Custance, see Pollen to Custance, 10 July 1908. There are no other references to this paper in the Pollen Papers.
17 Jellicoe to Pollen, December 1907.
18 Wilson to Pollen, 11 December 1907.
19 William Shakespeare (1564–1616), *The Tempest*, Act 1, Scene 2.
20 Sir Francis Drake (?1542–96), sea explorer and admiral; 'Drake's flagship' could have referred to any one of four vessels that carried Drake's flag: the *Pelican* (later renamed the *Golden Hind*), the *Elizabeth Bonaventure*, the *Revenge*, or the *Defiance*.
21 John Clerk of Eldin (1728–1812), successful Edinburgh merchant, who in 1782 privately printed his 'Essay on Naval Tactics', a work that appears to have contributed to Nelson's formulation of the tactics used at the Battle of Trafalgar.
22 Galileo Galilei (1564–1642), Italian mathematician, astronomer, and physicist, whose observations with a telescope of the movement of sun spots led him to publish a treatise in 1613 in which he supported the Copernican theory that the earth revolved around the sun against the orthodox view that had been propounded by Ptolomy, which held the reverse. The objections of leading scholars to Galileo's publication resulted in the Church's condemnation of Copernicanism in 1616.
23 HMS *Victory*, 104-gun ship-of-the-line; flagship of Admiral Horatio Nelson at the Battle of Trafalgar, 21 October 1805.
24 (Captain) R(eginald) H. S. Bacon, 'Some Notes on Naval Strategy', in T. A. Brassey, ed., *The Naval Annual, 1901* (Portsmouth: J. Griffin and Co., 1901), pp. 233–52.
25 Rear-Admiral Henry John May (1853–1904), not to be confused with Vice-Admiral Sir William Henry May.
26 Battle of the Yellow Sea, 10 August 1904.
27 Lewis Carroll (Charles L. Dodgson [1832–98]) wrote in the 'Bellman's Speech' ('Fit the Second') of his nonsense poem, *The Hunting of the Snark* (1876):

> He had bought a large map representing the sea,
> Without the least vestige of land:
> And the crew were much pleased when they found
>   it to be
> A map they could all understand.

> 'Other maps are such shapes, with their islands
>   and capes!
> But we've got our brave Captain to thank'
> (So the crew would protest) 'That he's bought *us*
>   the best –
> A perfect and absolute blank!'

NOTES TO 'NOTES, CORRESPONDENCE, ETC., ON THE POLLEN A.C. SYSTEM, INSTALLED AND TRIED IN HMS *ARIADNE*'

1. 'Notes on Charts, Made Before Christmas, Sent to Admiral Wilson', part of 'Notes, Correspondence, Etc., on the Pollen A.C. System, installed and tried in HMS *Ariadne*', but not included in this volume.
2. These conditions, which had been defined in the Admiralty letter of 21 September 1906, C.P. 14079, were as follows:

   1. The system can be worked at sea in a ship of war under moderately bad conditions of weather, the ship having a fair motion of pitch, roll, and yaw.
   2. The system to give the change of range of one moving ship observed from another with such degree of accuracy that with an initial range of 8,000 yards, the error does not exceed 80 yards in 3 minutes.
   3. This result to be obtained within two minutes from the commencement of observations, both ships steering a steady course (excluding ordinary yaw), speed not to exceed 16 knots.

3. See Appendix VII, 'Virtual Course'.
4. For the Wilson-Dreyer system, see 'Report of Admiral of the Fleet Sir A. K. Wilson on Rate of Change Experiments', in Great Britain, Admiralty, Gunnery Branch, Fire Control, G. 4023/08 (1908), bound together with other reports in *Miscellaneous Gunnery Experiments*, 1901–1913, in the Naval Library of the Ministry of Defence, London; 'Battle Practice' and 'Plotting', in Captain C. Hughes-Onslow, *Fire Control* (Royal Naval War College, Portsmouth, 1909) in the Pollen Papers; and the testimony of Frederic C. Dreyer, in Royal Commission on Awards to Inventors, *Minutes*, vol. 11 (4 August 1925), p. 18, in the Pollen Papers.
5. 'Report of Admiral of the Fleet Sir A. K. Wilson on Rate of Change Experiments', in *Fire Control*, G. 4023/08.
6. Testimony of Arthur Hungerford Pollen, in Royal Commission on Awards to Inventors, *Minutes*, vol. 9 (1 August 1925), p. 51, in the Pollen Papers.
7. Untitled, undated printed history, probably prepared for the Royal Commission on Awards to Inventors hearings by either Pollen or Greene, paragraph 34, in the Pollen Papers.
8. For the trials of 11, 13, and 15 January, see the log of the HMS *Ariadne*, Adm. 53/17318, Public Record Office, Kew; 'Notes on the Torbay Trials of the A.C. System', part of 'Notes, Correspondence, Etc., on the Pollen A.C. System, installed and tried in HMS *Ariadne*', but not included in this volume; Wilson to Pollen, 4 December 1907, which gives the programme of experiments; Admiralty Permanent Secretary to Pollen, 5 December 1907, G. 18915/27562, which made additions to the programme given the day before; Wilfrid Greene, in Royal Commission on Awards to Inventors, *Minutes*, vol. 3 (6 July 1925), pp. 3–12; and testimony of Frederic C. Dreyer, in Royal Commission on Awards to Inventors, *Minutes*, vol. 11 (4 August 1925), pp. 18, 22–4, and vol. 12 (5 August 1925), pp. 5–6; all except the first citation in the Pollen Papers.
9. Pollen to Wilson, 21 January 1908.
10. This material has been summarised in the editor's introduction.
11. No copy of Wilson's report of 31 January 1908, has yet been found, but see Pollen's summary of Wilson's criticisms, in Pollen to the Admiralty Permanent Secretary, 25 March 1908, in 'Notes, Etc., on the *Ariadne* Trials', included in this volume, but not to be mistaken with the present paper.
12. 'The best is the enemy of the good.' Not Lafontaine, but rather Francois-Marie Arouet (1694–1778), better known as Voltaire, in *Dictionnaire philosophique*, s.v. 'Art dramatique'.
13. See editor's introduction to 'An Apology for the A.C. Battle System: Being Notes for a Lecture to the War Course College, Portsmouth', included in this volume.

14 For Jellicoe's favourable reaction, see Jellicoe to Pollen, December 1907; for the favourable views of Captain Frederick Tower Hamilton of the HMS *Excellent*, and his Commander, Alfred Chatfield, see Pollen to his wife, late December 1907 or early January 1908.

NOTES TO 'EXTRACTS FROM A LETTER TO CAPT. REGINALD H. S. BACON, CVO, DSO, ROYAL NAVY. DATED 27 FEBRUARY 1908'

1 Pollen to Bacon, 25 February 1908.
2 Bacon to Pollen, 26 February 1908.
3 Pollen to Sir C. Inigo Thomas, 28 February 1908, and Pollen to Tweedmouth, 2 March 1908. The complete text of Pollen's letter to Bacon of 27 February 1908, is to be found in the Pollen Papers.
4 The description of the fire control system tested in the *Vengeance* given by Admiral of the Fleet Sir Arthur Wilson in his official report specified only two men for the range finder; see 'Report of Admiral of the Fleet Sir A. K. Wilson on Rate of Change Experiments', in Great Britain, Admiralty, Gunnery Branch, *Fire Control*, G. 4023/08 (1908), bound together with other reports in *Miscellaneous Gunnery Experiments 1901–1913*, in the Naval Library of the Ministry of Defence, London.
5 The old turret ram HMS *Hero* was sunk on 18 February 1908 off the Kentish Knock in gunnery trials.
6 On 20 November 1907, Admiral of the Fleet Sir Arthur Wilson had professed to being 'horrified' during his inspection of Pollen's instruments on board the *Ariadne* to find that no provision had been made to get computed data to the guns; see editor's introduction to 'An Apology for the A.C. Battle System' in this volume.
7 For Galileo, see 'An Apology for the A.C. Battle System: Being Notes for a Lecture to the War Course College, Portsmouth', Note 22.
8 William Harvey (1578–1657) stated the concept of blood circulation in a lecture in 1616, and published his findings in 1628. Harvey's views on blood circulation, while they did not go uncriticised, were generally accepted by leading physicians in his lifetime.
9 Sir Henry Bessemer (1813–98) invented a process that greatly reduced the cost of producing steel, which was patented in 1855. The failure of early commercial applications of the process, which was due to its inability to produce good results from phosphoric ores, in large part delayed the full acceptance of Bessemer's methods until after the expiration of the master patents.
10 Sir William Henry Perkin (1838–1907) discovered the first chemical as opposed to vegetable based dye, which became known as aniline purple, in 1856. Perkin's discovery was not, however, successfully exploited by British firms, and by the end of the century German manufacturers were in possession of 90 per cent of the chemical dye world market.
11 Sir William Thomson, first Baron Kelvin of Largs (1824–1907), the eminent scientist and inventor, had played a major role in the laying of transatlantic telegraph cables, which may in part explain the scepticism with regard to wireless telegraphy attributed to him by Pollen.
12 Robert Whitehead (1823–1905) invented the first torpedo in 1866. In 1868, Whitehead offered the Royal Navy monopoly rights to his invention at a price of £80,000, but was refused. After trials in 1870, the Admiralty was finally convinced of the necessity of equipping the fleet with torpedoes, and in 1871 the manufacturing rights were purchased for £165,000, although the right was not an exclusive one.
13 Sir Henry Joseph Wood (1869–1944) conducted the first English orchestra that was permanently constituted and that consisted of fully trained musicians.
14 Opera by Richard Wagner (1813–83).

15 Bacon was not convinced. 'The flexibility of the powers of a man either cerebral or mechanical,' he wrote to Pollen on 2 March, 1908, 'has to be balanced against the rapidity of operation of a machine. The liability to error of both from different causes is a variation of the problem and the available spare men must be balanced against the available spare machines. It is as I have previously pointed out the knowledge of the adaptability of men and matter at sea which draws the only distinctive line between sea and land experience.' Bacon to Pollen, 2 March 1908.

NOTES TO 'REFLECTIONS ON AN ERROR OF THE DAY'

1. See 'Notes, Etc., on the *Ariadne* Trials', included in this volume.
2. 'Report of Admiral of the Fleet Sir A. K. Wilson on Rate of Change Experiments', in Great Britain, Admiralty, Gunnery Branch, *Fire Control*, G. 4023/08 (9108), bound together with other reports in *Miscellaneous Gunnery Experiments, 1901–1913*, in the Naval Library of the Ministry of Defence, London.
3. For the 'time curve and plotting methods', see 'Notes, Correspondence, Etc., on the Pollen A.C. System installed and tried in HMS *Ariadne*', included in this volume.
4. Bacon Memorandum and Minute, March 1908, G. 4229/08, Adm. 1/8010, Public Record Office, Kew.
5. Bacon Memorandum and Minute, March 1908, G. 4229/08, Adm. 1/8010.
6. Bacon Memorandum and Minute, March 1908, G. 4229/08, Adm. 1/8010.
7. 'Proof of Evidence of Colonel E. W. Harding, RMA' (January 1920), p. 13, in the Pollen Papers.
8. Account of the Dreyer lecture in a letter to Lieutenant George Gipps, undated, from the HMS *Excellent*, by an officer who signed his nickname, which was 'Hooligan' (no officer by that name is to be found in the Navy Lists), in the Pollen Papers.
9. Untitled, typescript memorandum, probably by Dreyer, dated May 1908, in the Dreyer Papers, DRYR 2/1, Churchill College, Cambridge, and Dreyer to Captain Constantine Hughes-Onslow, 18 October 1908, in the Pollen Papers.
10. For the Admiralty view of the unsatisfactory results of the 1908 battle practices, see Great Britain, Admiralty, Gunnery Branch, *Information Regarding Fire Control, Range-Finding, and Plotting*, No. 464, G. 2826/09 (1909), p. 30, in *Miscellaneous Gunnery Experiments, 1901–1913*, in the Naval Library of the Ministry of Defence, London. For the disgust of the fleet over the failure of the time-and-range plotting method, see Lieutenant George Gipps to Captain Constantine Hughes-Onslow, 12 December 1908, in the Pollen Papers.
11. A list of recipients of this paper, the only one of its kind to be found in the Pollen Papers, gives the following names:

Lieutenant Ernest K. Arbuthnot, HMS *Duke of Edinburgh*
Lieutenant Oliver Backhouse, HMS *Excellent*
Lieutenant Roger Backhouse, HMS *Dreadnought*
Lieutenant Sidney R. Bailey, HMS *Revenge*
Lieutenant George R. B. Blount, HMS *Albion*
Lieutenant John Evelyn Bray, unassigned
Captain Sir Douglas E. R. Brownrigg, Baronet, HMS *Theseus*
Commander Henry C. Carr, HMS *Jupiter*
Lieutenant Bernard St George Collard, HMS *Glory*
Captain Bernard Currey, Assistant Director of Torpedoes
Lieutenant Kenneth Dewar, unassigned
Lieutenant Ralph Eliot, HMS *Jupiter*
Captain Seymour Erskine, HMS *Bedford*
Commander W. W. Fisher, HMS *Albermarle*
Captain William E. Goodenough, HMS *Albermarle*
Rear-Admiral Frederick Tower Hamilton, Inspector of Target Practice
Captain Arthur Limpus, HMS *Albion*

Lieutenant David T. Norris, Royal Naval War College
Captain Henry F. Oliver, HMS *Achilles*
Rear-Admiral Sir Percy Scott, HMS *Good Hope*
Captain Morgan Singer, HMS *Roxburgh*
Commander Bertram H. Smith, HMS *Duke of Edinburgh*
Commander Arthur Vyell Vyvyan, HMS *King Alfred*
Admiral of the Fleet Sir Arthur Knyvet Wilson, retired

12 For the favourable response to Pollen's paper, see Lieutenant Sidney R. Bailey to Pollen, 27 September 1908; Captain Sir Douglas E. R. Brownrigg to Pollen, 2 October 1908; Lieutenant Roger Backhouse to Pollen, 11 October 1908; Lieutenant Ernest K. Arbuthnot to Pollen, 17 October 1908; and Captain Seymour Erskine to Pollen, 16 January 1909.
13 During the trials of January 1908 in the HMS *Vengeance*, Dreyer had plotted ranges against time on a chart that had been placed on a deal kitchen table.
14 Dreyer to Hughes-Onslow, 18 October 1908, in the Pollen Papers.
15 Pollen probably refers to Scott's lecture to the United Service Institution of Hong Kong, 'Remarks on that portion of the Fighting Efficiency of a Fleet which is dependent on the Straight Shooting of the Guns', which was delivered on 28 February 1902, and later reprinted in Percy Scott, *Gunnery* (London: by the author, 1905), pp. 2–15, in the Arnold White Papers, WH1/65, National Maritime Museum, Greenwich.
16 The old turret ram HMS *Hero* was sunk on 18 February 1908 off the Kentish Knock in gunnery trials.
17 *The Mikado*, Act II, by Sir William Schwenk Gilbert (1836–1911) and Sir Arthur Seymour Sullivan (1842–1900).
18 St Barbara, 'Fool Gunnery in the Navy' (Part I), *Blackwood's Magazine*, 83 (February 1908), p. 310. The identity of 'St Barbara' remains a mystery, but see 'St Barbara' to Pollen, 6 March 1908.

NOTES TO 'NOTES, ETC., ON THE *ARIADNE* TRIALS'

1 Board of Admiralty Minutes, 1908, Adm. 167/42, Public Record Office, Kew, and Pollen to the Admiralty Permanent Secretary, 1 April 1908, in the Pollen Papers.
2 Admiralty Permanent Secretary to Pollen, 23 April 1908, C.P. 9389.
3 Pollen to the Admiralty Permanent Secretary, 5 January 1909.
4 See editor's introduction to 'Some Aspects of the Tactical Value of Speed in Capital Ships' in this volume.
5 See the Pollen-Jellicoe correspondence of 1908–9 in the Pollen Papers.
6 Admiralty Permanent Secretary to the Argo Company, 21 April 1909, C.P. 4854/09. 80S.
7 Date of printing printed on the title page, and reference made in the text to the Admiralty letter of 21 April 1909.
8 Pollen to Gipps, 17 May 1909.
9 See the Pollen-Jellicoe corespondence from April 1909.
10 For Captain Reginald Bacon's opposition to the Pollen system as Director of Naval Ordnance, see editor's introduction to 'An Apology for the A.C. Battle System: Being Notes for a Lecture to the War Course College, Portsmouth' in this volume.
11 For the Navy scare of the spring of 1909 and the apparent abandonment of the 'two-power standard,' see Arthur J. Marder, *From the Dreadnought to Scapa Flow*, 5 vols. (London: Oxford University Press, 1961–70), I, pp. 151–85.
12 For the 'time and range method', see editor's introduction to 'Notes, Correspondence, Etc., on the Pollen A.C. System, Installed and Tried in HMS *Ariadne*', and editor's introduction to and text of 'Reflections on an Error of the Day', both in this volume.
13 See the text to 'The Pollen System of Telemetry' in this volume.
14 In 1892, Colonel Henry Samuel Spiller Watkin (1843–1905), submitted a two-observer range finding and virtual-course plotting scheme to the Admiralty for trials. Watkin's system did not prove to be practicable, and was rejected. Pollen's reference to Watkin here and below stemmed from Commander Frederic Dreyer's accusations of Pollen

having plagiarised his system from Watkin. For the Watkin system and a later statement of Dreyer's accusation, see Great Britain, Admiralty, Gunnery Branch (Commanders Frederic C. Dreyer and C. V. Usborne), *Pollen Aim Corrector System, Part I: Technical History and Technical Comparison with Commander F. C. Dreyer's Fire Control System* (1913), G.O. 92/13, in the Naval Library of the Ministry of Defence, London.

15 See the editor's introduction to and text from 'Note on the Possibility of Demonstrating the Principle of Aim Correction Without the Use of Instruments Designed for the Purpose' in this volume.

16 For the tactical machine, see the editor's introduction to and text from 'Notes on a Proposed Method of Studying Naval Tactics', and the editor's introduction to and text from 'Memoranda and Instructions Introductory to the Use of Pollen's Tactical Instruments', both in this volume.

17 The relevant passage from the Appropriation Account is given in Pollen's appendix to this paper. This action was made necessary by a crisis in naval finance that had arisen in the fall of 1907, and was probably not directed against Pollen in particular. For the naval financial crisis of the fall of 1907, see Ruddock Mackay, *Fisher of Kilverstone* (Oxford: Clarendon Press, 1973), pp. 387–92.

18 These members were Rear-Admiral Sir Henry Jackson, the Third Sea Lord and Controller, and probably Lord Tweedmouth, the First Lord. For the reference to Jackson, see 'Note to Letters of August and September, 1906' in the appendix, below. Tweedmouth corresponded with Pollen in December 1907 (see Tweedmouth to Pollen, 18 December 1907) about the preliminary trials, and undoubtedly expressed his congratulations upon hearing of their success.

19 For Admiral of the Fleet Sir Arthur Wilson's conduct of the trials, see the editor's introduction to 'Notes, Correspondence, etc., on the Pollen A.C. System, installed and tried in HMS *Ariadne*' in this volume.

20 'The verdict of the world is conclusive', Saint Augustine of Hippo (354–430) in *Contra Epistulam Parmeniani*. 'Respecting the Donatist schism in N. Africa of the fifth century', noted W. Francis H. King,

The world (says St Augustine) is of opinion that their separation cannot be defended on its own grounds, much less when referred to the principle of unity which is of the Church's essence. Its judgment is too wide to admit of partiality, and too unanimous to allow of doubt. The decision is absolute. The passage owes its celebrity to Newman's employment of it, and the weight that it had in undermining his faith in the Anglican position will be remembered by all who have read his *Apologia*.

W. Francis H. King, ed., *Classical and Foreign Quotations: A Polygot Manual of Historical and Literary Sayings, Noted Passages in Poetry and Prose Phrases, Proverbs, and Bon Mots*, 3rd ed. (London: J. Whitaker, 1904), p. 311.

21 For the opposition of Captain Reginald Bacon, Lieutenant Frederic Dreyer, and Admiral of the Fleet Sir Arthur Wilson to the Pollen system, see the editor's introduction to 'An Apology for the A.C. Battle System: Being Notes for a Lecture to the War Course College, Portsmouth' in this volume.

22 The Guelphs and Hohenstaufens (Ghibellines), two German princely houses, competed bitterly for the leadership of the Holy Roman Empire from 1125 to 1268.

23 For Wilson's patronage of Dreyer, see Admiral Sir Frederic Charles Dreyer, *The Sea Heritage: A Study of Maritime Warfare* (London: Museum Press, 1955), pp. 52–3.

24 That is, the manual time-and-range plotting method.

25 That is, the manual virtual-course plotting method, which the Admiralty continued to recommend after the failure of the manual time-and-range plotting method in battle practice. Pollen had discussed the possibilities of automatic virtual-course plotting in the spring of 1906 with Captain (RMA) Edward Harding, but had rejected it as being a less satisfactory variant of the true-course plotting method, while manual plotting had been found to be impracticable in the *Jupiter* experiments. For the discussions of automatic virtual-course plotting in 1906, see 'Proof of Evidence of Colonel E. W. Harding, RMA' (January 1920), in the Pollen Papers.

26 In the spring of 1904, fire control experiments were conducted in the battleship *Venerable* off the coast of Greece near Prasa Island, which demonstrated that hits could be made at much longer ranges than had been previously thought possible through the use of

the salvo system. For Captain (RMA) Edward Harding's evaluation of the results, see Great Britain, Admiralty, Gunnery Branch, *Fire Control: A Summary of the Present Position of the Subject* (October 1904), in the Naval Library of the Ministry of Defence, London.
27 Included in this volume.
28 Included in this volume.
29 St Barbara, 'Fool Gunnery in the Navy' (Part I), *Blackwood's Magazine*, 83 (February 1908), p. 310.
30 For the submission of the Boys report, see editor's introduction to 'Note on the Possibility of Demonstrating the Principle of Aim Correction without the use of Instruments designed for the Purpose' in this volume.
31 Krupp armour was nearly a third more efficient than Harvey armour — that is, 5¾ inches of Krupp steel was equal in protection to 7½ inches of Harvey steel. Krupp armour replaced Harvey armour in all British battleships from the *Canopus* class, which were built under the 1896–7 estimates, and in all British first-class cruisers from the *Cressy* class, which were built under the 1897–8 estimates. For the efficiency of Krupp and Harvey armour, see Oscar Parkes, *British Battleships* (Hamden, Conn.: Archon, 1972; f.p. 1956), p. 396.
32 Great Britain, Laws, Statutes, etc. *Official Secrets Act, 1889*, 52 & 53 Vict., ch. 52.
33 Included in this volume.
34 See Note 25, above.

NOTES TO 'MEMORANDA AND INSTRUCTIONS INTRODUCTORY TO THE USE OF POLLEN'S TACTICAL INSTRUMENT'

1 Pollen to Rear-Admiral Richard Peirse, 30 March 1909.
2 Pollen to Rear-Admiral Lewis Bayly, 20 January 1909.
3 Date given on the title page of the print.
4 The only reference to this paper in the Pollen Papers is in a letter from Pollen to Vice-Admiral Sir Reginald Custance of 16 February 1909, in which Pollen asked Custance to read a draft of the 'Memoranda'.

NOTES TO 'TO REAR-ADMIRAL THE HON. STANLEY C. J. COLVILLE, CVO, CB'

1 British Patent 23,872/1904 (submitted 4 November 1904; accepted 5 October 1905).
2 Pollen to Rear-Admiral Stanley Colville, 1 July 1910, below, and Pollen to an anonymous correspondent, 2 November 1910, in 'The Quest of a Rate Finder' in this volume.
3 Pollen to the Admiralty Permanent Secretary, 16 November 1909, and Pollen to the Admiralty Permanent Secretary, 17 June 1910.
4 Pollen to the Admiralty Permanent Secretary, 16 November 1909.
5 Pollen to Captain Arthur Craig, 2 December 1909.
6 'Conference at Admiralty 10 Dec. 1909. Points for Discussion', in the Pollen Papers.
7 Vice-Admiral Sir (Arthur) Francis Pridham to Anthony Pollen, 14 May 1973, courtesy of Anthony Pollen. In 1909, Pridham was a lieutenant assigned to the *Natal*.
8 For Ogilvy's support of the Pollen system, see Pollen to Ogilvy, 16 November 1909, in the Pollen Papers, and Pollen to an anonymous correspondent, 2 November 1910, in 'The Quest of a Rate Finder' in this volume.
9 Admiralty Permanent Secretary to the Argo Company, 18 January 1910, C.P. 10321/15S.
10 Admiralty Permanent Secretary to the Argo Company, 11 April 1910, C.P. 13198/84S.
11 Admiralty Permanent Secretary to the Argo Company, 29 April 1910, C.P. 14084/93S.
12 Pollen to the Admiralty Permanent Secretary, 29 April 1910.
13 Pollen to an anonymous correspondent, 2 November 1910, in 'The Quest of a Rate Finder' in this volume.

14 The members of the committee were Rear-Admiral Stanley Colville, Captain A. P. Stoddart, Captain C. M. De Bartolome, and Commander W. W. Fisher. For the members of the committee and their instructions, see 'Memorandum: Aim Corrector Trials', dated 31 May 1910, in the Drax Papers, DRAX 3/3, Churchill College, Cambridge.
15 Pollen to Colville, 1 July 1910, below. For the dates of the trials in the *Natal* in 1909–10, see the logs of the *Natal*, Adm. 53/23981, and Adm. 53/23982, Public Record Office, Kew.
16 Admiralty Permanent Secretary to Pollen, 20 June 1910, G. 0381/14164.
17 On the basis of correspondence in the Pollen Papers, it is known that Prince Louis of Battenberg received a copy of this paper, and there may have been others.
18 Committee to the Commander-in-Chief, Home Fleet (Admiral Sir William May), 1 July 1910, Drax Papers, DRAX 3/3, Churchill College, Cambridge.
19 For the 'follow-the-pointer' system of sight-setting, see Vickers Sons & Maxim Ltd, *Vickers' Fire Control System, 1910 (The Vickers Mark III Follow-the-Pointer Instruments)* (1910), in the Naval Library of the Ministry of Defence, London.
20 Lieutenants Gerard B. Riley and Reginald Plunkett.
21 See the report by Lieutenant Reginald Plunkett on the Pollen system's performance in trials of 28 June 1910, especially his remarks with respect to the electrical system, in the Drax Papers, DRAX 3/3, Churchill College, Cambridge.

NOTES TO 'THE QUEST OF A RATE FINDER'

1 For the agreement of April, see the Admiralty Permanent Secretary to the Argo Company, 29 April 1910, C.P. 14084/93S.
2 Admiralty Permanent Secretary to the Argo Company, 19 August 1910, C.P.G. 0419/10/175S.
3 Pollen (for the Argo Company) to the Admiralty Permanent Secretary, 25 August 1910.
4 Royal Commission on Awards to Inventors, 'Claimants' Correspondence Bundle', Claim No. 1451 of The Argo Company Limited and Mr A. H. Pollen (1925), p. 117, in the Pollen Papers.
5 On the basis of correspondence contained in the Pollen Papers, the following are known to have received drafts or copies of this paper:

   Captain Montague Browning, Inspector of Target Practice (from January 1911)
   Vice-Admiral Sir Reginald Custance, retired
   Rear-Admiral Richard Peirse, Inspector of Target Practice (until December 1910)
   Rear-Admiral Edmond Slade, Commander-in-Chief, East Indies

   There were undoubtedly many others.
6 'Note on the Possibility of Demonstrating the Principle of Aim Correction without the use of Instruments designed for the Purpose' in this volume.
7 For the passage from the Appropriation Account, see the appendix to 'Notes, Etc., on the *Ariadne* Trials' in this volume.
8 Captain Reginald Bacon had replaced Rear-Admiral Sir John Jellicoe in September 1907 and shortly afterwards (perhaps six weeks before 25 November) Bacon relieved Captain (RMA) Edward Harding of duties related to fire control and replaced him with Lieutenant Frederic Dreyer.
9 For the question of the letter of 25 November 1907, see Note 17 of 'Notes, Etc., on the *Ariadne* Trials' in this volume.
10 'N.O.', that is, 'Naval Ordnance'.
11 Tactical exercises were known as 'P.Z. exercises' after the flag signal 'P.Z.', which indicated that they were to be performed; see Lord Chatfield, *The Navy and Defence: The Autobiography of Admiral of the Fleet Lord Chatfield* (London: William Heinemann, 1942), p. 106.
12 The secondary batteries of German dreadnought battleships and battle cruisers consisted of 5.9-inch guns, which were much more powerful than the 4-inch guns that composed the secondary batteries of British capital ships. British naval officers believed that the

NOTES TO PAGES 264–287   391

4-inch gun provided an adequate defence against attack by enemy destroyers or torpedo boats, and thus reasoned that the Germans intended to use the secondary batteries of their capital ships against other capital ships, which indicated that they would seek a close-range engagement. For the British view of German tactical intentions, see Captain Henry G. Thursfield, 'Development of Tactics in the Grand Fleet: Three Lectures', delivered 2, 3 and 7 February 1922 (third lecture missing), Thursfield Papers, THU 107, National Maritime Museum, Greenwich.

13 Ogilvy had been captain of the battleship *Revenge* when Lieutenant Frederic Dreyer's method of manually plotting ranges against time had been tested in that ship in the autumn of 1907. For the testing of Dreyer's gear in the *Revenge*, see Great Britain, Admiralty, Gunnery Branch (Commanders Frederic C. Dreyer and C. V. Usborne), *Pollen Aim Corrector System, Part I: Technical History and Technical Comparison with Commander F. C. Dreyer's Fire Control System* (1913) G.O. 92/13, in the Naval Library of the Ministry of Defence, London.

14 In six of the eight capital ships of the 1909–10 progamme, the 12-inch gun calibre that had been standard since the *Majestic* class battleships of 1893–4 and 1894–5 (the so-called Spencer Programme) gave way to a 13.5-inch gun armament, which was mounted in all subsequent British capital ships until the appearance of the 15-inch gun calibre in the vessels of the 1912–13 programme.

15 For the 'Natal Clock', that is, the Argo Clock Mark I, see British Patent 360/1911 (submitted 5 January 1911; accepted 1 August 1911). For an earlier experimental clock, see British Patent 2497/1908 (submitted 4 August 1908); accepted 22 December 1908).

16 See Appendix VIII, Trigonometric Computation of Change of Range and Bearing Rates, and Deflection, on making a 'cross-cut'.

NOTES TO 'OF WAR AND THE RATE OF CHANGE'

1   Pollen's preface to his essay, which was entitled an 'Apology', was dated 1 December 1910 (see text, below).
2   See 'Postscriptum', in the text, below.
3   See 'The Quest of a Rate Finder', editor's introduction, Note 5.
4   Reference to Pollen's query in Custance to Pollen, 9 January 1911.
5   Custance to Pollen, 9 January 1911.
6   For the 'Fleet Committees on Fire Control', see 'Reports of Committees on Control of Fire' (2 July 1904), Adm. 1/7758, in the Public Record Office, Kew, and Great Britain, Admiralty, Gunnery Branch (Captain (RMA) Edward Harding), *Fire Control: A Summary of the Present Position of the Subject* (October 1904), G. 10740/04, in the Naval Library of the Ministry of Defence, London.
7   'Whitehall, Whale Island, and Victoria Street' were the locations of, respectively, the Naval Ordnance Department, the Royal Navy's main gunnery school, and the offices of the Inspector of Target Practice.
8   Nestor, the King of Pylos, described by Homer in *The Iliad*, and Solomon, the King of Israel and Judah, described in the *Bible*, were particularly distinguished for their wisdom.
9   The reference is, of course, to the bow that only Ulysses was strong enough to draw, from *The Odyssey*, by Homer.
10 For the 13.5-inch gun, see 'The Quest of a Rate Finder', Note 14.
11 Probably a misquotation of the closing phrase of 'Observations of the Earl of St Vincent on "Clark's Naval Tactics" ', 2 June 1806, given in Jedediah Stephens Tucker, *Memoir of Admiral the Right Hon^e. the Earl of St. Vincent*, 2 vols. (London: Richard Bentley, 1844), ii, p. 283.
12 William James, *The Naval History of Great Britain from the Declaration of War by France in 1793 to the Accession of George IV*, 6 vols. (London: Macmillan, 1902; first published, 1822–24), ii, p. 52.

NOTES TO 'THE GUN IN BATTLE'

1. See Appendix VII, 'Virtual Course'.
2. For the Argo Clock Mark I, see British Patent 360/1911 (submitted 5 January 1911; accepted 1 August 1911).
3. For the arrangement of the disc, ball, and rollers, see Appendix IX, 'Disc-Ball-Roller Mechanism'.
4. For the dumaresq, see Appendix VIII, 'Trigonometric Computation of Change of Range and Bearing Rates and Deflection'.
5. British Patent 19,627/1911, which describes the Argo Clock Mark II, is still restricted. For a complete technical description of the Argo Clock Mark II, however, see Argo Company memorandum, 'The A.C. Range and Bearing Clock, Mark II', H.I./M.456/1 (8 May 1911), in the Pollen Papers. For a detailed description of the disc-ball-roller mechanism, see British Patent 17,441/1912 (submitted 4 April 1912; accepted 24 October 1912).
6. James Thomson, 'On an Integrating Machine Having a New Kinematic Principle', *Proceedings of the Royal Society*, 24 (1876), pp. 262–5; and William Thomson (Lord Kelvin from 1892), 'On an Instrument for Calculating $[\int \phi(x)\psi(x)dx]$, the Integral of the Product of Two Given Functions', *Proceedings of the Royal Society*, 24 (1876), pp. 266–8.
7. See editor's introduction to the 'Memorandum on a Proposed System for Finding Ranges at Sea and Ascertaining the Speed and Course of Any Vessel in Sight', included in this volume, and the text of 'The Gun in Battle'.
8. See editor's introduction to 'Note on the Possibility of Demonstrating the Principle of Aim Correction Without the Use of Instruments Designed for the Purpose' included in this volume.
9. For Boys's interest in differential analysers, see his untitled report to the Argo Company of 1 November 1912, in the Pollen Papers.
10. Argo Company memorandum, 'The A.C. Range and Bearing Clock, Mark II', and Wilfrid Arthur Greene, brief: 'The Claim of the Argo Company and Mr Pollen before the Royal Commission' (1925), part 2: paragraph 40, both in the Pollen Papers.
11. Greene, brief, part 2: paragraph 41.
12. Greene, brief, part 2: paragraph 41.
13. Argo Company memorandum, 'The A.C. Range and Bearing Clock, Mark II'.
14. For the bearings calculator, see British Patent 16,373/1913 (submitted 16 July 1913; accepted 16 July 1914).
15. Pollen to Rear-Admiral Richard Peirse, 24 November 1912.
16. Argo Company, 'Memorandum' (6 May 1913), in the Pollen Papers.
17. 'The Mark IV Clock', according to the 'Memorandum' of 6 May 1913, 'was exhibited in the Company's Office at 14, Buckingham Street, in February, March and April, 1912, and was there seen by the First Lord of the Admiralty, by three of the four Sea Lords, by the Directors of Ordnance and Contracts and their principal advisers and assistants, by the Inspector of Target Practice and his staff, by the Captain and officers of HMS *Excellent*, by the officer commanding the War College, by the representatives of the various intelligence branches, and many other expert officers.'
18. Argo Company to the Admiralty Permanent Secretary, 18 March 1912, A. 1064/8.
19. Admiralty Permanent Secretary to the Argo Company, 7 June 1912, C.P.N.S. 4847, and Argo Company, 'Memorandum' (6 May 1913).
20. For the dates of the trials in the *Orion* in 1912, see the log of the *Orion*, Adm. 53/24312, Public Record Office, Kew.
21. For the Bantry Bay trials, see Anthony Pollen, *The Great Gunnery Scandal: The Mystery of Jutland* (London: Collins, 1980), p. 92.
22. Director of Navy Contracts to the Argo Company, 26 October 1912, C.P. 5677.F 3/13/10905.
23. Pollen to Boys, 27 September 1912.
24. Pollen to Boys, 13 November 1912.
25. Charles Vernon Boys' untitled report to the Argo Company of 1 November 1912, in the Pollen Papers.

26 Pollen to Peirse, 19 November 1912.
27 'Copy Final Specification of Commander Frederic Charles Dreyer's Invention for "Improvements in and relating to the methods of controlling the fire of guns" ' (23 September 1910), in DRYR 2/1, Dreyer Papers, Churchill College, Cambridge.
28 See Appendix VIII, 'Trigonometric Computation of Change of Range and Bearing Rates, and Deflection'.
29 Great Britain, Royal Commission on Awards to Inventors, *Minutes of Proceedings*, vol. 13 (6 August 1925), pp. 101–5. For detailed descriptions of the several versions of Dreyer's apparatus, see Great Britain, Admiralty, Gunnery Branch (Commanders Frederic C. Dreyer and C. V. Usborne), *Pollen Aim Corrector System, Part I: Technical History and Technical Comparison with Commander F. C. Dreyer's Fire Control System* (1913) GO 92/13, in the Naval Library of the Ministry of Defence, London.
30 Prince Louis of Battenberg to Pollen in a telephone conversation, 18 December 1912, in 'Memorandum of Conversation with Prince Louis', 18 December 1912, in the Pollen Papers.
31 Prince Louis of Battenberg to Pollen in a telephone conversation, 18 December 1912.
32 Arthur Hungerford Pollen, 'Memorandum' (6 December 1912), in the Pollen Papers.
33 Admiralty Permanent Secretary to the Argo Company, 19 December 1912, C.P. 2262/S.
34 On the basis of correspondence contained in the Pollen Papers, the following are known to have received copies of the 'The Gun in Battle':

Admiral Sir Reginald Custance
The Admiralty

John Walter, the owner of *The Times* and a stockholder in the Argo Company, loaned his copy of the paper to Captain Henry George Thursfield, according to correspondence enclosed in the copy held by the National Maritime Museum. This paper was probably circulated widely.
35 Admiral Sir Reginald Custance, *The Ship of the Line in Battle* (London: Blackwood, 1912).
36 Gabriel Darrieus, *War on the Sea, Strategy and Tactics: Basic Principles*, translated by Philip R. Alger (Annapolis, Maryland: United States Naval Institute, 1908).
37 René Daveluy, *The Genius of Naval Warfare*, 2 volumes, translated by Philip R. Alger (Annapolis, Maryland: United States Naval Institute, 1910–11).
38 Romeo Bernotti, *The Fundamentals of Naval Tactics*, translated by H. P. McIntosh (Annapolis, Maryland: United States Naval Institute, 1912).
39 See Note 35.
40 Rear-Admiral Bradley A. Fiske, USN, 'The Mean Point of Impact', *United States Naval Institute Proceedings*, 38 (September, 1912), p. 1,003.
41 Rear-Admiral R. H. S. Bacon, 'The Battleship of the Future', *Transactions of the Institution of Naval Architects*, 52 (1910), pp. 1–21.
42 Vladimir Semenoff, *The Battle of Tsu-shima between the Japanese and Russian fleets, fought on 27th May 1905*, translated by A. B. Lindsay (London: John Murray, 1906).
43 John White, *Naval Tactical Problems* (Portsmouth: J. Griffin, 1911), 'Preface'.
44 Probably *A Treatise on Naval Gunnery,* first edition (1820).

NOTES TO 'THE NECESSITY OF FIRE CONTROL'

1 For Pollen's account of this period, see the text of the 'Untitled Paper', included in this volume, below. For the Admiralty's account of their action, see Great Britain, Admiralty, *Pollen Aim Correction System: General Grounds of Admiralty Policy and Historical Record of Business Negotiations* (February 1913), C.P. 15248/13, in the Naval Library of the Ministry of Defence, London.
2 Pollen to Peirse, 14 January 1913.
3 Date given on the title page.

NOTES TO '[UNTITLED PAPER]'

1. Pollen to Commander Gerard Riley, 15 October 1914, and Anthony Pollen, *The Great Gunnery Scandal: The Mystery of Jutland* (London: Collins, 1980), pp. 113–14, 117, 146–7.
2. Great Britain, Admiralty, *Pollen Aim Correction System: General Grounds of Admiralty Policy and Historical Record of Business Negotiations* (February 1913), CP 15248/13, in the Naval Library of the Ministry of Defence.
3. Sir Reginald Hall's draft for a letter to Admiral David Beatty, written probably in 1922, in the Pollen Papers.
4. For Pollen's wartime career as a journalist, see Anthony Pollen, *The Great Gunnery Scandal*, pp. 146–68, 203–24.
5. The approximate date of composition may be deduced from internal evidence: the first letter is dated 17 April 1916 and no mention is made of the Battle of Jutland, which took place on 31 May 1916, whose outcome illustrated Pollen's arguments perfectly.
6. On 11 July 1915, the German light cruiser *Königsberg*, which had sought refuge in the shallows of the Rufigi river delta in German East Africa, was sunk by the fire of two British monitors, the *Severn* and *Mersey*. The opposing warships were stationary throughout the engagement.
7. On 8 December 1914, the German armoured cruisers *Scharnhorst* and *Gneisenau*, and light cruisers *Nürnberg* and *Leipzig*, were sunk in an engagement off the Falkland Islands with the British battle cruisers *Invincible* and *Inflexible*, and armoured cruisers *Glasgow*, *Cornwall*, and *Kent*.
8. The Forbes Speed Log was the invention of William Charles Forbes (see British Patent 2351/1909).
9. On 24 January 1915, a German force of battle cruisers succeeded in escaping from a superior force of British battle cruisers, although the Germans suffered the loss of the armoured cruiser *Blücher*.
10. Probably *A Treatise on Naval Gunnery*, first edition (1820).
11. The battleships *Ocean* and *Irresistible* were sunk at the Dardanelles by Turkish mines (and in the case of the *Irresistible*, gunfire as well) on 18 March 1916. The *Triumph* and *Majestic* were sunk at the Dardanelles by torpedoes fired from the German submarine U.21 on separate days in late May 1916.
12. For the optical design of the Cooke-Pollen range finder, see British Patent 30,090/1912 (submitted 31 December 1912; accepted 31 December 1913). For an evaluation of the capabilities of the instrument, see Charles Vernon Boys's untitled report to the Argo Company of 1 November 1912 in the Pollen Papers.
13. 'Emeritus' was the pseudonym of Sir Charles Inigo Thomas (1846–1929), the Admiralty Permanent Secretary from 1907 to 1911. For the identity of 'Emeritus', see Pollen to Thomas, 22 October 1913.
14. Robert Bannatyne Finlay (1842–1929), later Viscount of Nairn. For the advice given to the Argo Company by Finlay, see R. B. Finlay and E. Russell Clarke, 'Opinion' (23 July 1913), in the Pollen Papers.

Addendum: In September 1913, the Argo Clock Mark IV was superseded by the Argo Clock Mark V in which the hand-setting of bearings during a turn was replaced by the automatic action of gyroscope control. For the Agro Clock Mark V, see Pollen to the Admiralty Permanent Secretary 10, 1913.

INDEX

Covering the Preface, General Introduction, Editorial Introductions and Texts, and Appendices

Aim Correction (A.C.), defined 55-6, 144
Argo Company ix, x, 194, 244, 252, 253, 255, 293, 294, 326, 329, 330, 331, 333, 347, 358

Battle Practice 68, 75, 77, 82, 86, 89, 97, 98, 100, 141, 142, 151, 173, 179, 182–4, 185–93, 196, 198, 200, 239, 259, 262, 273, 280, 282, 298, 329, 347–8, 349, 353
Battles: Agincourt 80, 317; Belgium Coast 340; Copenhagen 283; Dardanelles 340, 342, 358; Dogger Bank 338, 342; Falkland Islands 334, 335, 336, 338; Glorious First of June 283; Jutland x, 333; Khartoum 81; Lepanto 80; Lissa 125, 127, 290; Majuba 80; Manilla Bay 43, 81, 125, 126, 279; New Orleans 80; Nile 283; Quiberon Bay 287; Rufigi River 334, 335, 336, 340; St. Vincent 284, 286, 287; Trafalgar xii, 109, 134, 275, 283, 287, 299, 300; Tsushima xii, 83, 108, 125, 126, 127, 136, 139, 300, 304; Ulsan 300; Ushant 283; West Indies 283; Yalu 290; Yellow Sea (10 of August) 86, 139, 300
Board of Admiralty 105, 160, 161, 162, 164, 179, 194, 195, 198, 201, 202, 203, 204, 205, 206, 208, 215, 216, 219, 223, 224, 225, 226, 228, 229, 230, 231, 232, 233, 234, 235, 236, 243–4, 253, 257, 264, 267, 272, 293, 297, 350, 355

Calibration 22, 26, 30, 34, 55, 149, 151
Change of Range, variation in 6–7, 26, 40, 47, 48, 49, 82, 147, 151, 153, 156, 158, 166, 180–1, 291–2, 294, 299, 312, 315, 321–2, 327–8, 333, 337, 351, 369, 371–3
Coast Defence Fire Control 12, 199, 214–15
Committees on Naval Gunnery 14, 23, 70, 87, 89, 101, 105, 217, 226, 244, 245, 273, 359, 376–7
Continuous-aim, explained 1–2

Danger Space 4, 34, 35, 82, 91, 317, 326, 360
Differential Analysis x, 292
Director of Naval Intelligence 101, 233, 359
Director of Naval Ordnance 23, 101, 104, 105, 131, 159, 161, 164, 178, 194, 195, 201, 215, 233, 243, 244, 255, 256, 259, 274, 294, 304, 311–12, 333, 352, 357, 358, 359

Disc-Ball-Roller Mechanism x, 291, 291–2, 293, 373

Elliot Brothers, Ltd. 294
Error of the Day, defined 24

Fire Control Instruments (Pollen-Argo): Two-observer Range Finder 8–13, 14, 15–22, 23, 33–4, 35–45, 52–3, 56, 58, 70, 76, 87, 88, 89, 91, 199, 217, 278, 309, 311; Pollen-Cooke Single-observer Range Finder 293, 326, 342, 343–5, 347, 351; Gyroscope-controlled Range Finder Mounting and Bearing Indicator 71, 76, 102, 103, 105, 131, 132, 145, 152, 156, 157, 159–67, 167–71, 174–5, 183–5, 194–236, 210–11, 217, 243–54, 255, 256, 263, 321, 347; Plotting Tables 17–18, 20, 21, 23, 39–40, 58, 59, 70, 71, 76, 87, 88, 89, 91, 101, 102, 103, 104, 105, 114, 131, 132, 146, 152, 156, 157, 159–67, 167–71, 174–5, 183–5, 194–236, 243–254, 255, 256, 260, 264–70, 278, 293, 309–10, 312, 313, 314, 319, 321, 326, 342–5, 347, 351; Change of Range and Deflection Machine 23, 40, 41, 54, 59, 60, 61, 62, 70, 71, 76, 91, 105, 132, 146–7, 149–50, 152, 154, 157, 174, 175, 194, 310; Argo Clock Mark I (*Natal* Clock) 194, 244, 245, 249, 251, 255, 268–9, 291, 312, 314; Argo Clock Mark II 292, 293, 294, 373; Argo Clock Mark III 293, 373; Argo Clock Mark IV (A.C. Clock) 293, 294, 312, 314, 315, 319, 321–3, 326, 333, 336, 347–8, 349, 351, 353, 373; Argo Clock Mark V 330, 336, 339–41, 343–45, 373, 395; Sight-setting Gear 18, 132, 147, 149, 152, 154, 157, 174, 175, 194, 244, 245, 249–50, 255, 326, 343–4, 347
Fire Control Instruments (Service; a select list): Dumaresq 6, 7, 140, 157, 169, 186, 262, 292, 293–4, 371–3; Vickers Clock 6, 7, 77, 140, 157, 165, 169, 294; Wilson-Dreyer Manual Plotting System 156–9, 163–4, 173–4, 178–9, 180–2, 205–9, 232, 234; Dreyer Tables 293–4, 353; Director 3, 4, 348, 349; Follow-the Pointer Sights 29, 140

Fire Control Instruments, armour protection of 20, 27, 28–9, 30, 36, 42–3, 56, 76, 86, 144–5, 152, 163, 164, 170, 174, 178, 184–5, 213, 232–3, 236, 343
First Lord 8, 14, 172, 178, 179, 195, 203, 208, 215, 216, 222, 223, 224, 225, 228, 293, 294, 346, 347, 349, 350, 351, 352, 353, 354, 355, 357, 359
First Sea Lord 8, 14, 72, 105, 114, 179, 194, 244, 358, 359, 360
Fleets and Squadrons: Atlantic Fleet 357, 358, 359; Channel Fleet 14, 73, 76, 81, 131, 213, 357, 360; Channel Squadron 357, 360; East Indies Station 359; First Battle Cruiser Squadron 357; First Battle Squadron 359; First Cruiser Squadron 244, 357; First Light Cruiser Squadron 358; Grand Fleet 357. 358; Home Fleet 99, 262, 294, 357, 358, 359, 360; Mediterranean Fleet 2, 8, 357; Ninth Cruiser Squadron 359; Second Battle Cruiser Squadron 359; Second Battle Squadron 358; Second Cruiser Squadron 359; Sixth Cruiser Squadron 358; Third Battle Squadron 357; Third Cruiser Squadron 357, 358
Forbes Speed Log 336

Geometric Range, defined; *see also* True Range 81–2
Gun Layers Test 77, 95, 97, 98, 141 185–90, 192–3, 211, 329
Gun Range, defined 24
Gyroscopes 71, 83, 88, 101, 103, 158, 181, 201, 210, 211, 212, 213, 235, 248, 265, 292, 310, 311, 335, 336

Helm Free Fire Control, defined 243
Historical School 124, 287–8
Hitting Elevation 24–5

Inspector of Target Practice 101, 161, 233, 293, 349, 352, 359

Krupp Armour 92, 93–4, 123, 220

Linotype and Machinery, Ltd. ix, 8, 13, 14, 23, 70, 114, 131, 310

Material (Matériel) School 124, 135, 287–8

Navy Appropriation Account 202, 204, 223, 258

Official Secrets Act 222, 326, 346, 350, 353

PERSONS:
Alexander the Great 290
Bacon, Sir Reginald Hugh Spencer 131, 132, 159, 172, 178, 179, 194, 195, 205, 226, 243, 244, 259, 304, 357
Barr, Mark 14, 71, 357
Barry, Sir Henry Deacon 23, 357
Battenberg, Prince Louis Alexander of, *see* Mountbatten, Louis Alexander
Bayley, Sir Lewis 237, 357
Beatty, David, first Earl Beatty 338
Beresford, Lord Charles William De La Poer, Baron Beresford 8, 14, 357
Bernotti, Romeo 297
Bessemer, Sir Henry 175
Boys, Sir Charles Vernon 101, 217, 292, 293, 314, 315, 357
Broke, Sir Philip Bowes Vere 290
Bush, Vannevar x
Carr, Henry C. 92, 357
Churchill, Rt Hon Sir Winston Leonard Spencer 293, 294, 346, 347, 349, 350, 351, 352, 353, 355, 357
Clerk, John, of Eldin 134
Cochrane, Thomas, tenth Earl of Dundonald 284, 302
Colville, Hon Sir Stanley Cecil James 243, 244, 245, 357
Corbett, Sir Julian Stafford xi
Craig, Arthur, *see* Waller, Arthur Craig
Custance, Sir Reginald Neville (Barfleur, Maga) 97, 108, 125, 132, 271, 290, 295, 300, 301, 302, 317, 321, 323, 357
Darrieus, Gabriel 297
Daveluy, Rene 297
Dewey, George 43, 68, 81, 125, 279
Douglas, Sir Howard 318, 241
Drake, Sir Francis xii, 134, 290
Dreyer, Sir Frederic Charles 131, 132, 156, 157, 158, 161, 179, 205, 208, 209, 293, 294, 333, 358, 377
Dumaresq, John Saumarez 6, 358
Du Maurier, George Louis Palmella Busson 100
Elphinstone, Sir (George) Keith (Buller) 294, 358
Finlay, Robert Bannatyne 354
Fisher, John Arbuthnot, first Baron Fisher, of Kilverstone xi, 14, 72, 105, 114, 179, 194, 244, 358, 376
Fiske, Bradley 300, 301, 304, 323
Ford, Hannibal x
Frederick II, the Great, King of Prussia 80
Galileo, Galilei xii, 135, 175
Gipps, George 158, 159, 167, 195, 358
Goodenough, Sir William Edmund 8, 358
Harcourt, Robert Vernon 347, 349, 352, 358
Harding, Edward W. (Rapidan) 23, 77, 105, 131, 161, 179, 358, 376, 377
Harvey, William xii, 175

INDEX

Hawke, Edward 287, 316
Henley, Joseph C. W. 294, 358
Howe, Lord Richard 283
Hughes-Onslow, Constantine H. 179, 358
Isherwood, Harold x, 70, 71, 101, 158, 159, 179, 180, 225, 228, 253, 254, 290, 347, 358
Jackson, Sir Henry Bradwardine 101, 105, 131, 179, 358
James, William 286
Jellicoe, John Rushworth, first earl Jellicoe 23, 55, 101, 104, 105, 131, 132, 164, 174, 194, 195, 201, 215, 256, 274, 293, 294, 358
Jervis, Sir John; *see also* St. Vincent, earl of 284, 287
John of Austria (Don John) 80
Kelvin, Lord, *see* Thomson, Sir William
Keppel, Augustus 283
Kerr, Lord Walter Talbot 8, 14, 358
Kitchener, Horatio Herbert, Earl of 81

Lafone, Albert S. 166, 358
Lafontaine 160
Lamb, Charles 85
Landstad, D. H. x, 253, 254, 358
Lawrence, Sir Joseph, first Baronet ix, 14, 359, 376
Lock, William Henry 14, 71, 359
Locke, B. J. W. 376
Mahan, Alfred Thayer 97
Maria Theresa, Queen of Bohemia and Hungary, Archduchess of Austria 80
Marjoribanks, Edward, second Baron Tweedmouth 172, 178, 179, 203, 204, 215, 216, 222, 223, 224, 225, 228, 359
May, Henry John 138
May, Sir William Henry 14, 359
Moltke, Helmuth C. B. von 93
Moore, Sir Archibald Gordon Henry Wilson 244, 255, 294, 359
Mountbatten, Louis Alexander, first Marquess of Milford Haven, formerly styled Prince Louis Alexander of Battenberg 357, 359
McKenna, Reginald 195, 359
Napoleon I (Bonaparte), Emperor of France 93, 106, 286, 290, 301
Nelson, Lord Horatio xii, 80, 93, 109, 128, 181, 283, 284, 285, 287, 290, 299, 316
Nieman, H. W. x
Ogilvy, Frederick Charles Ashley 243, 244, 246, 250, 253, 264–5, 268, 269, 292, 359
Paderewski, Ignacy (Jan) 86
Pakenham, Sir Edward Michael (not Sir G. Pakenham) 80
Palmer, William Waldegrave, second Earl Selborne 8, 14, 359, 376

Parr, Alfred Arthur Chase 70, 87, 226, 359
Peirse, Sir Richard Henry 293, 326, 359
Perkin, Sir William Henry 175–6
Plunkett, Reginald (later Sir Reginald Aylmer Ranfurly Plunkett-Ernle-Erle-Drax) 250
Pollen, Arthur Joseph Hungerford: on tactics xi, 9, 11, 19–20, 43–4, 64–6, 72, 73–4, 77–81, 86, 94, 99, 106–10, 118–30, 142–4, 150–1, 154, 165, 202, 217–8, 238–42, 265, 273–90, 295–308, 315–25, 337–9, 341–2, 251; on strategy xi, 8, 11, 19, 154, 288; on capital ship design 55, 66–9, 72, 79–80, 92–5, 98–100, 110, 122–3, 217–18, 286, 337; on naval finance 11, 12, 68, 99–100, 175, 191, 196–7, 218, 285, 331–2, 345; on mechanization xi, 135–8, 162–3, 170–1, 175–7, 189–91, 200–1, 257–60, 310, 311, 312, 318–9, 340–1, 343; on cost of his fire control system 169–70, 175, 190–1, 217, 218–22, 231, 330–2, 341, 342, 345; on the creation of a staff system 197–9, 208, 215, 276–7, 352
Pollen, John Anthony xii, xiii
Pollen, John Hungerford ix
Raleigh, Sir Walter 290
Riley, Gerard B. 250
Rodney, Sir George Brydges 283, 290
St. Vincent, Earl of; see also Jervis, Admiral Sir John 283, 284
Scott, Sir Percy Moreton, first Baronet 1, 2, 3, 4, 14, 55, 101, 180, 243, 348, 349, 352, 353, 359, 376–7
Selborne, Earl of, see Palmer, William Waldegrave
Semenoff, Vladimir 304
Sims, William Snowden 24
Slade, Sir Edmond John Warre 132, 359
Smith, Bertram H. 92, 161, 359
Spee, Count Maximilian von 334, 335, 338
Sturdee, Sir Frederick Charles Doveton 334, 338
Thomas, Sir Charles Inigo (Emeritus) 172, 227, 230, 351, 353, 359
Thomson, James 292, 359
Thomson, Sir William, first Baron Kelvin of Largs 14, 176, 292, 310, 359
Togo, Heihichiro xii, 68, 86, 106, 108, 109, 125, 136, 300
Tweedmouth, Lord see Marjoribanks, Edward
Villeneuve, Pierre 287
Waller, Arthur Craig 161, 359
Warren, Herbert A. 92, 360
Watkin, Henry Samuel Spiller 199, 214–15
Wellesley, Arthur, first Duke of Wellington 301

Wellington, Duke of, see Wellesley, Arthur
White, John 307–8, 324
Whitehead, Robert 125, 176, 219, 220
Wilson, Sir Arthur Knyvet, third Baronet 131, 132, 156, 157, 158, 159, 167, 171, 178, 179, 194, 203, 204, 205–6, 208, 208–9, 223, 224, 225, 226, 228, 230, 231, 244, 360
Wood, Henry 177

Polyanthropy, defined 278
P. Z. Exercise 127, 262

Ram 25, 118, 122, 125, 137, 238
Rate Finder, defined 255
Rate Plotting (Time and Range, Time and Bearing) 158–9, 178–9, 180–1, 201, 255, 269, 271, 293–4, 373

REFERENCES:
1 Cross-references
Memorandum on a Proposed System for Finding Ranges at Sea and Ascertaining the Speed and Course of any Vessel in Sight 211–12, 214, 309
Fire Control and Long-Range Firing: An Essay to Define Certain Principia of Gunnery, and to Suggest Means for Their Application 55, 57, 58, 60, 87, 91, 212, 309–10
The *Jupiter* Letters: Extracts from Letters Addressed to Various Correspondents in The Royal Navy principally from HMS *Jupiter* 114, 121, 319, 321
Note on the Possibility of Demonstrating the Principle of Aim Correction Without the Use of Instruments Designed for the Purpose 257
An Apology for the A. C. Battle System: Being Notes for a Lecture to the War Course College, Portsmouth 319, 321
Extracts from a Letter to Capt. Reginald H. S. Bacon, CVO, DSO, Royal Navy 226
Reflections on an Error of the Day 201
2 Literary and Musical References (by author or composer)
Lewis Carroll [Charles L. Dodgeson], *The Hunting of the Snark* 146
Sir William Schwenk Gilbert and Sir Arthur Seymour Sullivan, *The Mikado* 186
Homer, *The Iliad* 184; *The Odyssey* 276
Charles Lamb, 'A Dissertation upon Roast Pig' 85
Frederick Marryat, *Peter Simple* 73
St. Augustine of Hippo, *Contra Epistulam Parmeniani* 205
William Shakespeare, *The Tempest* 134
Voltaire [Francois-Marie Arouet], *Dictionnaire philosophique* 160
Richard Wagner, *Parsifal* 177
3 Naval History and Theory (by author)
Reginald Hugh Spencer Bacon, 'Some Notes on Naval Strategy', *The Naval Annual, 1901* 138; 'The Battleship of the Future', *Transactions of the Institution of Naval Architects* 302
Barfleur (Sir Reginald Neville Custance), *Naval Policy: a Plea for the Study of War* 125
Romeo Bernotti, *The Fundamentals of the Naval Tactics* 297
Black Joke, 'Tactical Speed', *United Service Magazine* 105, 106, 108
Sir Reginald Neville Custance, *The Ship of the Line in Battle* 295, 300–1, 302, 321, 323
Gabriel Darrieus, *War on the Sea, Strategy and Tactics: Basic Principles* 297
Rene Daveluy, *The Genius of Naval Warfare* 297
Sir Howard Douglas, *A Treatise on Naval Gunnery* 318, 341
Bradley A. Fiske, 'The Mean Point of Impact', *United States Naval Institute Proceedings* 301, 304, 323
William James, *The Naval History of Great Britain from the declaration of War by France in 1793 to the Accession of George IV* 286
Maga (Sir Reginald Neville Custance), 'The Growth of the Capital Ship', *Blackwood's Magazine* 97; 'The Speed of the Capital Ship', *Blackwood's Magazine* 105, 106, 108
Alfred Thayer Mahan, 'Some Reflections upon the Far-Eastern War', *National Review* 97
Rapidan (Edward W. Harding), *The Tactical Employment of Naval Artillery* 77
St. Barbara, 'Fool Gunnery in the Navy', *Blackwood's Magazine* 186, 214
Sir Percy Moreton Scott, 'Remarks on that portion of the Fighting Efficiency of a Fleet which is dependent on the Straight Shooting of the Guns', in *Gunnery* 180

INDEX

Vladimir Semenoff, *The Battle of Tsushima between the Japanese and Russian Fleets* 304
Jedediah Stephens Tucker, *Memoir of Admiral the Right Hon. the Earl of St. Vincent* 284
John White, *Naval Tactical Problems* 307–8, 324

Royal Naval College, Greenwich 358
Royal Naval War College, Portsmouth 179, 237, 290, 357, 358, 359

Second Sea Lord 359

Tactical Machine 105, 106, 114, 115, 118, 122, 128, 129, 132, 150–1, 152, 154, 168, 202, 230, 237, 238–9, 241
Third Sea Lord (Controller) 14, 101, 105, 131, 161, 179, 194, 358, 359
Time of Flight of the Projectile 6, 26, 33, 57, 58, 71, 91, 147, 152, 268, 271, 308, 328
Torpedo 25, 71, 101, 118, 125, 176, 220, 242, 250, 275, 302, 303, 304, 307, 330, 337, 338, 352
True Range (Actual Range), defined 24

War Course College, Portsmouth 101, 131, 132, 133, 271, 359
WARSHIPS:
1 Modern British
   Agamemnon 80, 110
   Albemarle 359
   Amphion 358
   Ariadne 131, 132, 133, 156, 157, 158, 159, 163, 164, 165, 166, 168, 174, 175, 178, 179, 180, 183, 184, 185, 190, 194, 195, 198, 202, 203, 205, 213, 214, 215, 225, 228, 230, 231, 235, 258, 259, 261, 264, 266, 268, 269, 276, 314, 321, 350, 358
   Bellerophon 358
   Black Prince 110, 359
   Bulwark 77, 83
   Canopus 358
   Dido 308
   Drake 298, 308
   Dreadnought 3, 66–9, 80, 92, 94, 95, 100, 106, 110, 123, 134, 136–7, 138, 239, 262, 281, 299, 326, 329, 337
   Duke of Edinburgh 67, 80, 93, 359
   Duncan 75, 110
   Essex 186
   Excellent (Gunnery School at Whale Island, Portsmouth) 2, 101, 161, 167, 179, 233, 358, 359
   Exmouth 68, 77, 95
   Grafton 358
   Hero 174, 185
   Hindustan 358
   Indomitable 239
   Invincible 3, 110, 123, 144, 253
   Iron Duke 358
   Irresistible 342
   Jupiter 55, 70, 71, 76, 81, 83, 87, 92, 101, 103, 104, 131, 153, 157, 168, 174, 189, 200, 212, 213, 217, 222, 226, 229, 252, 256, 257, 258, 264, 265, 278, 312, 357, 360
   King Edward VII 68, 77, 93, 358
   Lord Nelson 80, 110, 123
   Magnificent 358
   Majestic 50, 52, 53, 138, 342
   Marlborough 358
   Narcissus 23
   Natal 67, 110, 243, 244, 246, 249, 250, 251, 252, 254, 255, 256, 260, 261, 264, 266, 267, 268, 269, 314, 321, 359
   Nelson x
   Newcastle 358
   Nile 357
   Nith 358
   Ocean 342
   Orion 293, 294, 347–8, 349, 351, 358, 359
   Pelorus 359
   Prince of Wales 186, 358
   Psyche 357
   Renown 170
   Revenge 264, 358
   Rodney x
   Royal Sovereign 67, 99, 317
   Severn 334, 336
   Shannon 156, 358
   Swiftsure 67, 97
   Temeraire 170
   Tenedos 357
   Terrible 359
   Thunderer 348
   Torpedo Boat 116 358
   Triumph 67, 97, 342
   Vanguard 358
   Venerable 211
   Vengeance 157, 158, 159, 163, 164, 166, 173, 178, 179, 203, 224, 276, 359
   Vernon 275
2 Modern Foreign
   Derfflinger (German Empire) 342
   Deutschland (German Empire) 94
   Emden (German Empire) 342
   Gneisenau (German Empire) 334
   Kaiser (German Empire) 93
   Kashima (Japan) 42
   Katori (Japan) 42
   Koenigsberg (German Empire) 334, 335, 336

Mikasa (Japan) 126
Moltke (German Empire) 334
Scharnhorst (German Empire) 334
Seydlitz (German Empire) 342
Worth (German Empire) 93
3 Sailing Era British and Foreign
Bucentaur (France) 124
Captain (Britain) 284

Chesapeake (United States) 298, 317
Shannon (Britain) 29, 317
Speedy (Britain) 285
Victory (Britain) 136, 137

Yaw 2, 36, 70–1, 83, 102, 103, 104, 145, 156, 157, 158, 199, 201, 213, 217, 221, 257

NAVY RECORDS SOCIETY
(FOUNDED 1893)

The Navy Records Society was established for the purpose of printing rare or unpublished works of naval interest. The Society is open to all who are interested in naval history and any person wishing to become a member should apply to the Hon. Secretary, c/o The Royal Naval College, Greenwich, London, SE10 9NN. The annual subscription for individuals is £10, the payment of which entitles the member to receive one copy of each work issued by the Society in that year. For Libraries and Institutions the annual subscription is £12.

The prices to members and non-members respectively are given after each volume, and orders should be sent, enclosing no money, to the Hon. Treasurer, c/o Barclays Bank, 54 Lombard Street, London, EC3P 3AH. Those volumes against which the letters 'A & U' are set after the price to non-members are available to them only through bookshops or, in case of difficulty, direct from George Allen & Unwin (Publishers) Ltd, PO Box 18, Park Lane, Hemel Hempstead, Herts HP2 4TE. Prices are correct at the time of going to press.

The Society has already issued:

Vols. 1 and 2. *State Papers relating to the Defeat of the Spanish Armada, Anno 1588.* Edited by Professor J. K. Laughton. (Vols. I and II.). (*£12.00/£15.00 ea.*)

Vol. 3. *Letters of Lord Hood,* 1781–82. Edited by Mr David Hannay. (*Out of Print.*)

Vol. 4. *Index to James's Naval History,* by C. G. Toogood. Edited by the Hon. T. A. Brassey. (*Out of Print.*)

Vol. 5. *Life of Captain Stephen Martin,* 1666–1740. Edited by Sir Clements R. Markham. (*Out of Print.*)

Vol. 6. *Journal of Rear-Admiral Bartholomew James,* 1725–1828. Edited by Professor J. K. Laughton and Commander J. Y. F. Sulivan. (*Out of Print.*)

Vol. 7. *Hollond's Discourses of the Navy,* 1638 and 1658. Edited by J. R. Tanner. (*Out of Print.*)

Vol. 8. *Naval Accounts and Inventories in the Reign of Henry VII.* Edited by Mr. M. Oppenheim. (*Out of Print.*)

Vol. 9. *Journal of Sir George Rooke.* Edited by Mr Oscar Browning. (*Out of Print.*)

Vol. 10. *Letters and Papers relating to the War with France,* 1512–13. Edited by M. Alfred Spont. (*£6.50/£12.00.*)

Vol. 11. *Papers relating to the Spanish War,* 1585–87. Edited by Mr Julian S. Corbett. (*Out of Print.*)

Vol. 12. *Journals and Letters of Admiral of the Fleet Sir Thomas Byam Martin,* 1773–1854 (Vol. II.). Edited by Admiral Sir R. Vesey Hamilton. (*See* 24.) (*Out of Print.*)

Vol. 13. *Papers relating to the First Dutch War,* 1652–54 (Vol. I.). Edited by Dr S. R. Gardiner. (*Out of Print.*)

Vol. 14. *Papers relating to the Blockade of Brest*, 1803–5 (Vol. I.). Edited by Mr J. Leyland. (*Out of Print.*)

Vol. 15. *History of the Russian Fleet during the Reign of Peter the Great, By a Contemporary Englishman.* Edited by Admiral Sir Cyprian Bridge. (*Out of Print.*)

Vol. 16. *Logs of the Great Sea Fights,* 1794–1805 (Vol. I.). Edited by Vice-Admiral Sir T. Sturges Jackson (£12.00/£15.00.)

Vol. 17. *Papers relating to the First Dutch War,* 1652–54 (Vol. II.). Edited by Dr S. R. Gardiner. (*Out of Print.*)

Vol. 18. *Logs of the Great Sea Fights* (Vol. II.). Edited by Vice-Admiral Sir T. Sturges Jackson. (£12.00/£15.00.)

Vol. 19. *Journals and Letters of Sir T. Byam Martin* (Vol. II.). Edited by Admiral Sir R. Vesey-Hamilton. (*See* 24). (£6.50/£12.00.)

Vol. 20. *The Naval Miscellany* (Vol. I.). Edited by Professor J. K. Laughton (*Out of Print.*)

Vol. 21. *Papers relating to the Blockade of Brest,* 1803–5 (Vol. II). Edited by Mr John Leyland. (*Out of Print.*)

Vols. 22 and 23. *The Naval Tracts of Sir William Monson* (Vols. I. and II.). Edited by Mr M. Oppenheim. (*Out of Print.*)

Vol. 24. *Journals and Letters of Sir T. Byam Martin* (Vol. I.). Edited by Admiral Sir R. Vesey Hamilton. (£6.50/£12.00.)

Vol. 25. *Nelson and the Neapolitan Jacobins.* Edited by Mr H. C. Gutteridge. (*Out of Print.*)

Vol. 26. *A Descriptive Catalogue of the Naval MSS. in the Pepysian Library* (Vol. I.). Edited by Mr J. R. Tanner. (*Out of Print.*)

Vol. 27. *A Descriptive Catalogue of the Naval MSS. in the Pepysian Library* (Vol. II.). Edited by Mr J. R. Tanner. (£6.50/£12.00.)

Vol. 28. *The Correspondence of Admiral John Markham,* 1801–7. Edited by Sir Clements R. Markham. (*Out of Print.*)

Vol. 29. *Fighting Instructions,* 1530–1816. Edited by Mr Julian S. Corbett. (*Out of Print.*)

Vol. 30. *Papers relating to the First Dutch War,* 1652–54 (Vol. III.). Edited by Dr S. R. Gardiner and Mr C. T. Atkinson. (*Out of Print.*)

Vol. 31. *The Recollections of Commander James Anthony Gardner,* 1775–1814. Edited by Admiral Sir R. Vesey Hamilton and Professor J. K. Laughton. (*Out of Print.*)

Vol. 32. *Letters and Papers of Charles, Lord Barham,* 1758–1813 (Vol. I.). Edited by Sir J. K. Laughton. (*Out of Print.*)

Vol. 33. *Naval Songs and Ballads.* Edited by Professor C. H. Firth. (*Out of Print.*)

Vol. 34. *Views of the Battles of the Third Dutch War.* Edited by Mr Julian S. Corbett. (*Out of Print.*)

Vol. 35. *Signals and Instructions,* 1776–94. Edited by Mr Julian S. Corbett. (*Out of Print.*)

Vol. 36. *A Descriptive Catalogue of the Naval MSS. in the Pepysian Library* (Vol. III.). Edited by Dr J. R. Tanner. (*Out of Print.*)

Vol. 37. *Papers relating to the First Dutch War,* 1652–1654 (Vol. IV.). Edited by Mr C. T. Atkinson. (*Out of Print.*)

Vol. 38. *Letters and Papers of Charles, Lord Barham,* 1758–1813 (Vol. II.). Edited by Sir J. K. Laughton. (*Out of Print.*)

Vol. 39. *Letters and Papers of Charles, Lord Barham,* 1758–1813 (Vol. III.). Edited by Sir J. K. Laughton. (*Out of Print.*)

Vol. 40. *The Naval Miscellany* (Vol. II.). Edited by Sir J. K. Laughton. (*Out of Print.*)

Vol. 41. *Papers relating to the First Dutch War,* 1652–54 (Vol. V.). Edited by Mr C. T. Atkinson. (£6.50/£12.00.)

Vol. 42. *Papers relating to the Loss of Minorca in 1756.* Edited by Capt. H. W. Richmond, R.N. (*£6.50/£12.00.*)
Vol. 43. *The Naval Tracts of Sir William Monson* (Vol. III.). Edited by Mr M. Oppenheim. (Rebinding.)
Vol. 44. *The Old Scots Navy,* 1689–1710. Edited by Mr James Grant. (*Out of Print.*)
Vol. 45. *The Naval Tracts of Sir William Monson* (Vol. IV.). Edited by Mr M. Oppenheim. (*£6.50/£12.00.*)
Vol. 46. *The Private Papers of George, second Earl Spencer* (Vol. I.). Edited by Mr Julian S. Corbett. (*£6.50/£12.00.*)
Vol. 47. *The Naval Tracts of Sir William Monson* (Vol. V.). Edited by Mr M. Oppenheim. (*£6.50/£12.00.*)
Vol. 48. *The Private Papers of George, second Earl Spencer* (Vol. II.). Edited by Mr Julian S. Corbett. (*Out of Print.*)
Vol. 49. *Documents relating to Law and Custom of the Sea* (Vol. I.). Edited by Mr R. G. Marsden. (*£6.50/£12.00.*)
Vol. 50. *Documents relating to Law and Custom of the Sea* (Vol. II.). Edited by Mr R. G. Marsden. (*£6.50/£12.00.)*
Vol. 51. *Autobiography of Phineas Pett.* Edited by Mr W. G. Perrin. (Rebinding.)
Vol. 52. *The Life of Admiral Sir John Leake* (Vol. I.). Edited by Mr G. A. R. Callender. (*£6.50/£12.00.*)
Vol. 53. *The Life of Admiral Sir John Leake* (Vol. II.). Edited by Mr G. A. R. Callender. (*£6.50/£12.00.*)
Vol. 54. *The Life and Works of Sir Henry Mainwaring* (Vol. I.). Edited by Mr G. E. Manwaring. (*£6.50/£12.00.*)
Vol. 55. *The Letters of Lord St. Vincent,* 1801–1804 (Vol. I.). Edited by Mr D. B. Smith. (*Out of Print.*)
Vol. 56. *The Life and Works of Sir Henry Mainwaring* (Vol. II.). Edited by Mr G. E. Manwaring and Mr W. G. Perrin. (*Out of Print.*)
Vol. 57. *A Descriptive Catalogue of the Naval MSS in the Pepysian Library* (Vol. IV.). Edited by Dr J. R. Tanner. (*Out of Print.*)
Vol. 58. *The Private Papers of George, second Earl Spencer* (Vol. III.). Edited by Rear-Admiral H. W. Richmond. (*Out of Print.*)
Vol. 59. *The Private Papers of George, second Earl Spencer* (Vol. IV.). Edited by Rear-Admiral H. W. Richmond. (*Out of Print.*)
Vol. 60. *Samuel Pepys's Naval Minutes.* Edited by Dr J. R. Tanner. (*Out of Print.*)
Vol. 61. *The Letters of Lord St. Vincent,* 1801–1804 (Vol. II.). Edited by Mr D. B. Smith. (*Out of Print.*)
Vol. 62. *Letters and Papers of Admiral Viscount Keith* (Vol. I.). Edited by Mr W. G. Perrin. (*Out of Print.*)
Vol. 63. *The Naval Miscellany* (Vol. III.). Edited by Mr W. G. Perrin. (*Out of Print.*)
Vol. 64. *The Journal of the First Earl of Sandwich.* Edited by Mr R. C. Anderson. (*£6.50/£12.00.*)
Vol. 65. *Boteler's Dialogues.* Edited by Mr W. G. Perrin. (*£6.50/£12.00.)*
Vol. 66. *Papers relating to the First Dutch War,* 1652–54 (Vol. VI.; with index). Edited by Mr C. T. Atkinson. (*£6.50/£12.00.*)
Vol. 67. *The Byng Papers* (Vol. I.). Edited by Mr W. C. B. Tunstall. (*£6.50/£12.00.*)
Vol. 68. *The Byng Papers* (Vol. II.). Edited by Mr W. C. B. Tunstall. (*£6.50/£12.00.*)
Vol. 69. *The Private Papers of John, Earl of Sandwich* (Vol. I.). Edited by Mr G. R. Barnes and Lieut-Commander J. H. Owen, R.N. (*£6.50/£12.00.*)

Corrigenda to *Papers relating to the First Dutch War,* 1652–54 (Vols. I. to VI.). Edited by Captain A. C. Dewar, R.N. (Free.)

Vol. 70. *The Byng Papers* (Vol. III.). Edited by Mr W. C. B. Tunstall. (£6.50/£12.00.)

Vol. 71. *The Private Papers of John, Earl of Sandwich* (Vol. II.). Edited by Mr G. R. Barnes and Lieut-Commander J. H. Owen, R.N. *(£6.50/£12.00.)*

Vol. 72. *Piracy in the Levant,* 1827–8. Edited by Lieut-Commander C. G. Pitcairn Jones, R.N. (£6.50/£12.00.)

Vol. 73. *The Tangier Papers of Samuel Pepys.* Edited by Mr Edwin Chappell. (*Out of Print.*)

Vol. 74. *The Tomlinson Papers.* Edited by Mr J. G. Bullocke. (£6.50/£12.00.)

Vol. 75. *The Private Papers of John, Earl of Sandwich* (Vol. III.). Edited by Mr G. R. Barnes and Commander J. H. Owen, R.N. (*Out of Print.*)

Vol. 76. *The Letters of Robert Blake.* Edited by the Rev. J. R. Powell. (£6.50/£12.00.)

Vol. 77. *Letters and Papers of Admiral the Hon. Samuel Barrington* (Vol. I.). Edited by Mr D. Bonner-Smith. (£6.50/£12.00.)

Vol. 78. *The Private Papers of John, Earl of Sandwich* (Vol. IV.). Edited by Mr G. R. Barnes and Commander J. H. Owen, R.N. (*Out of Print.*)

Vol. 79. *The Journals of Sir Thomas Allin,* 1660–1678 (Vol. I. 1660–66). Edited by Mr R. C. Anderson. (£6.50/£12.00.)

Vol. 80. *The Journals of Sir Thomas Allin,* 1660–1678 (Vol. II. 1667–78). Edited by Mr R. C. Anderson. (£6.50/£12.00.)

Vol. 81. *Letters and Papers of Admiral the Hon. Samuel Barrington* (Vol. II.). Edited by Mr D. Bonner-Smith. (*Out of Print.*)

Vol. 82. *Captain Boteler's Recollections* (1808 to 1830). Edited by Mr D. Bonner-Smith. (*Out of Print.*)

Vol. 83. *Russian War,* 1854. *Baltic and Black Sea: Official Correspondence.* Edited by Mr D. Bonner-Smith and Captain A. C. Dewar, R.N. (*Out of Print.*)

Vol. 84. *Russian War,* 1855. *Baltic: Official Correspondence.* Edited by Mr D. Bonner-Smith. (*Out of Print.*)

Vol. 85. *Russian War,* 1855. *Black Sea: Official Correspondence.* Edited by Captain A. C. Dewar, R.N. (*Out of Print.*)

Vol. 86. *Journals and Narratives of the Third Dutch War.* Edited by Mr R. C. Anderson. (*Out of Print.*)

Vol. 87. *The Naval Brigades in the Indian Mutiny,* 1857–58. Edited by Commander W. B. Rowbotham, R.N. (*Out of Print.*)

Vol. 88. *Patee Byng's Journal.* Edited by Mr J. L. Cranmer-Byng. (*Out of Print.*)

Vol. 89. *The Sergison Papers* (1688–1702). Edited by Commander R. D. Merriman, R.I.N. (£6.50/£12.00.)

Vol. 90. *The Keith Papers* (Vol. II.). Edited by Mr C. C. Lloyd. (£6.50/£12.00.)

Vol. 91. *Five Naval Journals,* 1789–1817. Edited by Rear-Admiral H. G. Thursfield. (£6.50/£12.00.)

Vol. 92. *The Naval Miscellany* (Vol. IV.). Edited by Mr C. C. Lloyd. (*Out of Print.*)

Vol. 93. *Sir William Dillon's Narrative of Professional Adventures (1790–1839)* (Vol. I. 1790–1802). Edited by Professor Michael A. Lewis. (*Out of Print.*)

Vol. 94. *The Walker Expedition to Quebec, 1711.* Edited by Professor Gerald S. Graham. (*Out of Print.*)

Vol. 95. *The Second China War,* 1856–60. Edited by Mr D. Bonner-Smith and Mr E. W. R. Lumby. (*Out of Print.*)

Vol. 96. *The Keith Papers,* 1803–1815 (Vol. III.). Edited by Professor C. C. Lloyd. (£6.50/£12.00.)

Vol. 97. *Sir William Dillon's Narrative of Professional Adventures (1790–1839)* (Vol. II. 1802–1839). Edited by Professor Michael A. Lewis. (*Out of Print.*)

Vol. 98. *The Private Correspondence of Admiral Lord Collingwood.* Edited by Professor Edward Hughes. (*Out of Print.*)
Vol. 99. *The Vernon Papers* (1739–1745). Edited by Mr B. McL. Ranft. (*Out of Print.*)
Vol. 100. *Nelson's Letters to his Wife and Other Documents.* Edited by Lieut-Commander G. P. B. Naish, R.N.V.R. (*£6.50/£12.00.*)
Vol. 101. *A Memoir of James Trevenen* (1760–1790). Edited by Professor C. C. Lloyd and Dr R. C. Anderson. (*£6.50/£12.00.*)
Vol. 102. *The Papers of Admiral Sir John Fisher* (Vol. I.). Edited by Lieut-Commander P. K. Kemp, R.N. (*Out of Print.*)
Vol. 103. *Queen Anne's Navy.* Edited by Commander R. D. Merriman, R.I.N. (*Out of Print.*)
Vol. 104. *The Navy and South America, 1807–1823.* Edited by Professor G. S. Graham and Professor R. A. Humphreys. (*£6.50/£12.00.*)
Vol. 105. *Documents relating to the Civil War, 1642–1648.* Edited by the Rev. J. R. Powell and Mr E. K. Timings. (*Out of Print.*)
Vol. 106. *The Papers of Admiral Sir John Fisher* (Vol. II.). Edited by Lieut-Commander P. K. Kemp, R.N. (*£6.50/£12.00.*)
Vol. 107. *The Health of Seamen.* Edited by Professor C. C. Lloyd. (*£6.50/£12.00.*)
Vol. 108. *The Jellicoe Papers* (Vol. I: 1893–1916). Edited by Mr A. Temple Patterson. (*£6.50/£12.00.*)
Vol. 109. *Documents relating to Anson's Voyage round the World, 1740–1744.* Edited by Dr Glyndwr Williams. (*£6.50/£12.00.*)
Vol. 110. *The Saumarez Papers: The Baltic, 1808–1812.* Edited by Mr A. N. Ryan (*£6.50/£12.00.*)
Vol. 111. *The Jellicoe Papers* (Vol. II: 1916–1935). Edited by Professor A. Temple Patterson. (*£6.50/£12.00.*)
Vol. 112. *The Rupert and Monck Letterbook, 1666.* Edited by the Rev. J. R. Powell and Mr E. K. Timings. (*£6.50/£12.00.*)
Vol. 113. *Documents relating to the Royal Naval Air Service* (Vol. I: 1908–1918). Edited by Captain S. W. Roskill, R.N. (*£6.50/£12.00.*)
Vol. 114. *The Siege and Capture of Havana: 1762.* Edited by Assistant-Professor David Syrett. (*£6.50/£12.00.*)
Vol. 115. *Policy and Operations in the Mediterranean: 1912–14.* Edited by Mr E. W. R. Lumby. (*£6.50/£12.00.*)
Vol. 116. *The Jacobean Commissions of Enquiry: 1608 and* 1618. Edited by Dr A. P. McGowan. (*£6.50/£12.00.*)
Vol. 117. *The Keyes Papers.* (Vol. I: 1914–1918). Edited by Dr Paul G. Halpern. (*£7.50/£17.50—*A & U.)
Vol. 118. *The Royal Navy and North America: The Warren Papers, 1736–1752.* Edited by Dr Julian Gwyn. (*£6.50/£12.00.*)
Vol. 119. *The Manning of the Royal Navy: Selected Public Pamphlets* 1693–1873. Edited by Professor J. S. Bromley. (*£6.50/£12.00.*)
Vol. 120. *Naval Administration, 1715–1750.* Edited by Professor D. A. Baugh. (*£6.50/£12.00.*)
Vol. 121. *The Keyes Papers* (Vol. II: 1919–1938). Edited by Dr Paul G. Halpern. (*£6.50/£17.50—*A & U.)
Vol. 122. *The Keyes Papers* (Vol. III: 1939–1945). Edited by Dr Paul G. Halpern. (*£6.50/£17.50—*A & U.)
Vol. 123. *The Navy of the Lancastrian Kings: Accounts and Inventories of William Soper, Keeper of the King's Ships, 1422–1427.* Edited by Dr Susan Rose. (*£6.50/£17.50—*A & U.)
Vol. 124. *The Pollen Papers: The Privately Circulated Printed Works of Arthur Hungerford Pollen, 1901–1916.* Edited by Dr Jon T. Sumida. (*£6.50/£17.50—*A & U.)